Procedure Manual for the

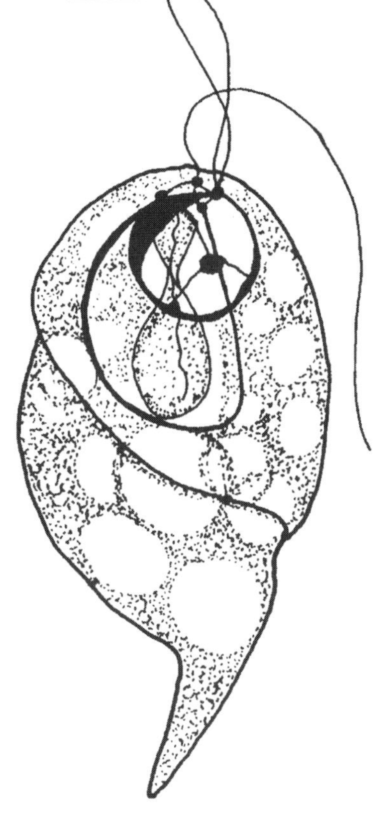

Diagnosis of Intestinal Parasites

Procedure Manual for the

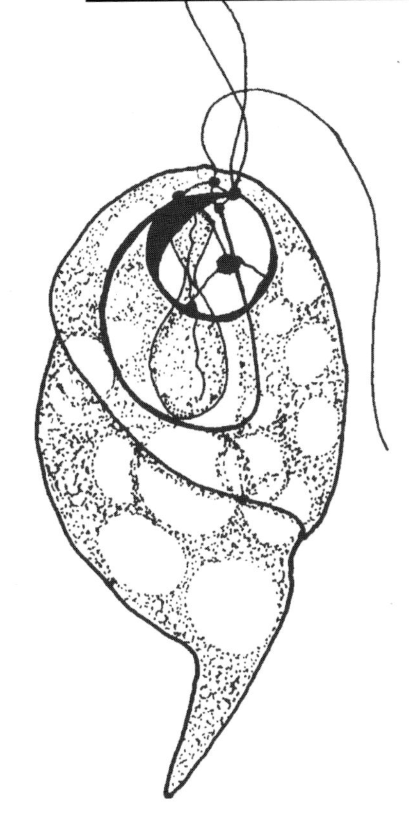

Diagnosis of Intestinal Parasites

Donald L. Price

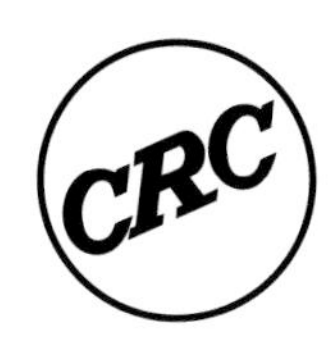

CRC Press
Boca Raton Ann Arbor London Tokyo

Library of Congress Cataloging-in-Publication Data

Price, Donald L.
Procedure manual for the diagnosis of intestinal parasites / Donald L. Price
p. cm.
Includes bibliographical references and index.
ISBN 0-8493-8654-3
1. Intestines—Parasites—Identification. I. Title.
[DNLM: 1. Intestinal Diseases, Parasitic—diagnosis—laboratory manuals. 2. Feces—parasitology—laboratory manuals.
WC 25 P945p 1993]
RC862.E47P75 1993
616.9′6—dc20
DNLM/DLC
for Library of Congress

93-7450
CIP

International Standard Book Number 0-8493-8654-3
Library of Congress Card Number 93-7450
Printed in the United States of America 1 2 3 4 5 6 7 8 9 0
Printed on acid-free paper

To my dear wife Ellyn, whose love, support, patience, constructive criticism, and encouragement enabled completion of this work;

and

to those who follow who will take whatever they find good in this book and will improve upon it, to those who will find ways to do what cannot be done, and to those who find "No" unacceptable.

Preface

The overall objective of this manual is to provide detailed, practical information on the microscopic diagnosis of intestinal parasites, i.e., of those parasites diagnosed by fecal examination. It deals principally with the parasites that infect (or infest) the gastrointestinal tract and associated organs of man. The intention is to provide information that will enable those directly involved in the diagnosis of intestinal parasitic infections to realize the greatest potential possible in the application of the closely related methods developed for finding and identifying parasites and eggs.

The manual has several secondary objectives. These include expanding detailed information relating to some methods, rather like a laboratory procedure manual with helpful hints; presenting how-to techniques that usually would not be addressed in a standard text; and presenting information on the types of materials and equipment that may make performing certain functions easier. Most of the information presented on fecal specimens applies to all methods used for collecting and handling them, preparing slides for microscopic examination, and examining the prepared slides. Emphasis is placed on procedures that are applied to preserved rather than unpreserved specimens, since preserved specimens are more practical for diagnosis of intestinal parasites in the clinical laboratory environment. The need for an effective, single-vial collection system has been expressed. The SAF collection/preservation product and the MIF-type products (described in the Introduction) that require the addition of iodine immediately before use are single-vial systems. The expanding use of the MIF technique and the present availability of the newer, upgraded MIF-type products have increased the need for more precise information on such products. Many text and reference books contain sections on methods but only a few superficially deal with these products and related procedures. Because of the lack of available, accurate information, the products and procedures are presented in greater detail in the manual.

In preparing the manual, the author has brought together in this single source a variety of published and unpublished information. Some of the material selected for inclusion in the manual was based on questions asked during continuing education classes, seminars, and lectures; and on research and testing performed in the laboratory. Some material originally written when the MIF Procedure Kit was placed on the market is also referenced. The author has attempted to cover information pertinent to the diagnosis of intestinal parasites, especially that which applies to examining wet preparations prepared from the specimen both before and after a concentration method is applied. Tables, drawings, and photomicrographs are used extensively throughout the manual to supplement and support written information.

Consideration has been given to the need for a system that includes methods for direct examination, concentration, and preparation of permanently stained fecal films. A more thorough understanding of these methods and how to use them effectively will ensure successful results.

The concerns include how to collect a fecal specimen, how to prepare the specimen for examination (including the application of various procedures), what parasites and eggs to look for, how to find them, and how to identify what is found. These same concerns apply regardless of the collection/preservation method employed, as does other information addressed in each part of the manual. For additional detailed information on laboratory materials and methods other than those included in this manual, most of which are applied to preserved specimens, the reader should refer to more general texts, such as Ash and Orihel, 1987; Garcia and Ash, 1979; Garcia and Bruckner, 1988; Isenberg et al., 1992; Markel et al., 1942; or Melvin and Brooke, 1982.

The material included in the manual is organized so that the users can readily find the information needed. Following a brief introduction, the manual is divided into four main parts and an appendix.

Part 1 deals with collecting and preserving the fecal specimen for the laboratory, and the various factors that have an impact on the quality of the specimen collected.

Part 2 includes a discussion of the microscope, the basic tool used in the parasitology section of a laboratory to make diagnoses of parasitic infections, and a discussion of methods for preparing the specimen for microscopic examination and for examining the specimen.

Part 3 consists of four chapters and considers additional methods and procedures that may be applied to fecal specimens. Chapter 3 deals with methods for concentrating the parasites and eggs present in fecal specimens. Chapter 4 gives methods for preparing permanently stained slides of protozoa in fecal films. Chapter 5 addresses the application of the various methods. Chapter 6 describes acid-fast stains, some special methods for egg counts, preparing semipermanent wet mounts, and posttreatment examination.

Part 4 presents the organisms involved in intestinal infections in four chapters. Chapter 7 introduces the reader to the parasites that are diagnosed by examining fecal specimens. Chaapter 8 deals with the helminths and Chapter 9 with the protozoa. Keys for the identification of helminth eggs and the intestinal amoebae are presented in the appropriate chapters. Chapter 10 presents a few of the objects and organisms found in fecal specimens that appear similar to and may be wrongly identified as parasites of man.

The Appendix adds more detail to technical information and terminology, gives special instructions on preparing stains and solutions, and provides a glossary of terms that apply to the examination of fecal specimens and to the parasites involved.

In considering the overall objective and who might use the manual, the background, training, and experience of individuals who signed up for continuing education courses and the institutions and organizations from where they came was a major factor in determining the scope of the manual. Individuals attending courses included practicing physicians with special interests in infectious diseases, pathologists responsible for clinical laboratories, laboratory directors, supervisory technologists responsible for diagnostic parasitology, bench technologists, individuals involved in general education in microbiology, and teachers involved in teaching medical technologists. Some of those attending the courses had been newly introduced to diagnostic parasitology while others had many years of experience.

An attempt has been made to provide the kind and extent of information that meets the special needs of individuals with such varied experience and backgrounds. By covering the subject from the time a fecal specimen is passed, through the pathways and steps to the final identification of the parasites and eggs involved, the individual should gain an overall understanding of the subject and become more familiar with the methods involved to accomplish the task of diagnosing intestinal parasitic infections.

Only through a clear understanding of the most effective use of a method can one expect to obtain the best results. One achieves efficiency through knowledge and understanding, and through the application of that knowledge.

"...Who, with a natural instinct to discern
What knowledge can perform, is diligent to learn."
—William Wordsworth
Character of a Happy Warrior

Introduction

The methods for collecting and submitting a fecal specimen for examination for eggs and parasites vary from laboratory to laboratory. Either a preserved or an unpreserved specimen may be received, depending on the operating procedures that are established by the laboratory director to serve the physicians and patients using the laboratory. Usually, a portion of a fecal specimen mixed in one of the accepted collecting/preserving solutions is received by the laboratory, either with or without an accompanying unpreserved portion of the specimen. If only an unpreserved fecal specimen is delivered or mailed to a diagnostic laboratory, the usual accepted protocol is to place a portion of the specimen in an appropriate collecting/preserving solution reserved for that purpose as soon as possible. It is this preserved portion of the fecal specimen that is most often processed and examined for intestinal parasites.

The various methods for handling, processing, and examining unpreserved fecal specimens have appeared in many textbooks, manuals, and scientific publications. These include methods for preparing wet mounts of feces mixed with saline and iodine, concentration methods, the preparation and fixation of smears for permanently stained fecal films, special procedures for collecting juveniles (juvenile nematodes are referred to as larvae in most texts), and for egg counts. There are special methods that are used in field studies for specific parasite groups. The Kato thick-smear method has been used extensively in epidemiologic studies on schistosomiasis. The original procedure has been modified several times (Martin and Beaver, 1968); probably the most important modification is the use of sodium azide preservation, which allows delayed examination of the specimen (Ash and Orihel, 1987; Bundy et al., 1985). For information on procedures for unpreserved specimens that are not covered in this manual, refer to texts by Ash and Orihel (1987); Garcia and Ash (1979); Garcia and Bruckner (1988); Isenberg (1992); Markel et al. (1986); and Melvin and Brooke (1982).

SURVIVAL OF TROPHOZOITES AND CYSTS IN UNPRESERVED FECAL SPECIMENS

The question of survival of parasites and eggs in fecal specimens arises when dealing with unpreserved specimens. Many factors may have an impact on the survival of protozoa in a fecal mass after it has been passed from the body, but only a few of these have been well investigated. Most studies correlate survival with time and temperature (Burrows, 1959). Markel and Quinn (1977) compared immediate fixation using PVA with delayed fixation using Schaudinn's fixative after the specimens reached the laboratory. In specimens from 100 positive patients, more were found to be positive in the specimens fixed immediately in PVA than in those fixed in Schaudinn's after arriving at the laboratory.

Where a species was present in both specimen groups, the numbers were greater in those fixed in PVA. If Schaudinn's fixative had been substituted for PVA for immediate fixation, the end result would have been the same.

Several studies have shown that trophozoites of *Entamoeba histolytica* may survive up to 5 hours in an incubator at 37°C and up to 16 hours at room temperature (Burrows, 1959). In observations made by the author, however, changes in trophozoites of *E. histolytica* began immediately in some freshly passed stools held at room temperature. Within 30 minutes, most of the trophozoites in these specimens had rounded up, many had become heavily vacuolated, and the number observed in the specimen had diminished. In some specimens, many active trophozoites of *E. coli* and *Endolimax nana* were seen on examination immediately after stool passage, but the numbers had markedly decreased after 4 hours, and only rounded up and mostly highly vacuolated trophozoites could be found. In another specimen that was held in the refrigerator over a weekend, trophozoites of *Entamoeba histolytica* were still active and appeared unaffected by the long delay before fixation.

In some specimens containing cysts of *E. nana* and *Giardia lamblia* that were held overnight at room temperature, the organisms within the cyst had begun to shrink away from the cyst wall; some appeared simply as blobs. Such cysts are usually referred to as "ghost cysts" (Plate 66:9). The quality of the cysts varied in different specimens held overnight. The numbers of apparently viable cysts in comparison to the numbers of ghost cysts varied from specimen to specimen held under similar conditions.

Since the relative numbers of ghost cysts found in different specimens held at either room temperature or in the refrigerator were inconsistent after a given period of time, some factors other than time and temperature must affect the survival of the organisms. Characteristics such as stool consistency, e.g., whether the specimen was watery, soft, or formed, did not appear to have a consistent relationship to survival time of the protozoa present. Specimens in which parasites did not survive well often had a putrid odor. Other investigators have reported similar findings.

Since it is probably something in the specimen itself rather than time or temperature alone that has the most significant impact on survival time of the organisms present in a specimen, how an unpreserved specimen is handled may have little effect on what protozoa are found and in what numbers. Fewer organisms are usually found where specimen fixation is delayed. It seems prudent, then, that a portion of the fecal specimen be placed in a suitable collecting/preserving solution as soon as possible after the specimen is passed, preferably immediately. There is no other way to be assured that what is present and recognizable when the specimen is passed is still present and recognizable when it is examined.

COLLECTING/PRESERVING SOLUTIONS

There are a number of collecting/preserving solutions available for preserving fecal specimens for parasitological examination, including formalin, PVA, SAF, PAF, MIF, and PIF, described below. PIF is a relatively new MIF-type solution, since it uses Lugol's iodine as an additive, but it contains no mercury. Usually, either the laboratory director or someone designated by the director selects the products and procedures to be used (see Chapter 1). The choices include:

- Formalin, which may be from 5 to 10% aqueous solution, may or may not be buffered, and/or have salt added (formol-saline).
- PVA (polyvinyl alcohol preservative), which contains polyvinyl alcohol powder, ethyl alcohol, glacial acetic acid, glycerin, and mercuric chloride. Some workers use a modified PVA in which cupric sulfate or zinc sulfate has been substituted for the mercuric chloride (Brooke and Goldman, 1949; Burrows, 1967b; Garcia et al., 1983; Garcia et al., 1993).
- SAF, which contains sodium acetate, acetic acid, and formaldehyde (Junod, 1962; Yang and Scholten, 1977).
- PAF, which contains phenol, alcohol, and formaldehyde (Burrows, 1967b).
- MIF-type solutions (except PIF), which have a stock solution containing tincture of merthiolate (thimerosal and alcohol), glycerin, and formaldehyde (MF stock). Ideally, Lugol's iodine should be added to the MF stock solution immediately before the sample of feces is added (Price, 1978; Sapero and Lawless, 1953).
- PIF, the newest of the MIF-type solutions, contains glycerine, alcohol, and formaldehyde, but is mercury-free.

The Use of Collecting/Preserving Solutions

Each of the collecting/preserving solutions listed below has advantages and disadvantages.

Formalin is the oldest and most used fixing and preserving solution. It is inexpensive, stable, and is used extensively as a fixative for tissues. Aqueous solutions containing from 5 to 10% formalin will kill and fix most helminth eggs and protozoan cysts, but an *Ascaris* egg with an intact cortex will not die; the embryo will continue to grow and develop if the specimen is held at controlled room temperature or kept in a refrigerator. Formalin-fixed specimens can be used for concentration methods and a number of concentration methods can be applied (see Chapter 3). Formalin-fixed specimens are inadequate and/or ineffective for fixing trophozoites of amoebae and other protozoa, usually rendering them unrecognizable. It is not a particularly good fixative for active juveniles (larvae) of nematodes because of the slow killing time. In most instances, if the specimen is fixed immediately after stool passage and the portion of the specimen is adequately mixed with the formalin, the juveniles are adequately fixed for identification. Specimens fixed in formalin can be stored at controlled room temperature for many years, although morphological detail of some cysts, eggs, and juveniles may diminish in time. Acid-fast staining procedures can be applied to dried fecal films prepared from the collection vial or after a concentration method has been performed. Formalin-fixed fecal specimens are unsatisfactory for preparing permanently stained fecal films by the usual methods.

PVA, on the other hand, is an excellent fixative for protozoan trophozoites and cysts, and it is especially suited to preparing permanently stained fecal films. It was developed by Brooke and Goldman in 1949 as a fixative for dysenteric stools, to help the fixed specimen adhere to the slide during staining. It proved to be an excellent collecting/preserving material for protozoa in specimens that were to be used for preparing stained fecal films. The PVA formula was modified and improved by Burrows (1967a). Generally, PVA-fixed fecal specimens are considered unsatisfactory for finding and identifying eggs and juveniles. Although the formalin-ether concentration method can be applied and eggs and cysts may remain recognizable, there appears to be no appreciable increase in the numbers found over direct examination of stained slides (Carroll et al., 1983). A formalin-fixed specimen should also be included for concentration for eggs and juveniles.

Because of the problems relating to disposal of materials containing mercury, PVA formulas have been introduced in which copper sulfate or zinc sulfate is substituted for mercuric chloride. The author obtained poor results with the copper sulfate product, in that protozoa fixed in other collecting/preserving solutions stained better and were more readily found and identified. Garcia et al. (1983) compared the two formulas of PVA and could not recommend the copper sulfate base product. PVA referred to in this manual is the mercuric chloride formulation.

The two fixatives, formalin and PVA, complement each other and are usually employed as a two-vial system, combining one vial of formalin and one of PVA, to ensure recovery of both helminth eggs and juveniles and protozoan trophozoites and cysts. A third empty vial is sometimes added for an unpreserved portion of the specimen to be transported to the laboratory.

SAF is a relatively good fixative for parasites and eggs found in fecal specimens. It was developed by Junod (1962) as a single-vial collecting/preserving solution. Specimens fixed in SAF can be examined directly by preparing saline or iodine wet mounts, concentrated by sedimentation methods, and permanently stained fecal films can be prepared (Scholten, 1972; Yang and Scholten, 1977). Iodine solutions are recommended for staining protozoa and eggs in wet preparations. According to Yang and Scholten (1977), juvenile nematodes were not recovered in SAF-fixed specimens and submission of an unpreserved specimen was necessary for finding and identifying these parasites.

Results of the application of sedimentation-type concentration methods on SAF-fixed specimens for cysts and eggs are comparable to the same method applied to formalin-fixed specimens. The quality of permanently stained fecal films stained with the usual trichrome-type stains is generally good in comparison with those made from PVA- or MIF-fixed specimens. Hematoxylin stains often yield very good results.

PAF was developed by Burrows (1967) especially for the identification of species of protozoa in wet preparations; when the appropriate dyes are used, the morphologic detail of amoebae is excellent. PAF-fixed fecal specimens are unsatisfactory for the preparation of permanently stained fecal films of protozoa with the usual trichrome and hematoxylin staining methods. Although preserved specimens can be concentrated by sedimentation, and eggs of helminths and juveniles of nematodes as well as protozoa present in the specimen are readily recovered, PAF is not considered by most workers in the field as satisfactory as a general collecting/preserving solution for routine work.

MIF was developed by Sapero and Lawless (1953) and modified by Price (1978) (see also Part 2). There have been several unpublished modifications since that date, but the 1978 procedure is generally used and is the one recommended by commercial companies that sell the product. The solution fixes, preserves, and stains in bulk all parasites and eggs usually found by fecal examination within 1 hour after the feces is well mixed with the solution (see the next section, Background). Properly preserved specimens are especially suitable for direct examination (wet mounts), concentration by sedimentation methods (Price, 1977), and preparing permanently stained fecal films (Price, poster session, American Society for Microbiology (ASM) meeting, 1980). Acid-fast staining can be performed on dried fecal smears made directly from the preserving solution or from the sediment plug after concentration (see Chapter 6). Parasites in fecal specimens fixed in MIF remain identifiable indefinitely. (Some of the photomicrographs that appear in this manual were taken of amoebae in fecal films made from specimens collected and preserved in 1960, which are therefore over 30 years old.)

PIF is a mercury-free collecting/preserving solution. The PIF Fecal Test is a system for effectively collecting, fixing, preserving, and staining trophozoites and cysts of protozoa; eggs of trematodes, cestodes, and nematodes; and juveniles of nematodes in bulk feces. Organisms are killed and preserved as they come into direct contact with the chemical mixture. Fixed fecal specimens may be examined after a minimum of 1 hour. Parasites and eggs in preserved specimens retain indefinitely the diagnostic features present at the time of fixation. Materials for the PIF Fecal Test are available from Alpha-Tec Systems, Inc., P.O. Box 17196, Irvine, CA 92713.

The Quality of Preserved Eggs and Parasites

The quality of the parasites in collecting/preserving solutions depends on the same factors regardless of the type of solution used, e.g., the consistency of specimens passed (formed, soft, or liquid), the time between passage and fixation, adequate mixing of the specimen with the solution, and also how well the specific directions for the product are followed (see Part 1). When the directions for a product are followed precisely, the quality of the specimen is usually ensured and the results of procedures are usually good. There is no way to get good results from a poor-quality specimen.

BACKGROUND OF THE MIF-TYPE COLLECTING/PRESERVING SOLUTIONS

For many years the author has worked with the MIF-type methods for collecting and preserving fecal specimens. Very little information regarding these methods is available to individuals who do diagnostic parasitology. To correct this omission from the literature, some detailed information concerning the history and use of these products and methods is included here.

The original MIF solution and technique were developed by Sapero and Lawless (1953) at the Naval Medical Research Unit Number 3 in Cairo, Egypt for collecting fecal samples in locations remote from the laboratory. The collecting/preserving solution consists of two components, an MF stock solution and Lugol's iodine (see Appendix for preparation). The iodine is added to the MF stock solution immediately before adding the fecal specimen. The MIF solution effectively kills, fixes, stains, and preserves trophozoites and cysts of protozoa; eggs of trematodes, cestodes, and nematodes; and juveniles (larvae) of nematodes present in fecal specimens. These properties allow the investigator to collect and preserve the specimen in the field, examine it at any time after collection, and still find and identify all stages of parasites present. The specimen can be examined over and over again, until it is exhausted. Because they are stained in the bulk specimen, protozoa can usually be identified without further procedures. In addition to being a good diagnostic tool, the specimen fixed in MIF can also be used effectively for field surveys, teaching, and proficiency testing.

The original procedure was modified by Price in 1958, at the 2nd U.S. Army Medical Laboratory. In 1960 it was introduced into East Africa, and from there to England and Europe, where it was modified slightly and designated TIF. The method was employed by a number of investigators but how the various procedures were performed may have varied. Blagg et al. (1955) developed a concentration procedure. Dunn (1968), working in Malaysia, found the method efficient for field studies of prevalence and for estimation of worm burdens applied to hookworm, *Trichuris*, and *Ascaris* infections. In 1977, Marion Laboratories, Inc., in collaboration with Price, produced a commercial product, the Culturette Brand MIF Procedure Kit (Price, 1978). With this development, the procedure to be employed was standardized. Price (1981) compared the commercial product with PVA/Formalin and reported a higher parasite yield and reduced working time. The commercial product has been successfully marketed since that time and is being used in many private, commercial, and hospital laboratories. In some laboratories, especially those having personnel with former U.S. Naval Laboratory experience, the stock solution is prepared in-house (see Appendix).

In 1977, a modified formalin-ether sedimentation method for MIF-preserved specimens was published (Price, 1977), and at the meetings of the ASM in 1980, a method for preparing permanently stained fecal films stained with trichrome stain was introduced. In 1980, the College of American Pathologists evaluated the MIF collecting/preserving solution and method and acknowledged its quality in "...providing a stained specimen with minimum preparation time," among other claims.

Over a period of several years, the author tested a variety of modifications of the original MIF stain/preservation product, with the objectives of reducing the mercury content while improving the overall properties of the product. In September 1987, a 510K was obtained from the Food and Drug Administration (FDA) for the TIF Fecal Test, a modification using the same principles that were the basis for the MIF procedure. Specific chemicals and dyes were substituted for Lilly's merthiolate, which resulted in a reduction in the mercury content by 40%. The product was modified again and a new substitute chemical formula was developed early in 1989, one that used no heavy metal. In initial studies, the product appeared to be comparable in all ways to the original MIF. It was tentatively designated mercury-free TIF, and later, PIF.

The principles on which the MIF-type products are based include quick fixation, gravity sedimentation, and staining parasites in small bulk fecal samples. None of the other presently available products has all these qualities. The PIF modification developed in 1989 is based on the same principles as its predecessors, but has the advantage that it contains no mercury or other heavy metal.

ENVIRONMENTAL CONSIDERATIONS

In selecting a product for use in the laboratory, the problem of disposal and its impact on the environment are important considerations. In certain parts of the country, especially Florida and California, mercury is considered to be one of the most important and dangerous pollutants and presents a major problem in fresh water supplies. The quality of the MIF product is not in question; however, commercial MIF kits and bulk packs purchased from a dealer and the MF stock solution prepared from the original formula contain thimerosal, which is 49.49% mercury. When the other chemicals and water are added, the final stock solution contains 0.2% mercury. To bring the 13 mL of solution in a single vial to the approved level of mercury in water requires dilution with 1500 mL of water.

Other products contain much larger amounts of heavy metals. For example, Schaudinn's fixative and PVA contain very large amounts of mercury, and PVA is made up with polyvinyl alcohol, which makes disposal a major concern.

Because of the mercury content, it is possible that PVA and the MIF-type products containing mercury, and possibly products using other heavy metals, will be restricted or possibly even excluded from general laboratory use. Although disposal has been poorly monitored in the past, it is gradually coming under more strict scrutiny.

Even though free of heavy metals, other collecting/preserving solutions (i.e., formalin, SAF, PAF, and PIF) contain hazardous chemicals; they also should be disposed of in a proper manner. Formalin, which is in all of the above-listed collecting/preserving solutions, is considered to be carcinogenic and should be handled appropriately. At this writing, there is a pressing need for clear, concise information on all aspects of the application of these products and techniques and procedures that relate to them.

APPLICATION OF COLLECTING/ PRESERVING METHODS

In hospitals, samples of fecal specimens may be collected in an appropriate container and a sample may be preserved in one of the collecting/preserving solutions on the hospital floor or, as an alternative, after the specimen reaches the laboratory. In outpatient clinics and physicians' offices, specimens may be preserved on site or kits may be given to the patient to use at home and bring or send to the laboratory (see Figure 1 in Chapter 1). When a preserved specimen reaches the laboratory, regardless of the locality where the specimen was preserved, personnel can fit the necessary processing and examining activities of a preserved specimen into a predetermined schedule. On the other hand, unpreserved specimens require immediate attention when they reach the laboratory in order to ensure accurate results; therefore, the work cannot be prescheduled. As a general rule, if the fecal specimen will not reach the laboratory within 1 hour after it is passed, provisions should be made to preserve a portion of the specimen in an appropriate fixative at the location where it is passed and at the time it is passed (see Chapter 5).

In field surveys, specimens may be fixed and preserved where the patient is located and the fixed specimen returned to the laboratory for subsequent examination. The collecting/preserving solution selected should be applicable for use in all circumstances where the procedure calls for a fecal specimen to be preserved for subsequent examination for intestinal parasites.

It is often not the quality of the product or the procedures that determines the quality of the results obtained but the experience, training, and conscientiousness of the individual employing the procedures. The individual who is comfortable with the procedure being performed usually does it better. The information included in this manual is intended to enhance the individual's ability to obtain good results.

The factors impacting on specimen quality that relate to collecting and preserving fecal specimens apply regardless of the collecting/preserving solution used. Optimum results can be expected only when the specimen is properly collected and processed, which includes dealing with the specimen as soon as possible, an appropriate ratio between the volume of the solution and the amount of feces added, and adequate mixing with the solution. When products are purchased commercially, directions for proper use of the product should be provided and the directions provided should be followed. When there are questions, the technical services of the supplier should be contacted for answers.

The Author

Donald L. Price, Ph.D., received his Doctorate in Zoology from the University of Maryland, with a specialty in Medical Zoology (Thesis Title: Epizootiological Studies on Some Filarioid Parasites of the Family Dipetalonematidae (Nematoda: Filarioidea) Found in Certain Small Mammals). He has broad experience as a scientist, consultant, and educator. For 22 years, Dr. Price served in the United States Army, first as a pilot in WW II, and for 18 years as a Medical Zoologist (parasitology) in the Medical Service Corps. He retired in 1968 as a Lieutenant Colonel.

During much of his military career, he was associated with Walter Reed Army Institute of Research and the Armed Forces Institute of Pathology. In-depth research activities included projects related to laboratory diagnosis and diagnostic techniques, disease transmission and ecology, host–parasite interrelationships, and epizootic diseases. As a part of his research, he discovered and named five new species of filarioid parasites. Studies concerned with human disease included the impact of physical, chemical, biological, and cultural factors on disease, particularly in third world countries; and were directed toward filariasis, malaria, onchocerciasis, schistosomiasis, and intestinal parasitic diseases.

Since retirement from the military, he has engaged in a variety of activities, including planning and implementing programs for service and research in science, health, and related fields. He participated in prearchitectual program planning for the University of Pennsylvania Medical School and for a number of other medical and laboratory facilities. He developed a new approach to primary health care delivery, including the development of a facility in Appalachian Pennsylvania.

Dr. Price has been a consultant to several corporations, organizations, and institutions. He worked as a consultant and provider for over eleven years with Pinellas County Health Department in St. Petersburg, Florida, initially in the development of the parasitology program for Southeast Asian refugees, and later in carrying out the diagnosis of parasitic diseases as a part of the program. Dr. Price was appointed as a consultant in parasitic diseases to the Bay Pines VA Medical Center, Bay Pines, Florida in 1980, and has continued in that capacity to the present time. While continuing his consulting activities, Dr. Price initiated and headed a small research and development company for several years.

Dr. Price has developed a number of new products for diagnostic parasitology used in the medical and veterinary fields. He organized and participated in many workshops and courses on diagnostic parasitology, especially for continuing education. He has lectured extensively in the United States and in several foreign countries, and has published many scientific papers and book chapters.

He continues his interest in parasitology, especially in the diagnosis of intestinal parasitic diseases and the development of new and/or modified diagnostic methods for parasitology. He has made available to other workers several quality control materials. He is a Biological Resource Specialist in the Natural Science Department, Bradenton Campus, Manatee Community College. He continues to be an active consultant to a number of area hospitals.

Contents

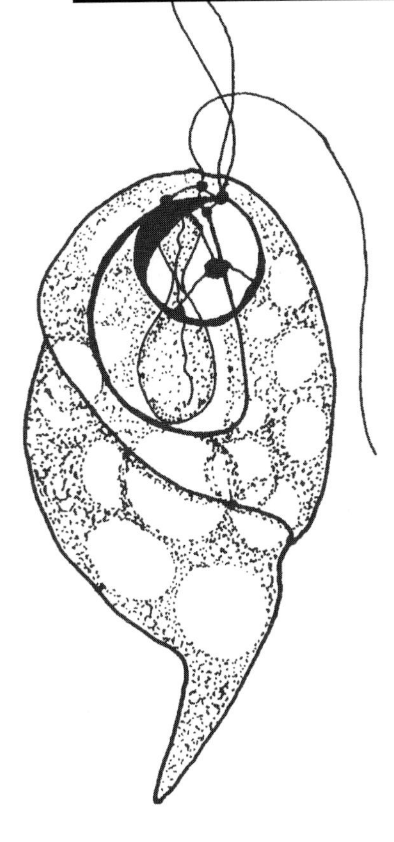

Procedure Manual for the

Diagnosis of Intestinal Parasites

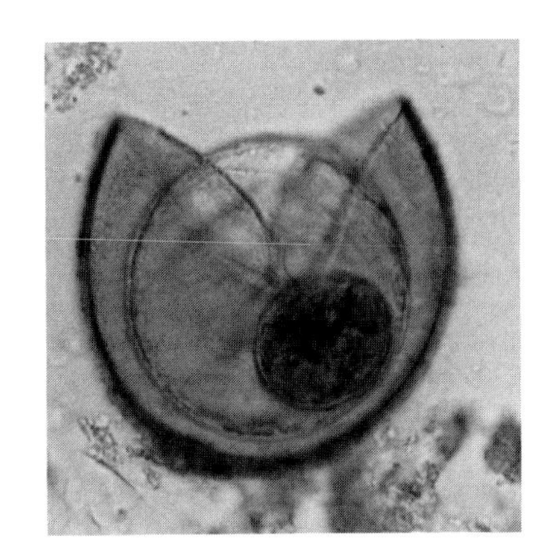

PART 1

Chapter 1

The Specimen

FECAL SPECIMENS IN THE CLINICAL LABORATORY

The clinical diagnostic laboratory performs an important and essential function in the diagnosis of clinical illness. Its work is primarily related to performing tests on various kinds of body fluids and excreta (the specimens), which include blood, urine, feces, pus, nasal discharge, gastric fluids, etc. Regardless of the kind of test to be performed in the clinical laboratory or the methods to be used, the first requirement is collecting the specimen.

With a few exceptions, specimens for laboratory examination or testing are obtained by medical or allied support personnel. The patient may provide a sputum or urine sample but usually support personnel are on hand to deal with the specimen. Other means may be used to collect the specimens, such as bronchial washings or collection with catheters. Feces, however, is unique among laboratory specimens in that neither the patient, physician, nor support personnel can control exactly when the specimen will be passed normally, its physical quality, and/or how much specimen will be available for laboratory tests. For clinical parasitology, normally passed stools are generally considered to be preferable (Stamm, 1957; Goldman, 1964; Melvin and Brooke, 1982). Procedures such as enemas can have a negative effect on specimen quality especially in patients with protozoan infections.

Visual examination of fresh feces immediately after it is passed can yield information that cannot be gained by other means. Form, color, consistency, quantity, and odor are easily observed and may be useful. One may observe mucus, frank blood, or foreign bodies in the specimen. It is good practice for the person collecting and processing the specimen that will be sent to the laboratory to make a note of such findings to be entered on the patient's chart and to accompany the specimen to the laboratory.

Chemical tests, such as total fat, bile pigments, nitrogen content, and trypsin, are performed on unpreserved fecal specimens. When the attending physician orders such tests, the entire stool, or what is called for in the protocol for the specific test, should be sent promptly to the laboratory.

FACTORS HAVING AN IMPACT ON SPECIMEN COLLECTION

Some Important Factors

Medications and Procedures

Many substances can render a fecal sample unsatisfactory for examination. Urine and tap water can destroy trophozoites of some protozoa. Many compounds and drugs are antagonistic to organisms that may be present and, although they may have no therapeutic value against the specific parasites, the drugs may temporarily reduce the numbers present in the feces to below a level that can be readily detected. This is particularly true of some antibiotics and antidiarrheal compounds. Whenever possible the specimen should be obtained before the patient has been given medication.

In addition to medication, a number of diagnostic procedures make the specimen unsuitable for examination. Radiologic examinations of the gastrointestinal tract or gall bladder where barium (Juniper, 1962) or radiopaque dyes are used (Melvin and Brooke, 1982) are examples of such procedures. The impact of antibiotics and/or diagnostic procedures may last up to 3 weeks after the course of treatment or the procedure has been completed.

Multiple Uses of Specimens

Methods used should provide the most effective use of the specimen and yield the most accurate results. For bacteriology, a two-swab transport system and for occult blood, the Smith–Kline occult card are commonly used. One of the generally accepted collection/preservation solutions should be used for collecting, fixing, and preserving parasites and/or eggs of parasites in fecal specimens.

Protection of Personnel

If the specimen is to be transported, it is especially important that it be processed (fixed) or sufficiently contained immediately after passage for the protection of those individuals who handle the container and could come into direct contact with the specimen. This is especially true with the ever-increasing risk of exposure to hepatitis B virus (HBV) and immunodeficiency virus (HIV). There is always the possibility of an accident that will bring individuals into contact with an unpreserved specimen. This can best be avoided by *immediately preparing transport materials, preserving a part of the specimen for parasitology, and discarding the remaining part of the specimen.* If the unpreserved specimen is to be transported, it must be properly contained (see Part 1, Precautions and Appendix, Safety in the Laboratory).

Immediate Transport and/or Preservation

The results obtained when performing some diagnostic procedures may be greatly affected if the specimen is allowed to sit at room temperature for even 30 minutes. Bacterial enzymes and other chemicals present in a fecal specimen may have an adverse impact on what is found on laboratory examination, especially for some bacteria and for parasites and eggs. Trophozoites of protozoa may be destroyed or rendered unidentifiable (Burrows, 1967b).

PROCEDURE FOR SPECIMEN COLLECTION FOR PARASITOLOGY

The conditions under which a stool specimen is passed largely determines who should collect and process the specimen. The routine established in the laboratory for intestinal parasites will be developed based on whether the laboratory receives preserved or unpreserved specimens. The laboratory staff should present a strong case for receiving preserved specimens.

Figure 1 shows the routing of the specimen from various points of origin (see Introduction). Only the patient knows when a specimen will be passed naturally without laxatives or enemas. In any situation, whether in a doctor's office or clinic, the patient should be instructed on how to collect the specimen and be given a proper container. A staff person should accept the container from the patient, note the physical characteristics of the specimen, and immediately process (preserve) a part of the specimen. If a fresh portion of the specimen is needed, it should be properly contained to prevent any contamination.

If the patient is to collect and process the specimen at home, a staff member should provide the patient with the necessary materials, thoroughly instruct the patient on how to collect and process it, and provide written instructions. Whenever possible, the patient should be asked to note any unusual physical characteristics seen in the fecal mass. If objects are seen in the feces, the preserved sample along with the remainder of the specimen properly contained should be brought or sent to the laboratory or physician's office.

On the hospital or nursing home floor, a trained staff person should accept a bed pan or container from the patient and immediately process the specimen. In such situations, the nursing staff is usually responsible for either collecting and processing the specimen or instructing other personnel to do so. Because of the constant change of nursing staff and support personnel in many hospitals, some members of the nursing staff are uninformed. This is an area where the system for training must be especially well planned if fecal specimens are to be properly prepared and reach the laboratory in satisfactory

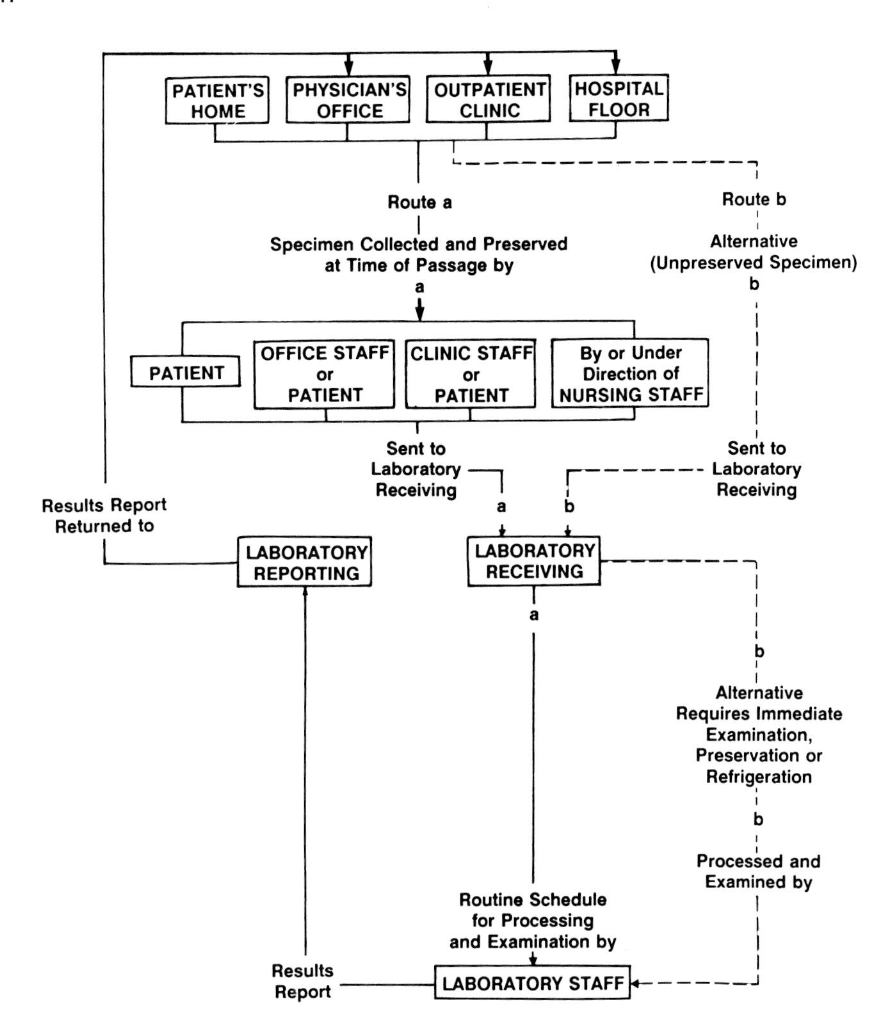

Figure 1. Sources and routing of patients' fecal specimens. The various types of localities that function as sources of fecal specimens, the localities where a collecting/preserving solution may be prepared and who may prepare it, and the routing to the laboratory where the specimens are processed, examined, and the results reported for both preserved and unpreserved specimens are shown in the diagram.

condition to perform the tests requested. In some hospitals, the nursing staff is very cooperative and quickly accepts responsibility for collecting and preserving fecal specimens while in others nurses find it difficult to accept the idea of handling feces as a part of their routine. The laboratory director should accept the final responsibility for developing the format of the system to be employed (see the chart depicted in Figure 1).

Multiple Specimens

Multiple specimens are recommended when fecal examinations are submitted for the diagnosis of parasitic infections (Sawitz and Faust, 1942; Swartzwelder, 1952; Alicna and Fadell, 1969). Some parasites, such as *Ascaris*, *Trichuris*, and hookworms, pass eggs on a relatively regular basis and usually eggs can be found in the feces of an infected individual on any day. Other parasites, especially the protozoa *Entamoeba histolytica* and *Giardia lamblia*, which are potentially pathogenic, are passed intermittently at intervals that may be as long as 7 days apart (Marsden, 1946; Rendtorff, 1954; Stamm, 1957) and prepatent periods may be even longer (Wolfe, 1978). Specimen collection should be made at 2- or 3-day intervals to increase the chances of finding parasites that are passed intermittently.

When a patient has multiple infections, depending on which species are present and their numbers, examination of a single, normally passed specimen may yield only about one third to one half of the species present (Sawitz and Faust, 1942; Melvin and Brooke, 1982). When only a single fecal specimen is ordered, the likelihood of a negative report is increased. A "No Parasites Found" report on a single specimen received by the requesting physician may lead him in another direction which could result in failure to diagnose an important parasitic infection. If one or more parasites are found on a subsequent examination performed at another laboratory, it could invite criticism. In the clinical medical environment, the emphasis should be on diagnosis and patient care.

Some investigators and health care providers disagree and suggest that the submission of multiple fecal specimens is unnecessary for the recovery of enteric parasites (Montessori and Bischoff, 1987; Senay and MacPherson, 1989). Because of cost containment, they suggest that subsequent specimens need to be submitted only when no parasites are found initially but the clinical episode persists. Reviews of their laboratory records showed that more than 90% of total species of parasites found were present on the first examination but the authors do not state which parasites were and which were not found on the first examination.

In a series of 670 positive cases of giardiasis, Wolfe (1978) reported that on examination of the first specimen, 76% of the positive cases were found, 90% after the second, and 97.6% after the third. From personal experience, *Entamoeba histolytica* and *Giardia lamblia* are the protozoan parasites most frequently undetected in the first specimen submitted. Among the helminths, the eggs of tapeworms *Taenia* spp. and especially *Hymenolepis nana*, are often undetected on examination of the first fecal specimen submitted but are found on examination of specimens submitted subsequently.

Other authors have suggested that multiple specimens should be collected but pooled for examination to reduce technician time (Peters et al., 1986). In cases of amoebiasis and giardiasis, parasites may reach numbers that are readily detectable only once in 5 to 7 days. Pooling 3 specimens taken over a 5-day period, of which only one may have numbers of parasites that were high enough to be detectable, can reduce the chances of finding the parasites. If specimens with numbers of parasites or eggs below the usually detectable level are added to those with low numbers but high enough to be detected, the number of parasites per unit may be reduced below the usual detectable level. Technician time may be increased if the patient's symptoms persist and it becomes necessary to request additional specimens. Improving efficiency through better training for personnel and using more effective methods is the better way to reduce technician time (see Part 3, Chapter 5).

When a patient presents with an intestinal upset and a clinical diagnosis is not evident, some physicians tend to initially treat the condition before any kind of fecal examination for either bacteria or parasites has been performed. In cases where the first therapeutic course of treatment is unsuccessful, subsequent examination of a fecal specimen for parasites is often negative even though the parasite(s) are still present. One case in point is of a *Hymenolepis nana* infection where the patient was treated before a fecal examination was performed; the infection was not diagnosed until 45 days after the initial visit to the physician's office, after three additional visits, three courses of medication, and three fecal examinations. In addition, the patient developed an intestinal yeast infection for which treatment was necessary. Such cases are not unusual when patients have a parasitic infection.

Only the patient's attending physician can request an examination of a fecal specimen for intestinal parasites and it is the attending physician who finally decides the number of specimens that will be sent to the laboratory for examination. More often than not, the specimen request for eggs and parasites accompanying the specimen has no patient information, such as foreign travel or possible exposure, that might be helpful in diagnosis. The more the physician knows about diagnostic parasitology and the greater the communication with the laboratory, the greater the possibility of obtaining optimal results.

Good practice might include collecting normally passed stools on Monday, Wednesday, and Friday, preferably before the patient has been treated, which should greatly improve the chances of finding parasites present over the single specimen collected, or one collected after some special procedure is performed on the patient. A general discussion of specimen collection is presented by Melvin and Brooke (1982).

Planning for Fecal Specimen Collection

Ideally, a plan that takes into consideration as many of the variables as possible will probably yield the best results and provide the best patient care which is the objective of any diagnostic program. The importance of establishing a procedure for all aspects of collecting and preparing the fecal specimen cannot be overemphasized. If optimal results are desired and expected, proper collection and preparation of the specimen is paramount. The selection of suitable methods for collecting the specimen is the first step in good planning.

Fecal specimens are rarely submitted as STAT specimens and it is relatively unusual for a hospital laboratory to get requests from the emergency room for a fecal examination for eggs and parasites. Laboratory personnel working at night or on weekends should know what collecting/preserving solution to use, where it is located, and how to properly deal with a fecal specimen that arrives at the laboratory during their shift.

Specimen samples should be preserved as soon as possible after passage since the main objective is to preserve the integrity of the parasites as close to their *in vivo* condition as possible. Although a portion of a fresh fecal specimen can be placed in a collecting/preserving solution after the specimen has reached the laboratory, any delay in preserving or examining the specimen may affect the results (see Introduction).

In a hospital environment, a special plan should be established for dealing with fecal specimens for parasitology. A protocol can be developed by the laboratory director and personnel in direct contact with the patient. The protocol should give uppermost thought to patient care and take into consideration safety, efficiency, and cost. It is especially useful to involve a gastroenterologist, an infectious-diseases specialist, and a nursing-staff representative in discussions with the laboratory director and key staff members when developing a protocol to insure that individuals from various disciplines have knowledge of the plan and that the information is adequately disseminated. When private-practice physicians are involved, consultations between the participating physicians, the laboratory director, and key laboratory staff members may be scheduled.

A plan should establish a standard operating procedure for collecting and immediately processing a fecal specimen and transporting it to the laboratory for the usual laboratory tests in bacteriology, parasitology, and the presence of occult blood, as well as for special tests less-frequently performed. If fecal specimens for intestinal parasites and eggs cannot be obtained at the site where they will be examined, and they cannot be preserved immediately, some special arrangements should be in place to insure immediate transport to the laboratory where trained personnel can process and examine them. Proper handling instructions and an outline of routing to the laboratory should be at those sites where fecal specimens are collected.

Simply having written instructions and procedures to follow is not adequate. There should be a way of checking to make certain the instructions are being followed properly along with a method for training individuals who are newly introduced into the network of people responsible for specimen collecting and handling. Follow-up and training are usually the weakest links in a good operating procedure, especially when more than one department and/or discipline is involved. Clear, concise, and complete instructions should be available at each station along the route to the laboratory where the final examination will be performed.

FACTORS HAVING AN IMPACT ON SPECIMEN PREPARATION

Collecting a Proper Specimen

As there are many factors that have an impact on the quality of the parasites and eggs found in fecal specimens, there are others that have an impact on preparing the specimen that is submitted to the laboratory for examination. How and where the specimen is prepared play a role in the final results obtained. Some of the most important factors are listed below.

- A clearly written set of instructions for collecting and preserving the specimen should be available.
- A sample of the fecal specimen should be added to the vial and thoroughly mixed with the preserving solution as soon as possible after passage. If MIF-type solutions are used, Lugol's iodine is added immediately before the fecal specimen is added to the vial (see Figures 2 and 3).
- If the fecal mass is formed, some softer, inner portions should be selected along with some of the specimen from the outer surface.
- When a stool specimen consists of only very hard lumps, several lumps should be crushed in the collecting container and portions of these should be added to the collection vial.

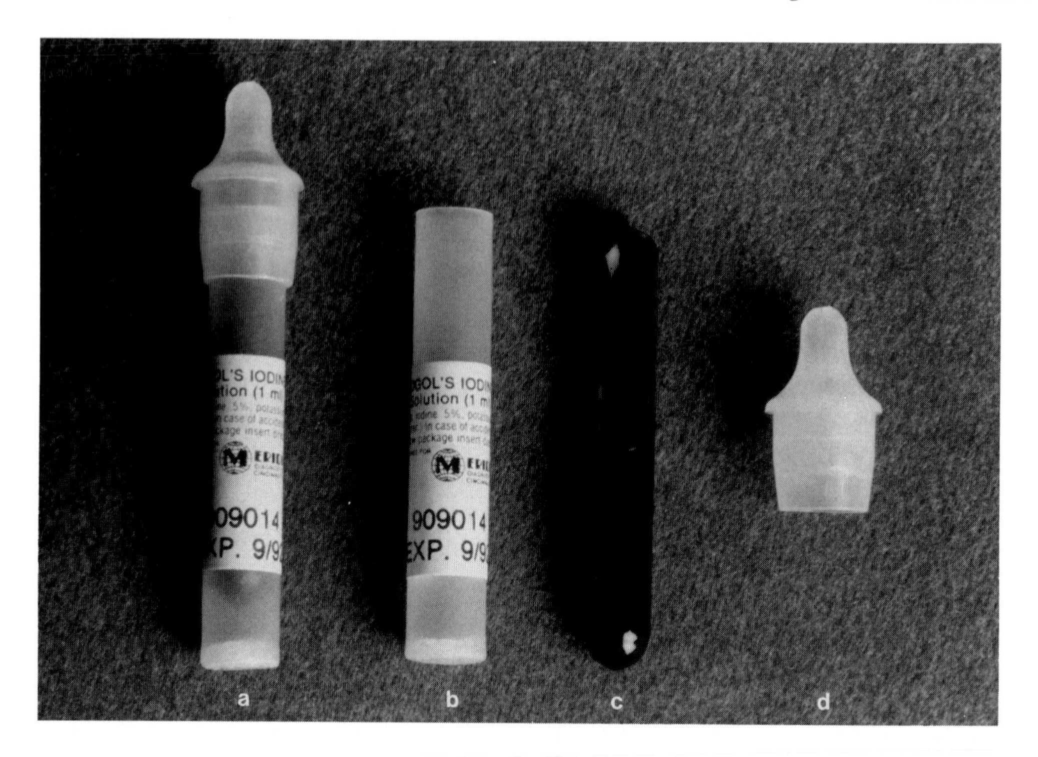

Figure 2. Lugol's iodine ampoule. The MIF-type methods (including PIF) require the addition of 1 mL of Lugol's iodine to the MF (merthiolate-formalin) stock solution immediately before the sample of feces is added to the vial. Since Lugol's iodine is unstable in air and light, a method was developed to seal the iodine under an inert environment in a glass ampoule, which stabilizes the iodine and gives it a long shelf life. The photograph shows a commercially made, intact Lugol's iodine ampoule and its components: (a) the assembled dropper for delivery of the iodine, (b) a plastic sleeve in which the ampoule is encased, (c) a glass ampoule containing Lugol's iodine, (d) a plastic tip with hole (pore opening) through which the iodine is dispensed.

- If the fecal mass is soft or unformed, the selection is less critical but the portion added to the vial should be the proper amount and representative of the entire stool specimen.
- A slightly larger portion of a liquid specimen (2 to 3 mL) may be poured directly into the vial since picking up a portion may be difficult.
- The portion added to the vial will vary according to the amount of the liquid in the collection vial, but is usually about 1 to 2 g, or about the size of a marble or first joint of the little finger.
- The maximum level of material in the vial after the fecal sample is added may be marked on the outer part of the vial and this level should not be exceeded. If the level is not marked on the vial, the portion to be added should not be greater than one-

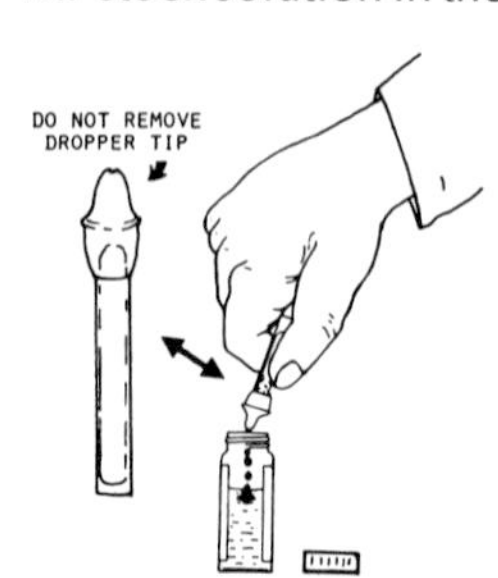

Figure 3. Adding Lugol's iodine to the specimen vial. Lugol's iodine should be added to the MF stock solution in the vial just before adding the feces when using the MIF-type collecting/preserving solutions. To add the iodine, invert the dropper so that the hole (pore opening) in the dropper tip is over the mouth of the specimen vial, squeeze the plastic sleeve of the dropper to break the glass ampoule inside, and gently squeeze the dropper several times to force out all of the iodine. As soon as the iodine is mixed with the stock solution, the proper amount of feces should be added to the vial.

fourth of the total volume after the specimen is thoroughly mixed with the solution. A ratio of one part feces to four or five parts fixing solution is optimal. Too much specimen added to the vial does not allow proper fixation. Too little feces does not provide a sample adequate for examination and parasites present in the fecal mass may not be present in the sample.

Mixing

The importance of thoroughly mixing the fecal specimen with the collecting/preserving solution cannot be overemphasized. Most protozoa are extremely small and the sooner they come in contact with the fixing solution the better they are fixed and preserved. If the solution has to penetrate a large fecal mass to reach the organisms, fixation may be delayed. More than 50 trophozoites of *Giardia lamblia* or *Entamoeba histolytica* can fit on the head of an ordinary pin, so several trophozoites in a fecal particle much larger than a pinhead might not come into immediate contact with the fixing solution. Therefore, the fecal mass should be vigorously broken up and mixed in the solution so that even the smallest organism comes into direct contact with the fixing solution. Breaking the particles to the minimum size possible, which is the objective of mixing, is usually possible by pressing lumps or larger pieces against the sides of the vial.

Usually, mixing is accomplished using three wooden applicator sticks. Any apparatus that enhances mixing (except sonic mixers that destroy the parasites) is helpful. If about 18 glass beads, 4 mm in diameter, are added to the vial, it greatly helps in mixing the feces with the solution when the vial is shaken. The best procedure is to select the portion of the fecal sample to be added, open the vial (add iodine if appropriate), add the fecal sample, and begin mixing immediately. The vial is then tightly capped and shaken vigorously (see Plate 1).

WRITTEN INSTRUCTIONS FOR SPECIMEN COLLECTION

What is the purpose of written instructions for collecting and preserving a fecal specimen? If it is important to have a specimen with well-preserved parasites, then written instructions are important, particularly when a patient will collect and preserve the specimen. Communication with patients can be very difficult especially with those who are ill. They may not pay close attention to the person giving instructions and may not understand or remember them. The only way to be certain that the person being instructed has the information needed is to provide written instructions. The instructions will vary with the type of collecting/preserving solution and the situation under which they are given.

A copy of instructions that were used for obtaining specimens from HIV-positive patients is given in Figure 4. In this case, the patient was given the instructions for collecting and preserving the specimen in MIF after an initial visit with a physician. Because the time of the visit with the physician did not always correspond with the time the specimen was available, most specimens were collected and preserved by the patient at home and later brought to the laboratory. Approximately 90% of the specimens were preserved correctly. It is advisable that instructions be prepared by personnel directly involved with the patients and with the examination of the specimen since conditions in which specimens are collected vary greatly. No matter how well the instructions are prepared, some of the specimens that reach the laboratory will not have been properly preserved.

Precautions

Avoid Direct Contact with the Specimen

Individuals handling fresh specimens should protect themselves from contamination by using gloves and protective clothing and should make certain that their hands, clothing, or surroundings do not become contaminated. Areas that become contaminated should be cleaned with an appropriate cleaning solution selected for that purpose and contaminated clothing should be properly handled and cleaned. Hands should be thoroughly washed with an appropriate soap before and after handling the collection container or preparing the specimen vial.

If the patient has prepared the specimen, the outside of the specimen vial may be contaminated. Good practice is for the individual who receives the specimen to wear gloves and a protective apron or gown to avoid unnecessary exposure.

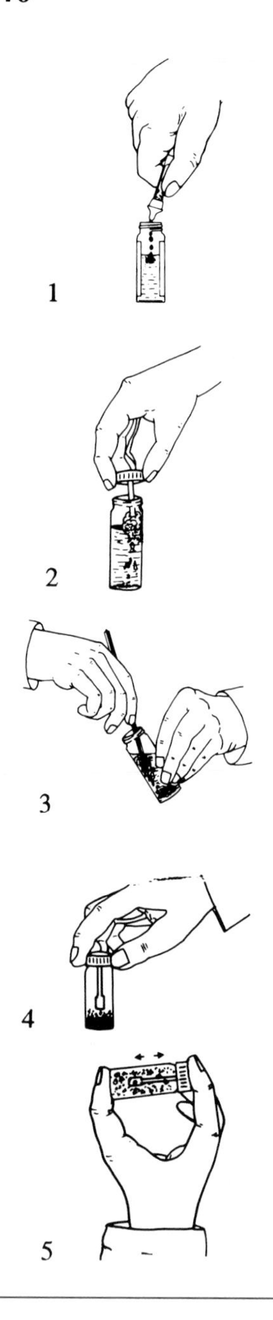

Plate 1. Important steps in collecting and preserving a fecal specimen. Before beginning, the vial should be properly labeled with the patient's identification and accompanied by a proper request to make certain that the correct procedures are run and that the results will be returned to the proper place. (1) If the collecting/preserving method is the MIF-type, add the Lugol's iodine using an ampoule as shown or by transferring 1 mL of Lugol's iodine to the specimen vial by pipette. (2) Using the spatula, transfer 1 to 2 g feces to the specimen vial (an amount about the size of a marble or the first joint of the little finger). (3) Use three wooden applicator sticks to thoroughly mix the specimen with the solution. Larger pieces of fecal material can be broken up by pressing them against the inside of the specimen vial. (4) Fasten the cap tightly on the vial. (5) Shake the vial vigorously to further break up and mix the fecal particles with the solution. The specimen, sealed in the vial, adequately mixed with the preserving solution, and accompanied by the appropriate information, should be sent to the laboratory.

After a specimen is in the vial for 1 to 2 hours, and is well mixed with the collecting/preserving solution, most of the organisms are killed and are no longer infective (see Collection/Preserving Solutions in the Introduction). In PVA-fixative and MIF-type solutions, all parasites and eggs in the vial should be killed, fixed, and preserved indefinitely; in formalin solutions, eggs of *Ascaris* are able to survive and continue development. Regardless of the fixing solution used, if the specimens are not adequately mixed with the fixing solution, some of the organisms may not have been in contact with the solution long enough to be killed and further mixing may be necessary.

Avoid Direct Contact with the Solutions

Certain precautions are advisable in handling the collecting solutions, especially when the patient will collect and prepare the specimen. All of the Collecting/Preserving solutions contain toxic chemicals and individuals who will handle the solutions should be properly instructed regarding their use. First aid instructions are usually printed on the vials and in the printed handout material accompanying the vials sold commercially. Those individuals that are involved in dealing with specimens and associated chemical solutions should know proper handling practices and emergency procedures. When appropriate, patients should be made aware of this information (see Appendix, Safety in the Laboratory).

Disposal of Solutions and Preserved Specimens

Disposal of the products is also an important consideration. Mercury is now considered one of the most hazardous of the environmental pollutants and solutions containing mercury should be discarded according to specific guidelines established by the U.S. Environmental Protection Agency (EPA). Other chemicals, especially formalin, may also require special disposal procedures. As a precaution, whenever disposal procedures for a product are not established by a laboratory, it is best to contact the local Environmental Control organization for disposal instructions.

▶

Figure 4. Instructions for collecting fecal specimens. The kind of instructions needed for collecting fecal specimens for the diagnosis of intestinal parasites varies depending on the collecting/preserving solution to be used and the conditions under which the specimens will be collected. After passage, the specimen may be added to the vial with the collecting/preserving solution by trained laboratory personnel, staff personnel in a clinic or doctors office, staff on a hospital floor, or by the patient. If the specimen is turned over to someone else for processing, the patient does not need detailed information. In each case, familiarity with the part of the process in which the individual is involved determines what information is needed. Instructions shown here were used by patients who had the responsibility of obtaining the specimen, adding it to the collecting/preserving solution, and returning it to the laboratory for examination. The collecting/preserving solution used in this case was MIF.

INSTRUCTIONS FOR COLLECTING FECAL SPECIMENS

This collection kit and instructions are designed to help you properly and conveniently collect a fecal specimen that is adequate for laboratory personnel to effectively perform the laboratory tests your doctor ordered, which can eliminate the need for costly and time-consuming retesting. The Kit consists of a screw-capped vial with spatula in the cap and containing a red/orange fluid, a plastic dropper of iodine, and three wooden sticks. **Please follow the instructions exactly.**

Collect the fecal specimen (bowel movement) in the cup provided or in a clean, dry container. A plastic cup, waxed carton, bottom half of a plastic or paper milk container, plastic bag, bed pan, or a piece of plastic stretched over the toilet bowl under the seat may be used to prevent the feces from falling into the water. **Avoid passing urine into the container.**

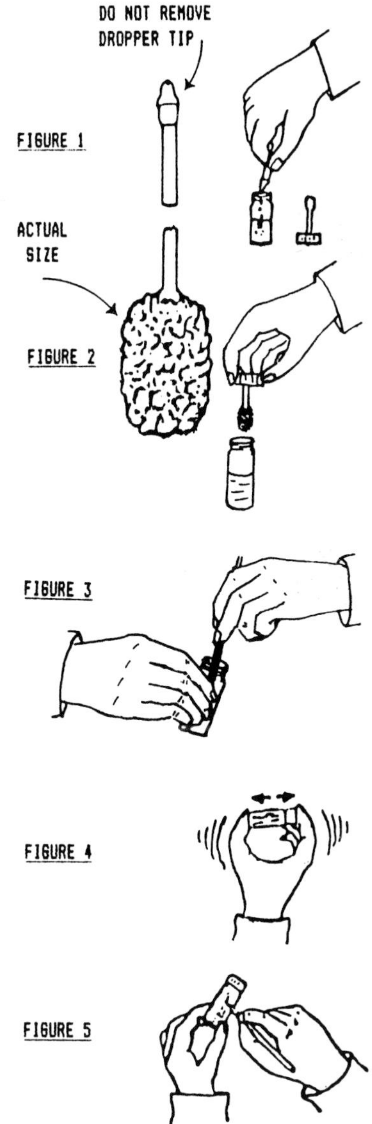

When you have the specimen, remove the cap from the vial containing the red/orange solution, hold the dropper with tip down over the mouth of the vial as in Figure 1 (**do not remove the dropper tip**), then squeeze the plastic sleeve to break the glass ampoule inside, and squeeze it several more times to force all of the iodine out of the dropper and into the vial.

Use the plastic spatula in the cap of the vial to pick up a portion of the fecal specimen. **A total portion equal to the size of the first joint of the little finger is sufficient**; see Figure 2. It is best to take portions from each end and the middle, especially if the specimen is hard. A half inch portion of a very runny specimen may be poured into the vial. **Too much feces** prevents adequate fixation and staining, and **too little feces** does not provide an adequate sample for examination. **Use the spatula to crush hard lumps in container before adding them to the vial.**

Add the fecal sample on the spatula to the fluid in the vial by pressing the spatula against the inside wall of the vial. Set the cap with the spatula aside, then use the applicator sticks to mix the feces with the solution until all particles are broken up; see Figure 3. Discard the applicator sticks with the fecal container.

Recap the vial, making certain that the cap is on tight, then hold the vial in a horizontal position, and **shake vigorously** 20 times or more to further mix the feces and the fluid; Figure 4. Write the patient's name, date, time the specimen was collected, and SSN on the label; Figure 5. After 10 minutes, shake the vial again in the same manner. **Collect no more than one specimen per day.**

Wash hands thoroughly, then fill in the blanks below. Please check the block that best describes your specimen and return the specimen and these instructions as directed.

Formed (hard)___ Soft (mushy)___ Loose (runny)___ Watery___

Note: For diagnostic use only. Do not take internally. Avoid contact with eyes or skin. Keep out of reach of children.

First aid: *Iodine*: Flush skin and eyes with water. If ingested (swallowed), give a teaspoon of Ipecac or Starch Paste (mixture of flour and water) to induce vomiting. *Red/orange solution*: Flush skin and eyes with water. If ingested (swallowed), give milk. Call your poison control center and your physician.

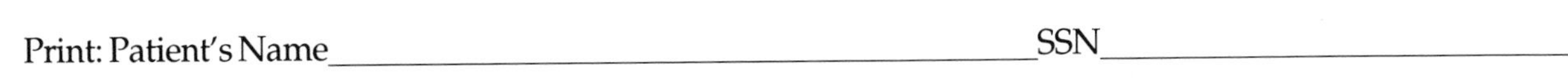

Print: Patient's Name______________________________SSN____________________

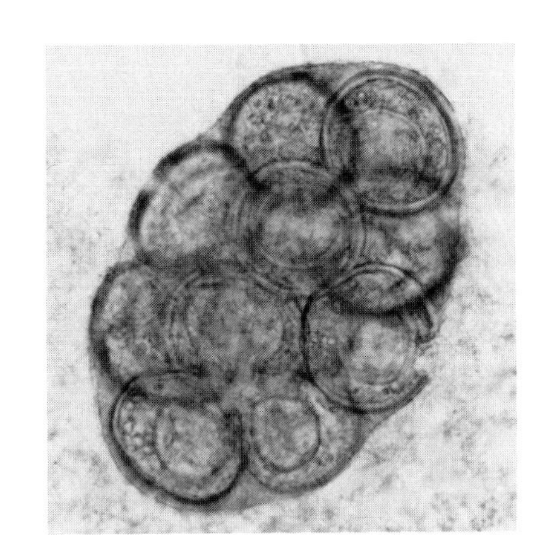

PART 2

Chapter 2

The Microscope and Examination of the Specimen

THE MICROSCOPE AND MICROSCOPY

The compound clinical microscope is an exacting instrument used to magnify and resolve fine detail within a transparent specimen (Figure 1). It is such an effective instrument that even when used improperly it can be relatively efficient, especially when used for examining tissue sections and other permanently mounted materials. Proper use of the microscope becomes more critical when it is used to examine wet mounts (see Appendix, Figure 1).

Use of the Microscope

Once the collected specimen reaches the laboratory, it will be prepared for microscopic examination by any of several methods. It is important to realize that most of the procedures carried out for parasitological examination, especially those for intestinal parasites, involve using the light microscope, making parasitology in the clinical diagnostic laboratory almost entirely subjective. The microscope, which is the most important instrument used for diagnostic parasitology, is more often than not used incorrectly.

In the past, critical illumination was used which results in focusing the light at the specimen. Critical illumination is discussed in older texts but is rarely used with modern microscopes. Modern microscopes are designed so that Köhler illumination can be used most effectively. In this system, light is focused at the aperture diaphragm of the condenser for even intensity at the specimen plane.

In 1893, August Köhler developed a system of illumination for the microscope that satisfied all the theoretical requirements for optimum light microscopy and provided homogeneous illumination from a nonhomogeneous light source. All quality light microscopes allow utilization of this system of illumination. Köhler illumination is the system that should be used for examination of all wet mounts that are examined under bright-field illumination; it is advisable to use the system for all bright-field work (see Appendix, Köhler illumination, Figure 1).

Auxiliary Condenser Lenses

Before adjusting a microscope for Köhler illumination, there are certain specific things one should know about the microscope being used. Microscope condensers often have auxiliary lenses that can be either turned into or out of the light path. Some of these are swing-out substage lenses and others are flip-top lenses. Whether an auxiliary condenser lens is to be used or not depends on the type and Numerical Aperture (N.A.) of the condenser and the N.A. of the objective being used. The manufacturers of microscopes provide instruction manuals that

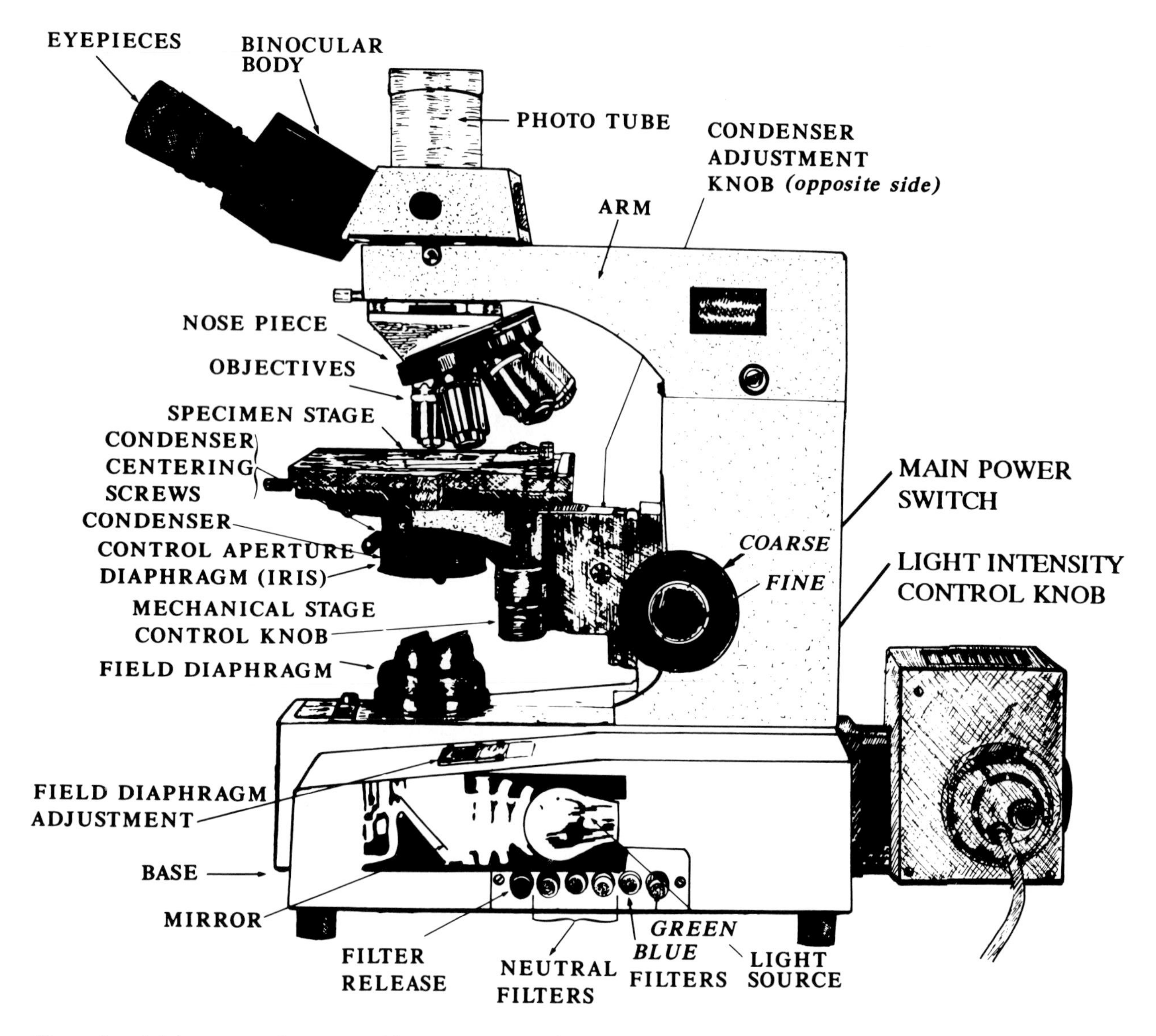

Figure 1. A laboratory microscope. The components of a microscope that are referred to in the text are named and located in the drawing. Some of the components may be located slightly differently on other microscopes but the function will be similar. Not shown is the flip-top auxiliary condenser lens at the top of the basic condenser. On the microscope shown, it should be in the light path when objectives of 10× magnification or higher are used. On other types of microscopes, an auxiliary condenser lens may be located differently. Knowing when to use the auxiliary condenser lens is especially important for the proper use of the microscope.

should be consulted to determine the condenser setting for each level of magnification (i.e., each objective). When information provided in the instruction manual is not clear, consult the microscope dealer representative to make certain when to use an auxiliary condenser lens.

Aperture Diaphragm and Numerical Aperture (N.A.)

In an optical system, the diameter of an opening through which light passes is called an aperture. The N.A. of a microscope objective is the diameter of the opening through which it allows light to reach the eyepiece. Generally, the N.A. is engraved on the objective and the condenser. It expresses mathematically the diameter of the solid cone of light passing through the condenser that reaches the specimen, or in the case of the objective, the diameter of the cone of light that passes through the front lens of the objective.

The aperture diaphragm on the condenser controls the size of the cone of light that passes through the upper lenses of the condenser, reaches the specimen, and enters the objective. The aperture diaphragm is adjustable and, to work effectively, it needs to be adjusted to (approximately) correspond to the N.A. of the objective. First, a wet mount slide should be placed on the microscope stage and the specimen brought into focus. To match the N.A. of the objective with the opening of the aperture diaphragm on the condenser, remove an eyepiece (usually the right one on binocular microscopes), look into the microscope tube, and close or open the aperture diaphragm so that the edge of its leaves are at the edge of the circle of light. This can be determined for each objective in the same manner. Most microscopes have a scale on the aperture diaphragm and it is appropriate to note and record the the number on the scale that corresponds to the N.A. for each objective.

For general use, the aperture diaphragm should not be closed more than one third of the diameter of the field of light entering the objective. The aperture diaphragm is used to control contrast. When there is too little contrast, the objects viewed lack definition. When contrast is too great, resolution is reduced. When very high contrast is needed, the one-third closure limit may need to be exceeded. The aperture diaphragm should not be used to control light intensity. The amount of light is controlled by the variable intensity control or with neutral density filters.

After using a microscope setup for Köhler illumination for a short time, recognizing when the aperture diaphragm of the condenser is properly set will be almost automatic. To check oneself when using the microscope, occasionally look at the reading on the aperture diaphragm and compare it with the number recorded to match the N.A. of the objective being used.

Light Source and Use of the Microscope

Most clinical microscopes have a fixed, precentered lamp (light source). If the microscope being used has an adjustable lamp, it must be centered before setting up the microscope for Köhler illumination (consult your microscope instruction manual).

The person using a microscope should become familiar with the components named in Figure 1 and how they function on each new microscope put into use. In some cases, the component listed may not be built into a particular microscope but can usually be supplied as an auxiliary component when needed to accomplish a certain task. For example, if there is no field diaphragm, a thin disk such as a washer, with a small opening, can be placed where the field diaphragm should be located and the edges of the central opening used to set the level of the condenser. Other innovations may be necessary to establish Köhler illumination, which should be accomplished for effective use of the microscope.

Microscope Setup for Wet Preparations (Köhler Illumination)

Very complicated procedures have been described for achieving Köhler illumination; however, with most of the modern microscopes a relatively simple procedure can be used. In setting up the microscope for Köhler illumination, the 10× objective should be used and the aperture diaphragm on the condenser should be set appropriately at the N.A. of that objective. Follow the step-by-step directions given below for adjusting the microscope. For more detailed information on Köhler illumination, see the Appendix, consult your microscope dealer, or the Eastman Kodak booklet, *Photography Through the Microscope* (Delly, 1980).

Step-By-Step Procedure (refer to Figure 1 and Appendix, Köhler Illumination, Figure 1)

1. Set the 10× objective into the light path and open the aperture diaphragm on the condenser to the appropriate diameter.
2. Set a wet-mount slide on the stage, turn on the light source, and bring an object in the microscopic field into focus.
3. Adjust the light using the variable intensity control on the microscope to the minimum comfortable level.
4. Close the field diaphragm on the microscope to almost its smallest opening. If the field diaphragm is not in focus, go directly to step 5.
5. While looking in the microscope, raise and/or lower the condenser until the field diaphragm comes into focus and the edges of the leaves on the field diaphragm are in sharp focus. (At least part of the circle of light must be visible.)
6. Use the centering screws on the condenser to bring the opening into the center of the field. Condensers on some microscopes have a sliding ring adjustment instead of screws.
7. Refocus on an object in the microscopic field and make a fine adjustment to the condenser to

center the field diaphragm and bring its leaves into sharp focus.

8. Open the field diaphragm to just beyond the field of view to provide a completely lighted field with no obstruction.

9. Adjust the contrast for your vision using the aperture diaphragm of the condenser. In most cases the diaphragm will be partially closed when using the 10× objective and reset as objectives of higher magnification are brought into the light path (see the information on the use of the aperture diaphragm presented earlier in this Part).

10. Adjust the intensity of the light using the variable intensity control on the microscope. If the range of light intensity is indicated (usually by a green band) with a limit (usually red), do not increase the intensity to beyond the recommended level (into the red area) or the filament of the bulb may be burned out. If the light is too bright, it may be necessary to use neutral density filters to obtain optimum light intensity.

Use the minimum light necessary to clearly see objects in the microscopic field. Such settings help to prevent eye strain when using the microscope for long periods of time. The microscope is now properly adjusted, ready for examination of wet mounts, and should require only small adjustments of the aperture diaphragm and light intensity control for proper illumination. Do not adjust the condenser to control the light but retain Köhler illumination for all microscopic work. Occasionally, the edges of the leaves of the field diaphragm should be brought into view to make certain that the condenser remains centered and in focus.

USING THEIR REFRACTIVE INDEX TO FIND PARASITES AND EGGS

Once the microscope is properly adjusted for Köhler illumination, the advantages can be seen. Each material (object) has a specific index of refraction (see Appendix, Index of Refraction). As the fine focus adjustment on the microscope is moved to bring the objects into and out of focus, a line of light moves through the objects in the field. If an object in the field has a higher index of refraction than the surrounding fluid, as light rays pass through it they are refracted so as to produce a slight convergence just above the object. When the microscope is focused slightly above the object, a band of concentrated light can be observed.

Place a wet mount containing cysts of a small protozoa such as *Endolimax nana* or *Giardia* on the microscope stage. Using the 10× objective and partially closing the aperture diaphragm (about ⅓ to ½ depending on the microscope) bring the object into focus. As the objective is raised slightly and the object gradually goes out of focus, a band of light can be seen moving from the edge of the object toward its center. This area of concentrated light is called the Becke Line. The higher the index of refraction of the object in relation to that of the surrounding fluid, the brighter the light (Ehlers, 1987) (see Appendix, Becke Line). As the light rays move up through a protozoan on a wet mount, it comes into focus, then the edges brighten (at the Becke Line), and light moves toward its center as the protozoa goes out of focus. So, as the light goes back and forth through it, an object brightens and darkens (dark to light to dark). When the fine-focusing adjustment is moved rapidly to bring various objects in and out of focus, some will appear to "flash" (brighten more than others) which is dependent on their index of refraction. This phenomenon allows the viewer to distinguish between different objects in the microscopic field at relatively low magnification. As the fine focusing adjustment is moved to bring the plane of light through them, most parasites with relatively high indexes of refraction, appear to brighten more than most plant or other materials with lower indexes of refraction. Thus, when examining a wet mount, move the fine adjustment rapidly, watch for the "flash", and find the parasites more readily.

Magnification needed to identify an object depends on its size and on the experience and training of the microscopist. Find the parasites at low power (scanning magnification, usually 100×) and use higher power to make the identification.

EYEPIECE MICROMETER

Measurement of a parasite or egg is often essential for accurate identification. Therefore, another component needed for microscopic examination of parasites and eggs is a calibrated eyepiece scale referred to as an eyepiece micrometer. The eyepiece micrometer is a flat, glass disk that has a line scale etched into the surface usually with 50 or 100 divi-

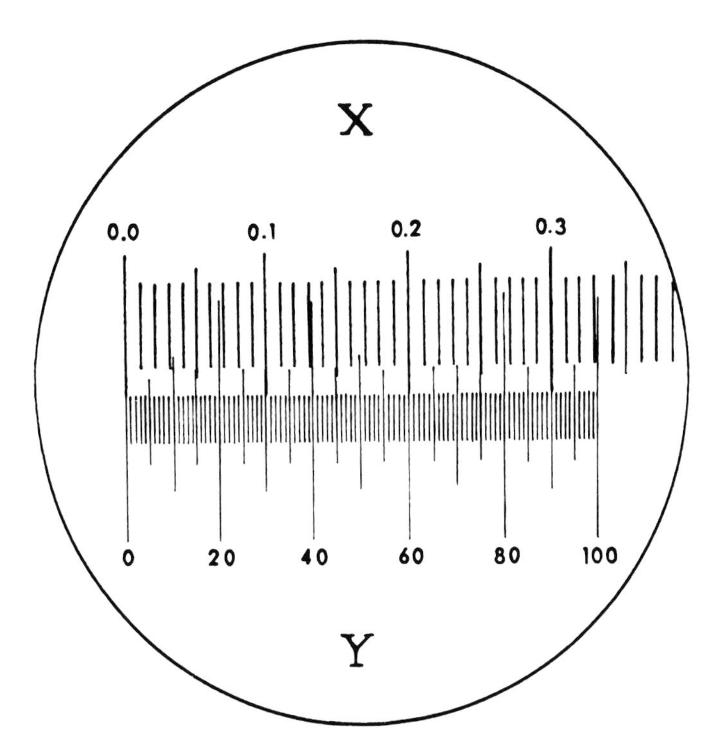

Figure 2. The eyepiece micrometer.The figure shows the stage (slide) micrometer (X at the top) and the eyepiece micrometer (Y at the bottom) lined up for calibration. The point furthest to the right (the highest numbers) where the two scales match determines the calibration factor. In this case, 90 eyepiece units are equal to 0.3 units on the stage micrometer scale. To determine the calibration factor, divide 0.3 by 90 to get millimeters (0.3/90 = 0.0033) and multiply by 1000 to get microns (1000 × 0.0033 = 3.3). With the eyepiece and objective used in the example, each unit of the eyepiece micrometer is 3.3 μm. The value of a unit of measurement of the eyepiece micrometer for each objective should be determined so that accurate measurements can be made at all magnifications (see text). [The illustration is from "Use and Care of the Microscope" (St 27-7/85) published by Reichert (now Leica, Inc.) and appears on page 16. Used by permission of Leica, Inc.]

sions. It fits into the eyepiece of the microscope. It is calibrated by using a stage (slide) micrometer placed on the stage of the microscope that has a calibrated scale of 0.1- and 0.01-mm divisions. In the calibration procedure, the scale of the eyepiece is superimposed over that of the stage micrometer so that the "0" points are aligned as shown in the diagram (Figure 2) with X representing the stage micrometer and Y representing the eyepiece scale. The object is to determine a calibration factor for each unit of the eyepiece scale for each objective, i.e., each level of magnification.

Using the diagram as an example, find a point farthest to the right where the eyepiece scale (Y) lines up exactly with the micrometer scale (X). In the diagram, 90 units on the eyepiece scale line up with 0.3 on the micrometer scale. The calibration factor becomes 0.3/90 = 0.0033 mm and multiply by 1000 to convert to microns (0.0033 mm × 1000 = 3.3 μm). Each unit of the eyepiece scale, using the diagram, equals 3.3 μm which is the calibration factor. Using the objective with this factor, if an egg measures 20 units in length on the eyepiece scale, its length is 20 × 3.3 = 66 μm.

The calibration factor, i.e., the size of each unit of the eyepiece scale at the plane of the specimen using each objective with the complementary eyepiece, should be checked and a chart of the calibrated values should be prepared and placed at a convenient location on or near the microscope. These values should be checked regularly, especially if several people use the microscope and in case an objective or eyepieces on the microscope are changed. Further details on how to calibrate your microscope can be obtained from your microscope manufacturer's representative.

SELECTION OF LENSES

For optimum effectiveness, a microscope used for parasitology may require objectives different from those usually found on a laboratory microscope. In most instances, 10× wide-field eyepieces are appropriate. For scanning wet mounts, a 10× objective is usually optimal but this power is usually not adequate for finding protozoa on permanently stained slides. If permanent slides are well prepared and protozoa are well stained, a 20× objective is often adequate for scanning. This power is also useful for identifying larger protozoa in wet mounts. Protozoa such as *Endolimax nana, Entamoeba hartmanni*, and *Dientamoeba fragilis* usually require a higher magnification for positive identification on either wet mounts or permanently stained slides; often the 100× oil immersion objective is needed. The selection of an objective that provides for magnifications between 200 and 1000× is most difficult. If the 40× dry objective is selected, it should be used only when there is no oil on the cover glass. If during the course of examining a slide the oil immersion objective is used, then the oil should be removed before using the 40× objective again. If the oil is not removed, the advantages of a flat field are lost and it is likely that oil will get on the 40× lens when it is used alternately while examining the same slide. Then, the 40× lens should be cleaned before examination of the slide is continued (see Appendix, Removing Immersion Oil From Slides).

If an oil immersion lens is selected for an intermediate lens, changing powers causes fewer problems and it offers better resolution of detail. Although resolution is not ideal, the 10 and 20× objectives can be used alternately when there is oil on the slide. Intermediate oil immersion lenses vary in magnification from about 45 to 65×. An intermediate lens from 45 to 50× is most practical. A 40× dry objective is often useful. If the laboratory can afford the extra expense, 50× oil and 40× dry lenses and a 5-place nosepiece can be selected. If there must be a choice between the two intermediate lenses, the 50× oil immersion objective is preferred over the 40× dry lens for diagnostic parasitology. The low-power (4×) lens is rarely used in routine parasitology but is occasionally needed. For the rare cases in which it is needed, a 4× lens should be available. It can be placed on the nosepiece when it is needed or a six-place nosepiece can be used.

The arrangement of the lenses on the nosepiece is also important. For parasitology, using the 10× objective as the pivot point, dry lenses may be placed on one side and oil immersion lenses on the opposite side i.e., 10×, 20×, 40× going one direction and 10×, 50×, and 100× going the opposite direction.

PREPARING WET-MOUNT SLIDES FROM THE PRESERVED SPECIMEN FOR MICROSCOPIC EXAMINATION

Before performing any procedure on a preserved specimen, it should be checked to make certain that the lumps in the specimen have been broken up and that the specimen has been thoroughly mixed with the fixing solution. If the specimen has not been properly mixed, it should be remixed after which it should be allowed to stand undisturbed for an hour so that fixation can be assured before proceeding further. Parasites or eggs that were embedded in fecal lumps may not be fixed equally to those that came directly into contact with the collecting/preserving solution earlier.

Specimens Preserved in Formalin or SAF

How a slide is prepared from a preserved specimen has a great impact on the results obtained. Wet mounts may be prepared from formalin- and SAF-fixed specimens either before or after a concentration procedure is run on the preserved specimen. A sample of the specimen may be mixed with saline, or with iodine to color organisms and make them easier to find and perhaps easier to identify. Iodine solutions that are recommended are D'Antoni's and Dobell and O'Connor's (see Appendix, Iodine Solutions).

Often, two samples are examined, one with and one without the addition of iodine. After a concentration procedure has been run, iodine may be added either to the sample on the slide or to the plug in the centrifuge tube before the sample is transferred to the slide which insures more even distribution of the iodine.

The thickness of the sample under the cover glass is extremely important. If it is too thick, use of the oil immersion objective may not be possible and/or there may be so much material on the slide that the organisms are hidden. If it is too thin, organisms present in the specimen may not be present on the slide. To insure proper thickness of the sample under the cover glass, the following recommendations are made in the published literature. A slide may be placed on a sheet of newsprint, a drop of saline (or iodine) placed on the slide, and a drop of specimen added and mixed with the saline. A cover glass is then added. The newsprint should be just legible through the sample. The author prefers the method recommended for preparing wet mounts from specimens fixed in PIF or MIF (see Preparing Wet Mounts from Specimens Preserved in PIF or MIF, and Figures 4 through 9).

Although many workers do not feel it necessary to ring the cover glass before examination (i.e., seal the specimen under the cover glass), the author recommends that all wet mounts be ringed with vaspar (equal parts of paraffin wax and Vaseline heated at 150 to 170°F) or another suitable sealant. Sealed samples under the cover glass will not dry out if examination is interrupted and can be reviewed by supervisors or other personnel at some later time. The edges of the cover glass should be freed of moisture before it is ringed. Occasionally, something found on one slide may not be found on another slide prepared later and, if the specimen on the slide is not sealed to prevent it from drying, confirmation may not otherwise be possible.

Some investigators believe that the examination of wet mounts prepared directly from preserved specimens yields very little. They suggest that only the concentrated specimen should be examined (see Part 3). Others recommend that protozoa should not be identified specifically when found in wet mounts but that identification should be confirmed only on permanently stained fecal films prepared from the specimen. The author does not agree with either position unless only the two-vial PVA/formalin system is being considered. The value of examination of wet mounts prepared directly from the fixed specimen and the accuracy of identification of protozoa in wet mounts is addressed in material that follows.

Specimens Preserved in PAF

A specimen fixed and preserved in PAF (Burrows, 1967b) can be examined directly (using special stains) or after it has been strained through gauze and washed in 0.85% saline. A saline/ethyl acetate concentration can be run for finding greater numbers of cysts, eggs, and juveniles of nematodes; however, trophozoites suitable for identification are not recovered by the concentration method.

To prepare a direct wet mount, the upper one third or one half of the excess fixative is poured off, the specimen is mixed thoroughly with the remaining fixative, and a drop of the suspension is transferred to a microscope slide. A drop of stain of equal size (Thionin or Azure A, see Appendix) is added to the slide and mixed with the fecal sample. After the stain has been able to act for a few minutes, a cover glass is added and sealed with vaspar (see Part 3, Concentration of PAF-Fixed Specimens).

Specimens Preserved in PIF or MIF

When the specimen is properly prepared in a flat-bottom vial, the specimen in the vial has two or three distinct layers. After being allowed to settle undisturbed for 1 hour or more, the specimen in a vial prepared from a formed stool will have three distinct layers: an upper supernatant layer; a fine, light-colored interface layer; and a darker, coarse base layer (usually, the more compact the fecal mass, the thicker the interface layer). A specimen in a vial prepared from an unformed or liquid stool will have only two layers: a supernatant layer and a base layer; an interface layer is either not produced or is very thin (see Figure 3).

The PIF/MIF Pipette

Because "gravity sedimentation" is used in the preparation of direct wet mounts, the pipette used to draw the specimen from the vial requires some consideration. For optimum results, a pipette with a specific bore size is especially important for transferring a portion of the specimen fixed in either PIF or MIF to a microscope slide for either examining the wet specimen directly or for preparing slides for staining. In selecting an optimum-sized pipette, consideration was given to the nature of the fixed specimen and to how the specimen would be used.

Gravity sedimentation is a method for concentrating materials of similar specific gravity at specific levels in a solution or sediment. When the proper amount of Lugol's iodine is added to the stock solution immediately before a sample of the specimen is added, an ideal density for gravity sedimentation of parasites and eggs is achieved. To take full advantage of the gravity sedimentation method, the fecal specimen should be thoroughly mixed so that all parasites and eggs come into direct contact with the solution and a pipette with the proper inside diameter should be used to draw the sample. Cysts and eggs of parasites fixed in formalin can be concentrated by gravity sedimentation; however, the specific gravity of the solution does not produce results equal to those seen using MIF-type fixatives.

The sample to be examined should be drawn with a pipette by capillary action. The pipette should draw most of the sample from the level where parasites

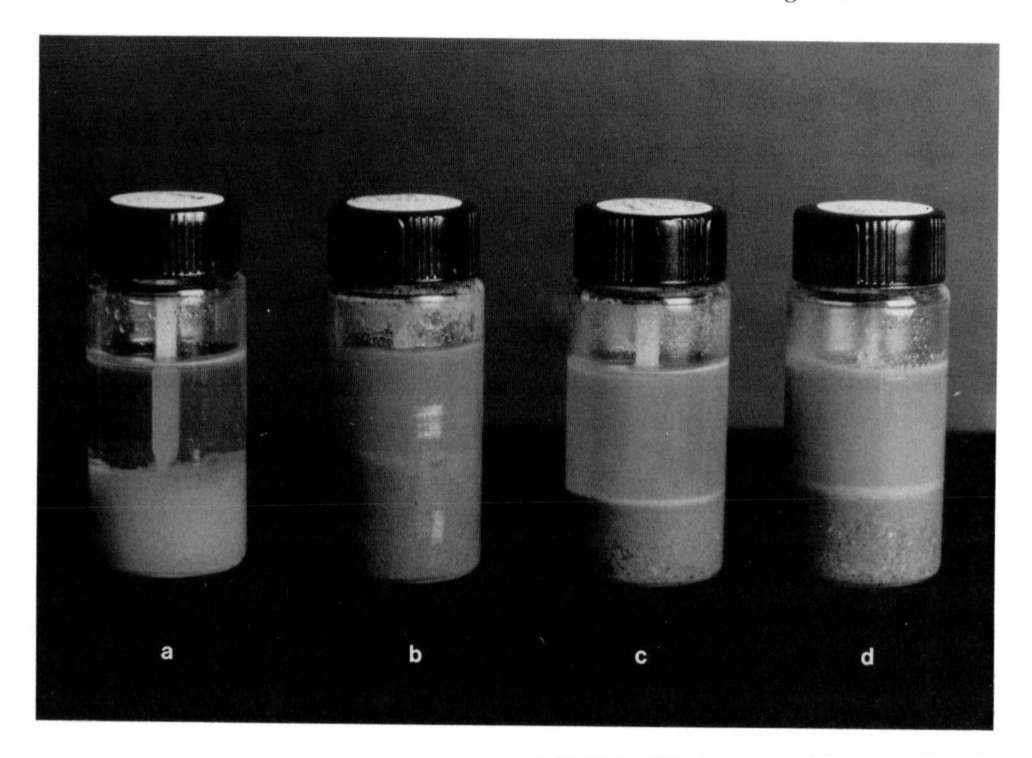

Figure 3. Vials containing fecal specimens prepared from formed and unformed stools in MIF collecting/preserving solution. Four vials with specimens added are shown to demonstrate the layering that may occur when fecal specimens of different consistencies are mixed with MIF-type collecting/preserving solutions. (a) When the specimen is very watery, a relatively uniform, amorphous base layer is formed with an upper fluid layer; (b) when the specimen is unformed and very soft only two layers result: a relatively even base layer slightly more dense at the bottom and an upper fluid layer; (c) when the specimen is semiformed and soft, three layers form, a base layer with more coarse material at the bottom, a very thin, pale interface layer, and a usually cloudy upper fluid layer; (d) three distinct layers are produced when the fecal specimen is formed: a dark base layer, a pale interface layer, and an upper, usually cloudy, fluid layer.

and eggs are most concentrated. A capillary pipette with an inside diameter of less than 2 mm will not draw a sufficient sample or draw from a broad enough area. A pipette with an inside diameter of 3 mm or greater draws too much sample and does not have the capillary action to lift and transfer the sample properly. A pipette with an inside diameter between 2.2 and 2.6 mm is ideal and can draw a sample of optimum size and draw most of it from the site of the greatest concentration of parasites and eggs. Either glass or clear plastic tubing may be used. Thin-walled glass tubing sold as 4-mm O.D. (outside diameter) usually has the correct inside diameter but the inside diameter should be checked.

Another consideration is that parasites and eggs tend to be drawn to the sides of the pipette, especially to glass. If the bottom opening at the pipette tip is constricted (slightly narrowed), many parasites and eggs will be trapped just inside the pipette tip. Therefore, a pipette with straight sides and an unconstricted bottom tip is essential for maximum efficiency. Use of a proper pipette will ensure the best results.

Obtaining a Specimen Sample

The depth of the fluid in relation to the amount of specimen in the vial affects the volume of the sample drawn up by the pipette by capillary action. The proper amount of sample is usually drawn when there is a ratio of one part specimen to four or five parts fluid. If the specimen portion is less than one fifth of the volume in the vial, the sample transferred to the slide may be thin. If the specimen portion is greater than one fourth of the volume, the specimen transferred to the slide may be thick. One must use care in transferring the proper amount of the specimen to be examined in order to be certain to find parasites and eggs that may be present.

Preparing Wet Mounts from Specimens Fixed in PIF or MIF

As stated, in a flat-bottom vial parasites sediment in the fixing/preserving solution according to weight, size, and stage (gravity sedimentation). Samples of specimens for direct examination (i.e., examination of wet mounts made directly from the fixed specimen) should be drawn only after the speciment has been allowed to settle for at least 1 hour to make certain that parasites and eggs have concentrated properly. Specimens allowed to stand for several days should be reshaken and allowed to stand undisturbed for an hour or more before reexamination. In specimens that form three layers, most of the parasites and eggs will be found in the top part of the base layer just below the interface layer. In specimens that form only two layers, most of the parasites and eggs will be found in the middle third of the base layer (see Figure 3).

Either too much or too little feces in the vial cancels the value of gravity sedimentation. Use the following procedure for preparing wet mounts (see Figures 4 through 8).

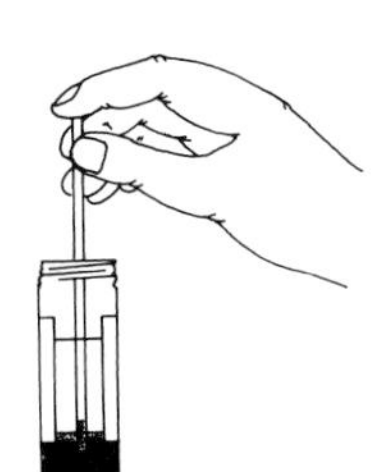

Figure 4. Obtaining a sample. To draw a sample for examination, place a finger over the upper tip of the pipette, insert the pipette along the side of the vial to the desired level in the specimen, lift the finger from the pipette tip to allow a sample of the specimen to flow into the pipette, replace the finger, and withdraw the pipette to transfer the sample to a slide.

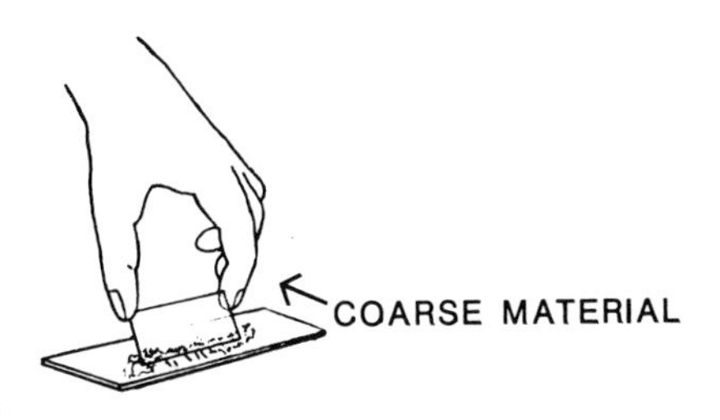

Figure 5. Adding the cover glass. Step 1. A 22 × 40-mm cover glass is recommended to cover the specimen to insure that a reasonable amount of sample is examined. The sample under the cover glass should be kept as thin as possible so that the oil immersion lens can be used readily for parasite identification. To keep coarse material from being under the cover glass, hold the cover glass by the edges using the thumb and forefinger, tilt the slide slightly and use the cover glass to pull all the material on the slide to one edge.

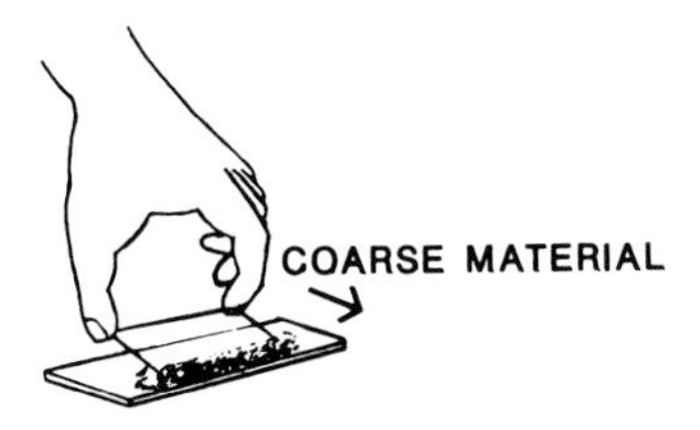

Figure 6. Adding cover glass. Step 2. Lift the cover glass and reverse the slide. Push the wet edge of the cover glass over the surface of the slide into the fluid part of the sample and, keeping the coarse material at the edge of the slide, lower the cover glass to allow the fluid portion of the sample to flow across the slide under the cover glass.

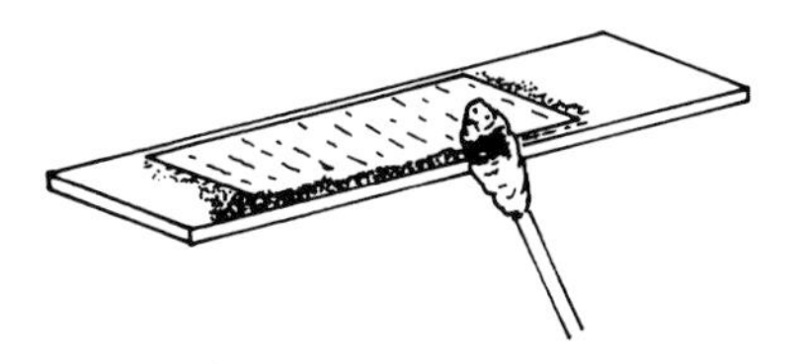

Figure 7. Removing coarse material. Touch the edge of the slide to a paper towel to remove some of the excess material, then remove the remaining coarse material from the edge of the slide with a cotton-tipped applicator stick. Any excess fluid remaining may be removed by touching the edge of the slide again to a towel or with the applicator stick to make the sample as thin as possible and free the edges of moisture before the vaspar is added.

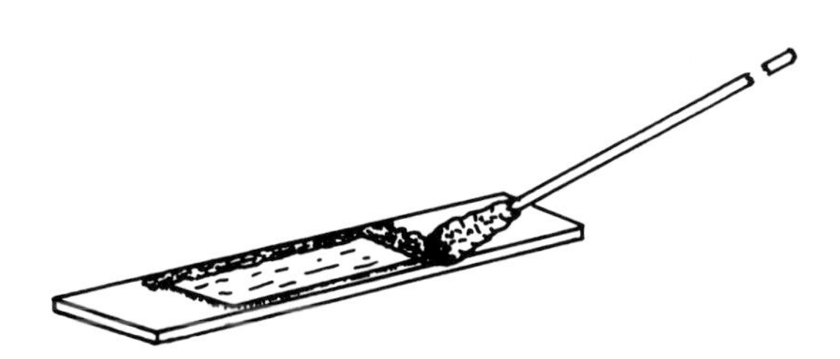

Figure 8. Sealing with vaspar. Sealing the sample under the cover glass allows repeated examination of the slide and temporarily preserves it so it can be examined at some later time by other laboratory personnel, supervisors, or trainees.

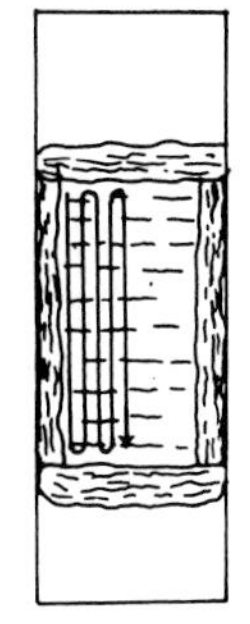

Figure 9. Scanning the slide. Good practice is to always begin in the same corner of the cover glass and scan either left and right or up and down, whichever is most comfortable for the examiner. Moving the fine adjustment as the sample is scanned will help the examiner find parasites and eggs as explained in the discussions on Köhler illumination, refractive index, and Becke Line (see also Appendix).

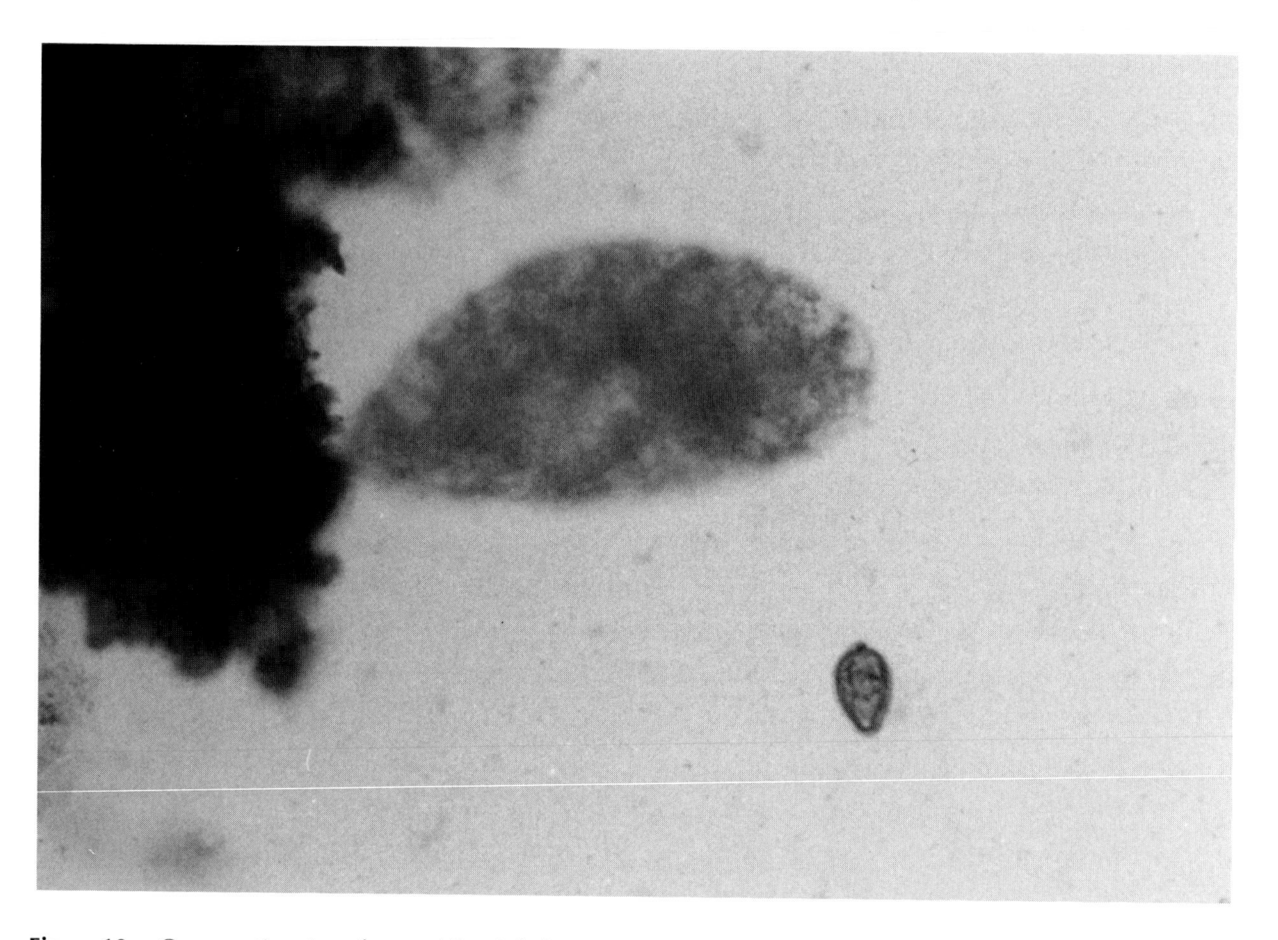

Figure 10. Comparative size of eggs of *Fasciola hepatica* and *Clonorchis sinensis* at 100× magnification, wet mount, MIF. The average length of *F. hepatica* eggs is about 140 μm whereas that of *C. sinensis* is about 31 μm. When scanning a wet mount at higher power, e.g., 400×, it is often possible to miss the larger egg especially if there is much debris in the sample being examined. The slide should be examined first using 100× and, if considered necessary, an additional 200 or more fields can be examined at higher power.

Finding parasites and eggs in wet preparations fixed in formalin, SAF, PIF, MIF, or TIF is a matter of technique and experience. Those familiar with the appropriate techniques and the identifying features of parasites, eggs, and juvenile nematodes should have no difficulty either finding or identifying them regardless of the method of fixation (see Part 4). Most workers began identifying parasites in wet preparations made from fresh specimens or those fixed in formalin.

Scanning the Slide

Wet preparations should be scanned using 100× magnification (usually a 10× wide-field eyepiece and 10× objective) and the entire sample under the cover glass should be scanned. Higher magnification should be used to check suspicious objects and to confirm identification.

Microscopic Examination of the Prepared Slide

Microscopic examination can reveal many things in addition to parasites and eggs. Leukocytes (white blood cells), epithelial cells, plant cells, fibers, pollen grains, undigested animal protein in the form of muscle fibers, etc. are seen in normal stools. In excessive numbers, muscle fibers, budding yeast cells, mucus, leukocytes, and erythrocytes should be reported.

One problem in finding and identifying parasites and eggs in fecal specimens relates to the variation

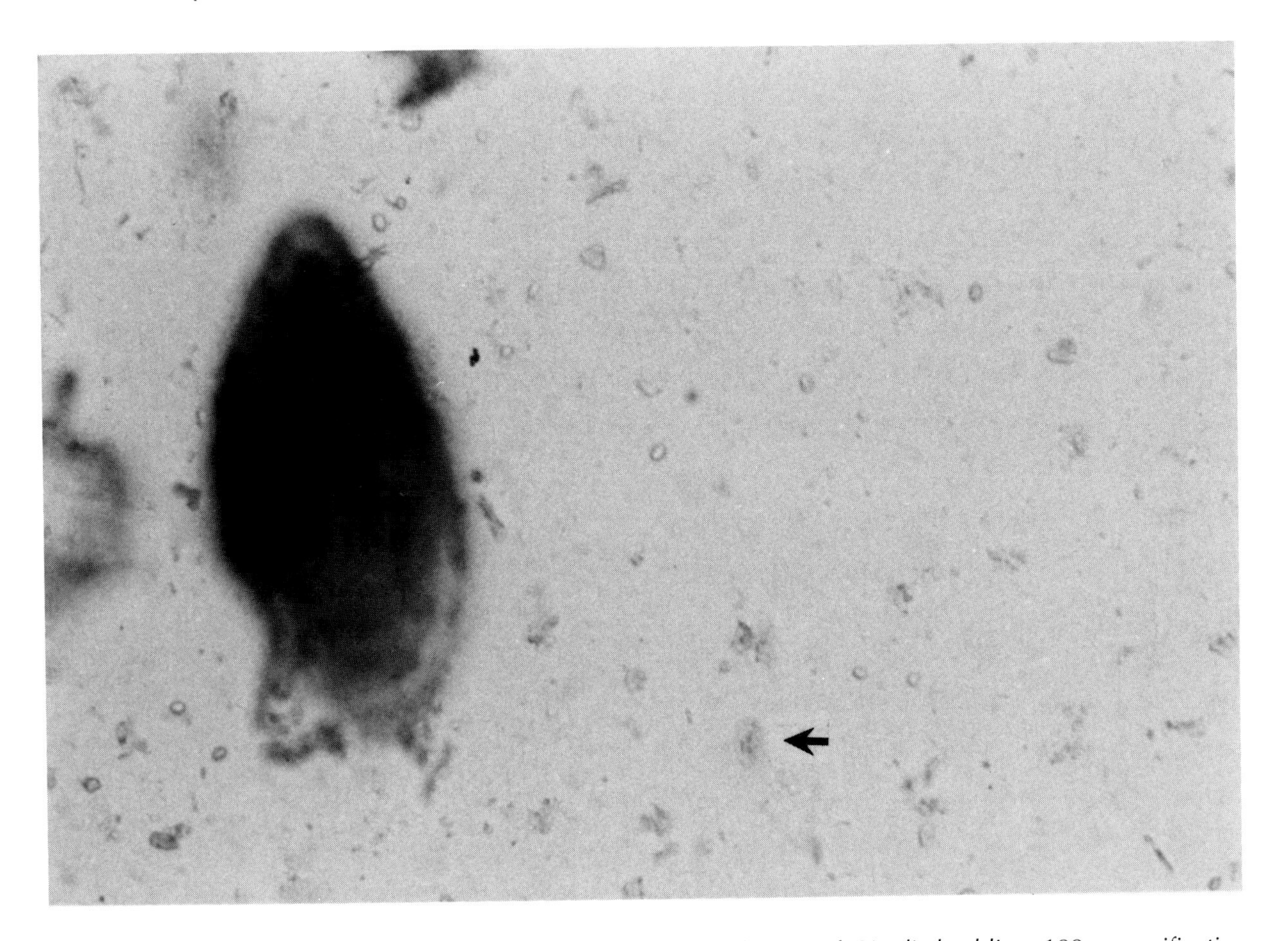

Figure 11. Comparative size of an egg, *Schistosoma mansoni,* and a cyst of *Giardia lamblia* at 100× magnification, wet mount , MIF. Even though the cyst of *G. lamblia* measures 8 to 15 μm in length, it can be seen even if not identified at 100× magnification (arrow). Its refractive index helps one find the small protozoa when scanning. One can quickly change the magnification to 400 or 500× to make an identification. The much larger egg of *S. mansoni* is readily picked out at this magnification.

in size between helminth eggs such as *Fasciola hepatica* and *Clonorchis sinensis* (Figure 10) and the vast difference in size between the larger helminth eggs and the small protozoa such as *Schistosoma mansoni* and *Giardia lamblia* (Figure 11). At 100× magnification, one can recognize the presence of both the smaller refractile bodies (as in Figure 11) as well as the larger eggs.

At 400× magnification, however, larger eggs, such as *S. mansoni*, can virtually disappear in the debris when a small protozoan, such as a cyst of *Giardia*, is brought into focus (Figure 12). After the larger egg found at scanning magnification is identified, the smaller refractile body can be examined at a higher magnification to make an identification.

When scanning at 100× magnification, using a 22 mm × 40 mm cover glass ringed with vaspar, and moving 1 field at a time, there are approximately 200 fields under the cover glass to be examined. One must learn to scan the slide using 100× and use higher magnification to identify parasites in order to cover the entire sample within a reasonable time. If higher magnification is used for scanning, the number of fields to be examined increases greatly. For example, at 400× magnification there are approximately 3200 microscopic fields using a vaspar-ringed cover glass (Figure 13). It is unreasonable to expect the microscopist to examine over 3000 fields on every prepared wet mount, and examination of a wet mount is an integral part of all systems used for diagnosing intestinal parasites.

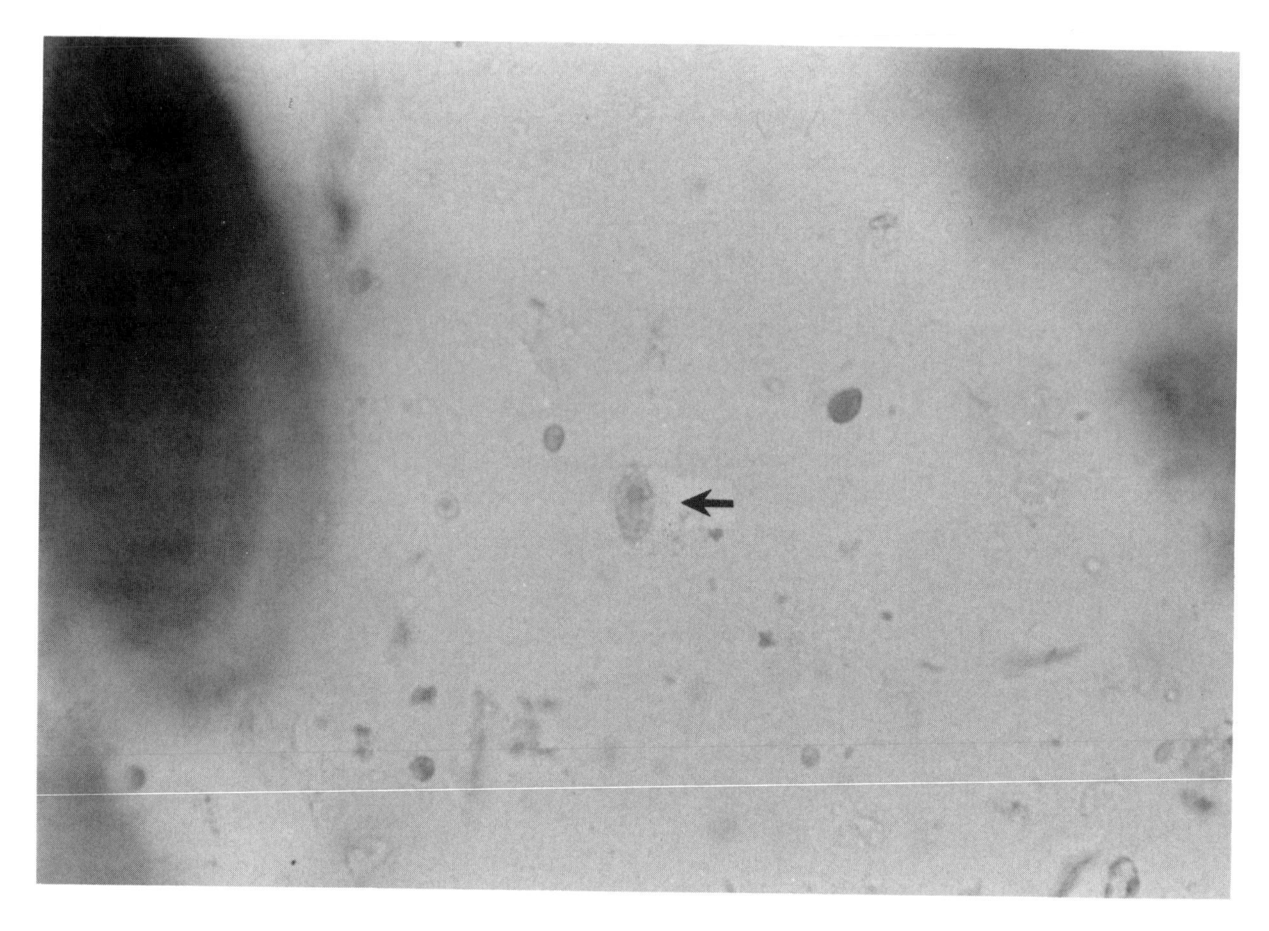

Figure 12. Focusing on a large egg of *Schistosoma mansoni* and a cyst of *Giardia lamblia* at 400× magnification, wet mount, MIF. The same field as shown in Figure 11 at 400× magnification. When the cyst of *G. lamblia* (arrow) is in focus, the egg of *S. mansoni* is nearly lost in the debris because of its large size.

Magnification	Fields Each Direction		Total Fields
100X	11 fields	20 fields	= 220 fields
400X	44 fields	80 fields	= 3520 fields
500X	55 fields	100 fields	= 5500 fields

Figure 13. The number of microscopic fields on a slide at different magnifications using a 22 × 40-mm cover glass. The number of visual microscopic fields that can be seen based on the magnification varies greatly. As shown in this figure, using a 22 × 40-mm cover glass there are 220 fields at a magnification of 100×, 3520 fields at 400×, and 5500 fields at 500× that one must see in order to cover the entire specimen. Organisms present on most well-prepared wet mounts can be found when the examination is carried out at 100×, scanning magnification (see Refractive Index). Identification of what is found, especially protozoa, may require higher magnification.

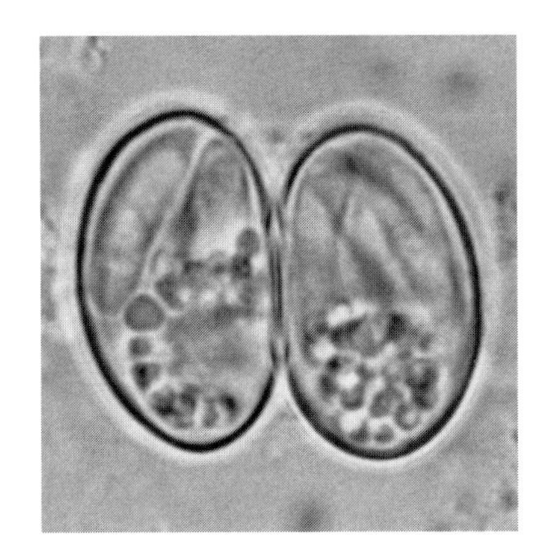

Part 3

Introduction to Part 3

The preparation and examination of direct wet mounts was presented in Chapter 2. There are a number of complementary methods performed on the preserved fecal specimen that enhance the individual's ability to find and identify parasites and eggs by microscopic examination. These include, but are not limited to, methods for concentrating eggs and juveniles (larvae) of helminths, and trophozoites and cysts of protozoa (Chapter 3), and methods for preparing permanently stained fecal films of protozoa (Chapter 4).

Concentration methods are applied to most fecal specimens sent to the laboratory for examination for eggs and parasites. The specimens received by the laboratory may be either unpreserved or preserved in SAF, MIF, PIF, or formalin, depending on the protocol of the laboratory. An unpreserved specimen may be fixed in one of the preserving solutions or processed without prior preservation. Results of a concentration method applied to an unpreserved specimen and to the same specimen after it has been preserved may not be the same. In fact, the same parasite or egg may concentrate differently in specimens supplied by the same individual on different days or in specimens from different individuals.

Different parasites and eggs also concentrate differently. The heavier eggs of trematodes concentrate in a way different from that of the lighter eggs of hookworms, *Trichuris*, and *Ascaris* in individual specimens, and protozoa concentrate in a manner different from that of eggs.

Variations in results also occur using different methods, and the procedure followed for a given method may differ from one laboratory to another, in fact even between individuals in the same laboratory. Variations include the volume of specimen used, the number of washing steps, if and how surfactants are used, the size of the plug, the amount of the plug transferred to a slide for examination, and how the plug is transferred. These variations can affect the results obtained.

Variations also occur in permanently stained fecal films — whether the specimen was unpreserved or fixed in a collecting/preserving solution, the way the film was prepared, the stain used, and the procedure followed, all have an impact on the results.

How the methods are grouped to form systems, and the effectiveness and efficiency of the different systems are discussed in Chapter 5. Special methods, such as acid-fast staining, preparation of temporary wet mounts, egg counts, and after-treatment examination are presented in Chapter 6.

All of the methods described can be applied to unpreserved specimens, but some preliminary steps may be necessary and there may be some differences in how the procedure for a particular method may be employed. Some methods are applied to unpreserved specimens and, in some cases, the use of a collecting/preserving solution becomes a part of the procedure.

The procedures for concentration methods and most of the special methods are part of the text; procedures for staining methods are given as tables. The instructions for preparing solutions used for a particular method are given in the Appendix.

Chapter 3

Concentration Methods

INTRODUCTION

Concentration methods that have been developed were designed to improve one's ability to find eggs and/or cysts of parasites. Over the years, some procedures or modifications of them have persisted. They fall into two general categories: flotation and sedimentation methods. In all procedures, specific gravity plays a role (see Appendix, Specific Gravity), and, more often than not, centrifugation is included as an integral part of the procedure. Where procedures are described in the text, some information, suggestions, and/or comments may be added that are not part of the originally described procedure. These will usually relate to technique and may be in parentheses.

USE OF THE CENTRIFUGE FOR CONCENTRATION METHODS

Most of the methods for concentration involve the use of the centrifuge as part of the procedure. The speed of the rotation of a centrifuge, i.e., the number of revolutions per minute (rpm), produces a "relative centrifugal force" (RCF) or g force (g = the force of gravity) at each speed. The g force is expressed as "xg", e.g. 500 xg = 500 times the force of gravity. Different centrifuges running at the same rpm may produce different g forces (RCF). Often, in older published literature where a method was described, only the rpm to be used was given for the procedure involving centrifugation. The rpm may be appropriate for the method using one type of centrifuge but not when using another type of centrifuge. Since types of centrifuges used in laboratories vary greatly, it is desirable to know the g force to use and the rpm required to achieve that g force. In performing centrifugal procedures for concentrating intestinal parasites, the most commonly used g forces are between 500 and 700 xg. Forces too low will not provide adequate concentration of parasites and eggs and those too high may collapse the organisms being concentrated. Use only centrifuges with swing-free heads.

A nomogram (nomograph) and the procedure for calculating the g force (RCF) produced at each rpm for the centrifuge being used is provided in the Appendix (see Figure 2). The results obtained may be recorded and a chart prepared for each centrifuge that gives the rpm required to produce the desired g forces. The chart should be placed conveniently near the centrifuge so that the users can readily determine the rpm needed for a particular procedure.

FLOTATION METHODS

Brine and Sugar Flotation

Brine Flotation, introduced by Bass in 1906, consists of floating eggs and cysts in a saturated salt solution and is still used in veterinary practice. It

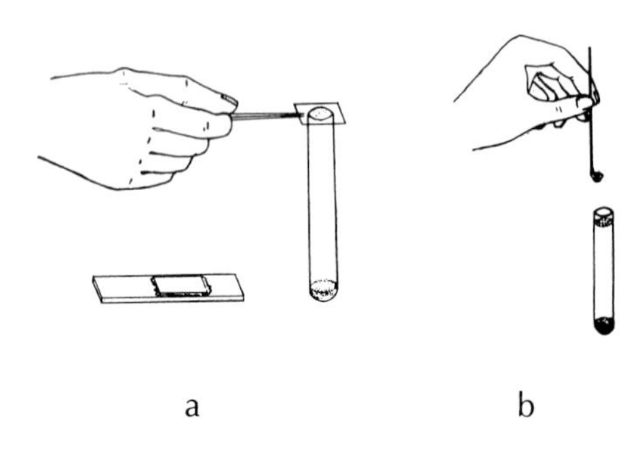

Figure 1. Flotation procedure—transferring a sample to a slide following centrifugation. (a) Flotation fluid is added to the tube to form a meniscus at the top. A cover glass is placed on the top of a flotation tube in contact with the fluid meniscus making certain that there are no bubbles. After an appropriate time, the cover glass is removed with forceps, bringing with it a drop of fluid, and is placed on a slide so that most of the fluid remains under the cover glass. The sample under the cover glass is examined for cysts and eggs. (b) Some fluid may be added to bring the level near the top of the flotation tube. After the specified time, a wire or plastic loop can be used to transfer several drops of the upper surface of the fluid to a microscope slide for examination for cysts and eggs.

can be used for concentrating various eggs and coccidia, including cryptosporidia, from fresh feces. The concentration method using a saturated sugar solution and centrifugation proposed by Sheather in 1923 is considered to be a better method by most workers. The two procedures are performed on fresh specimens.

In the brine method, about 1 g of feces is emulsified in 4 to 5 mL of the saturated salt solution (specific gravity 1.20) in a round-bottom tube. The tube is then filled to the brim with the salt solution so that a meniscus is formed, a cover glass (preferably 22 × 40 mm) is placed on the tube so that it is in contact with the salt solution and there are no air pockets. The tube is allowed to stand for 10 to 15 minutes. The cover glass is then carefully removed along with some adhering solution and placed with the solution down on a slide for immediate examination (Figure 1a). Protozoan cysts and some eggs may collapse, making them unrecognizable.

Sugar flotation uses sucrose preserved with phenol. A stock saturated sugar solution can be prepared by dissolving 500 g of sucrose and 6.5 g of phenol crystals in 320 mL of distilled water (7 mL of liquid phenol may be substituted for crystals). The sugar solution is usually heated gently until clear before adding the phenol. Use appropriate safety precautions for handling phenol.

The procedure is as follows:

- Mix about 1 g of feces with 7 to 8 mL of the saturated sugar solution in a round-bottom test tube, then add more sugar solution to almost fill the tube.
- Centrifuge the suspension at 500 to 550 xg for 5 minutes and allow the centrifuge to come to a complete stop.
- After about 1 minute, use a wire or plastic loop 5 to 7 mm in diameter to transfer several drops from the surface to a glass slide and cover with a cover glass for immediate examination (Figure 1b).

The cover glass method as described in the brine flotation procedure above may be used instead of the loop.

Some workers prefer to make a suspension of fresh feces in 0.9% saline or water, centrifuge it once, pour off the supernatant fluid to remove some of the debris, and use the resulting plug (about 1 g) for the flotation procedure. One or two drops of Dobell and O'Connor's iodine solution may be added to the material on the slide to enhance identification of cysts.

Zinc Sulfate Flotation

The zinc sulfate flotation method introduced by Faust et al. in 1938 for concentrating eggs and cysts in fresh fecal samples is still in regular use in many laboratories (Faust, 1949). A modification for use with formalin-fixed specimens, which requires a solution with a higher specific gravity (1.20 rather than 1.18) to float eggs and cysts was introduced by Bartlett et al. in 1978 and is used extensively.

Materials and Equipment

- Zinc sulfate solution (at the correct specific gravity)
- Round-bottom test tubes (usually 100 × 13 mm)
- Wooden applicator sticks for mixing
- Wire or plastic bacteriological loop (5 to 7 mm in diameter)
- Cover glasses, if the alternate method (Figure 1a) is used (22 × 40 mm or 22 × 22 mm)

- Centrifuge (with a swing-free head), calibrated

PROCEDURE

1 A portion of a well mixed specimen, usually about 4 to 5 mL of a fecal suspension, is strained through 2 layers of wet surgical gauze into a round-bottom tube. (The original procedure recommends using a small, 13 × 100-mm Wassermann tube, but larger tubes are usually satisfactory and specially designed screens presently available for concentration procedures can be substituted for surgical gauze.)

2 Formalin (the same strength as the fixing solution) is added to the tube and thoroughly mixed with the specimen.

3 The tube is spun in a centrifuge at 500 to 550 xg for 1 minute, and the supernatant fluid is carefully poured off making sure that none of the plug is lost (see Figure 5). The washing step may be repeated several times if the supernatant fluid is not clear. For the washing steps in most pro-cedures, 1% saline should be used for unpreserved specimens and formalin (the same strength used for fixing the specimen) should be used for formalin-fixed specimens.

4 The sediment is then mixed with about 10 mL of zinc sulfate suspending solution. After mixing, additional suspending solution is added to bring the level to about one-half inch from the top of the tube, and the tube is spun again at the same speed for 1 minute.

5 When the centrifuge has come to a complete stop, the tube is gently placed upright in a rack.

6 Using a wire or plastic bacteriological loop (5 to 7 mm), several drops of the surface material are carefully transferred to a microscope slide, covered with a cover glass, and examined immediately for eggs, cysts, and juvenile nematodes (see Figure 1).

An alternate method for transferring the specimen is to add additional suspending solution to the tube immediately after centrifugation until a slight meniscus is formed at the top of the tube. As described in the brine and sugar flotation procedures, a cover glass is placed on the tube in contact with the meniscus (without bubbles) for up to about 2 minutes but not longer. The cover glass is then carefully removed with forceps and placed on a slide so that most of the adhering fluid remains under the cover glass (see Figure 1). The sample of the specimen on the slide is examined for eggs, cysts, and juvenile nematodes. An iodine solution (Dobell and O'Connor's or other) may be added to the sample on the slide to enhance identification of what is found (see Appendix, Iodine Solutions).

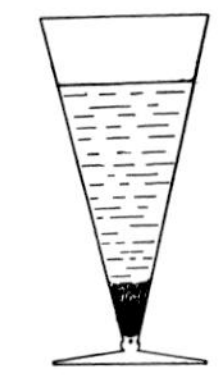

Note that trophozoites of protozoa, operculated eggs, and eggs of schistosomes are not recovered by flotation procedures.

Figure 2. Gravity sedimentation. The procedure is carried out using a conical flask, or a pharmaceutical graduate for larger volumes of feces. After washing the fecal specimen several times in an appropriate solution, it is poured into a conical flask, the flask is filled with sedimenting fluid, and the solids are allowed to settle. A gradient of solid materials is formed at the base of the flask with the coarser, heavier materials at the bottom and the finer, lighter materials at the top as shown in the drawing.

SEDIMENTATION METHODS

Gravity Sedimentation

Gravity sedimentation was briefly discussed earlier since the development of the MIF, TIF, and PIF methods are based partly on the principle of gravity sedimentation. The container used may be conical or cylindrical but, in either case, the sides should be straight to the base for best results (see Figure 2). A round or conical bottom will change the results of layering in the sediment. The bottom will have little effect when large amounts of fecal material are being sedimented so that the surface layers of the sediment are well above the curvature in the container. If an unpreserved specimen is being sedimented, a cold solution of 1% saline should be used. The flask should be placed in the refrigerator while sedimentation is taking place because of the greater possibility of deterioration of the parasites at room temperature. Because colder solutions are slightly more dense (water is most dense at 4°C), they are preferred when sedimenting preserved specimens by gravity sedimentation. Parasites and eggs cannot be concentrated unless they are separated from the fecal clumps. The importance of breaking up all particles to their smallest possible size so that concentration can be accomplished is therefore reemphasized. Miracidia may emerge from eggs of schistosomes and some other trematodes if pure water is used. The laboratory procedure consists of mixing the unpreserved fecal mass in a 1% saline solution, allowing the solids to settle, then carefully pouring off the supernatant fluid so as not to lose any of the sediment.

If the specimen is fixed in formalin, then the matching strength of formalin (or formol-saline) is used instead of saline. This washing procedure is usually repeated until the supernatant fluid is clear, after which a portion of the sediment is examined. A portion of the sediment of an unpreserved specimen can be placed in a collecting/preserving solution for future examination.

In the MIF, TIF, and PIF procedures, the specific gravity of the solution has been adjusted so that parasites and eggs concentrate at a specific level in a flat bottom vial at room temperature and therefore can be found more readily. The procedure for transferring a sample of the sediment from the appropriate level in the vial or flask to a slide for examination is described in Part 2.

Using the Centrifuge for Sedimentation Procedures

The use of the centrifuge improved sedimentation procedures and such procedures proved to be especially good for operculate and schistosome eggs. A variety of solutions were used for improving the sedimentation of various stages of parasites (see Ash and Orihel, 1987). In most procedures, ethyl ether was used in combination with the solution to help separate eggs, cysts, and juvenile worms from the fecal matter and to remove fatty material and some debris from the sediment. One or two drops of a surfactant, usually 20% Triton X-100, may be added to help separate organisms from mucus and fatty materials. Some methods (special solutions) were developed to concentrate eggs of certain species or types, such as schistosomes. These special methods are not usually appropriate for routine work.

In performing all concentration procedures certain details should remain constant. Centrifuge tubes made of polypropylene are preferable for most procedures involving the use of a centrifuge. Fecal material has a tendency to adhere to glass but not as much as to polypropylene. Some other plastics are affected by the ethyl acetate or ethyl ether used in the various procedures.

The amount of the specimen to be concentrated using a 15-mL centrifuge tube may vary with the procedure and the preservative. Generally, the amount should be about 0.5 to 1.5 mL of the fecal suspension after straining or 0.3 to 0.8 mL after several washing steps by centrifugation are completed. Too much of the specimen overloads the solutions and defeats the purpose of the concentration method. This is particularly true if several washing steps are used before the final centrifugation. If the amount used is less than 0.5 mL of the suspension (or a plug of less than 0.3 mL after washing), the possibility of finding the organisms in the sediment is reduced.

When pouring off the upper solutions and material in the tube after centrifugation, the tube should be held so that the plug is always visible and should be turned to a vertical position (see Figure 5). If the plug can be seen and begins to break up, the tube can be quickly turned to a horizontal position to retain all of the plug. If the tube is held at a sharp angle, e.g., 45°, while pouring off the supernatant materials, the upper surface of the plug containing the lighter organisms may be washed away with the waste fluids and some of the parasites and/or eggs may be lost.

The concentration methods used for specimens fixed in SAF, PIF, MIF, or TIF are all sedimentation methods and include formalin-ether sedimentation, formalin-ethyl acetate sedimentation, MIFC (Blagg et al., 1955, technique using MF stock solution), and CONSED (a relatively new solution and procedure, unpublished; see CONSED Sedimenting System).

Special Apparatus for Sedimentation Procedures

Special funnels and tubes, with screens as substitutes for the cotton gauze, have been introduced and are designed especially to remove coarse material from the suspended fecal sample before sedimentation. The effectiveness of these depends largely on the composition and consistency of the fecal mass and on the solution in which the feces is mixed and/or preserved. Fecal specimens with large amounts of fibrous material or clumps of mucus do not pass well through fine screens. Screens with an electrical charge, such as metal screens, tend to attract and hold organisms, especially those fixed in MIF-type solutions. Screens at least 1 inch diameter or breadth, with mesh openings of approximately 1 mm^2, made of a material that does not hold a charge, and does not absorb liquid appear to give the best results when tested with a variety of fecal specimens of different compositions and consistencies.

Special apparatus, e.g., closed systems in which the specimen is passed from one tube through a screen into an attached tube, and different types of kits are available from a number of major distributors and manufacturers.

Use of Surfactants in Sedimentation Procedures

In many applications of sedimentation methods a few drops of a surfactant, usually 20% solution of Triton X-100, are recommended especially when large amounts of mucus and/or fatty materials are present in the specimen. Directions may include a surfactant as a regular part of the procedure. These products help break up the mucus masses which allows the release of the parasites that may be trapped in them; however, if gauze or a fine-mesh screen is used for straining the specimen, mucus clumps may be held back and may not be present in the sample to be concentrated.

A product called Mucolex (polyoxyethyl, 3.1%; ethanol, 1.4%; formaldehyde, and distilled water; Lerner Laboratories) is used in some laboratories instead of Triton. Burrows (1968) suggests that three other surfactants may be substituted for Triton: Pluronic P-75, manufactured by Wyandotte Chemicals Corporation, Wyandotte, MI. (The solution is made by dissolving 10 g of the white paste in 100 mL of distilled water); Irium (sodium lauryl sulfate), Fisher Scientific Company; and Brij 35, manufactured by Atlas Powder Company, Wilmington, DE.

One drop of any of the above surfactants can be substituted for 1 drop of Triton NE (or 3 drops of a 20% solution of Triton X 100). Adding a surfactant to the centrifuge tube as a part of the procedure is usually necessary only when there is excessive mucus or fatty material in the specimen and it passes through the screen into the centrifuge tube. The author has found that the results usually are not improved by adding a surfactant when concentrating specimens with low amounts of fat and/or mucus. If a surfactant is used, it may be better to add it to the specimen in the vial to help break up the mucus clumps and free the parasites before straining the portion to be concentrated.

CONSED Sedimenting System

The CONSED System* was designed especially for use with MIF- and PIF-fixed fecal specimens but can be used with SAF-, PAF-, or formalin-fixed specimens. The system calls for a unique sedimenting solution, straining funnel, and procedure. The funnel screen should have a diameter or width of 1.0 in. or greater, with openings about 1.0 mm^2, and be made of a material that does not hold a charge. The procedure used for the CONSED Sedimenting System can be used with 10% formalin substituted for the CONSED sedimenting solution, but the yield is not as great (see Table 1).

CONSED Sedimenting System for Preserved Specimens:

PROCEDURE

1. Place an appropriate straining funnel in a 15-mL centrifuge tube, shake the specimen in the vial, and remove the vial cap.

2. Slowly pour most of the specimen through the funnel into the centrifuge tube. Do not overfill the centrifuge tube. If the screen clogs while straining, a spatula or an applicator stick may be brushed across the screen to help the fluid portion with its suspended materials to pass into tube. *Note*: Parasites fixed in PIF or MIF tend to stick to gauze and metal screens. A screen 1 inch or greater in diameter or breadth, with openings 1 mm^2 or slightly greater, gives the best results. If a proper screen is not available, it is sometimes better to eliminate the straining step.

3. Return all but 1 mL of specimen (slightly less if specimen is thick and more if thin) back into the original vial, and recap the vial.

* Materials for the CONSED Sedimenting System are available from Alpha-Tec Systems, Inc., PO Box 17196, Irvine, CA 92713.

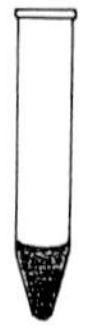

Figure 3. Portion of specimen to be sedimented.

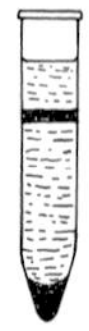

Figure 4. Centrifuge tube after sedimentation.

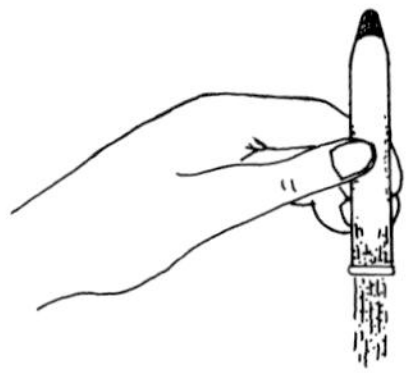

Figure 5. Holding the tube correctly.

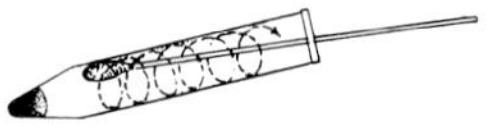

Figure 6. Removing debris from inside wall of tube.

Figure 7. Adding diluting solution to the plug.

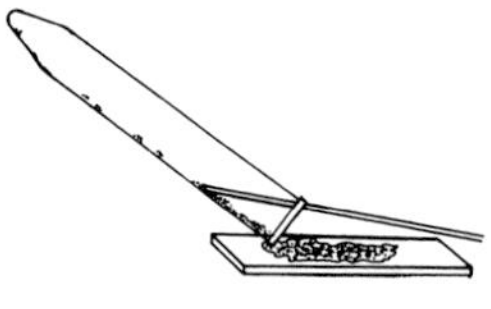

Figure 8. Transferring plug material onto a slide.

COARSE MATERIAL

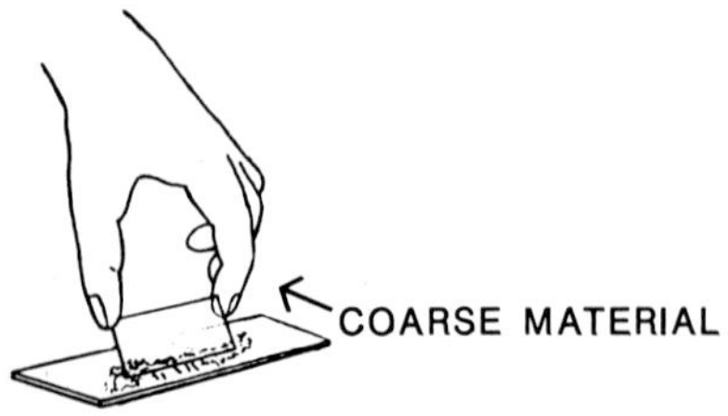

Figure 9. Adding cover glass, step 1.

COARSE MATERIAL

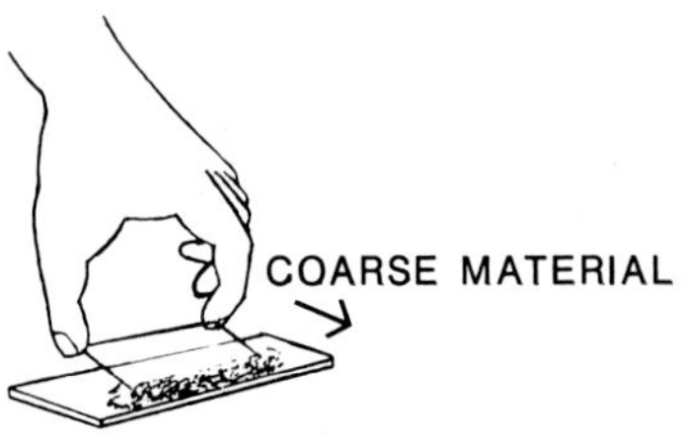

Figure 10. Adding cover glass, step 2.

4 Add 8 mL of CONSED solution to the portion of the specimen suspension in the tube (about 1 mL), cap the tube, and shake it well. Remove the cap, add 4 mL of ethyl acetate, recap tube, and shake vigorously. Place the tube in a centrifuge with a free swinging head, spin for 2 minutes at about 600 xg, and allow the centrifuge to come to a complete stop. The result should be a tube with four layers: a top layer of mostly ethyl acetate, an interface layer of fatty fecal debris, a lower solution layer, and a plug containing eggs and parasites that are present.

5 With the stick end of a cotton-tipped applicator, free the interface ring from the tube. Hold the tube so that the plug is always visible, invert the tube to a vertical position, pour out the three upper layers leaving the plug undisturbed. (If the plug begins to break up, immediately turn the tube to a horizontal position to save all of the plug.)

6 Keeping tube inverted (or horizontal), remove the debris adhering to the side of the tube using the cotton-tipped end of the applicator stick.

7 Turn the tube upright and add a few drops of diluting solution to the plug (an amount equal to the size of the plug), mix with a wooden applicator stick, and allow the tube to stand for 5 minutes. Do not remove the applicator stick.

8 Tip the tube to pour the material in the tube on to a glass slide using the applicator stick to assist in pulling all of the materials on to the slide.

9 As for preparing slides directly from the specimen vial, use a 22 × 40 mm (or 22 × 50 mm) cover glass. To keep the coarse material from under the cover glass, hold it by the edges with the thumb and forefinger, tilt the slide slightly, and use the cover glass to pull the material on the slide to one edge.

10 Lift the cover glass and reverse the slide. Push the wet edge of the cover glass across the slide into the fluid portion of the sample. Keeping the coarse material at the edge of the slide, lower the cover glass to allow the fluid portion of the sample to flow across the slide under the cover glass.

11 Touch the edge of the slide to a paper towel to remove some of the excess fluid and material. Remove the remaining coarse material from the edge of the slide with a cotton-tipped applicator stick. Any remaining excess fluid may be removed by touching the edge of the slide again to a towel or with an applicator stick to make the sample as thin as possible and free the edges of moisture before the vaspar is added.

Figure 11. Removing the coarse material.

12 The cover glass is ringed with vaspar to attach it to the slide and seal the sample under the cover glass. Equal parts of paraffin and vaseline are melted together at 150 to 170° F and applied to the edges of the cover glass with a cotton-tipped applicator stick.

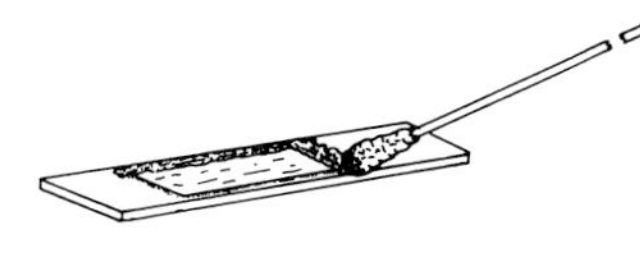

Figure 12. Ringing the cover glass with vaspar.

13 Wet mounts prepared from concentrated specimens should be scanned using 100× magnification to find parasites and eggs and higher magnification for identification when necessary. The entire specimen under the cover glass should be examined.

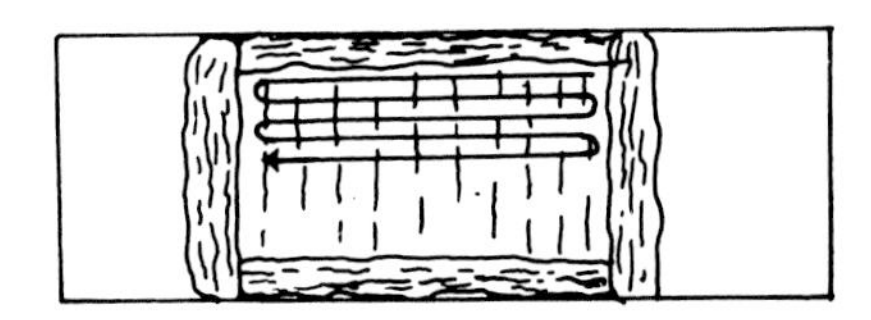

Figure 13. Scanning the prepared slide.

The author has found the procedure described here for transferring the plug to a slide after centrifugation and adding the cover glass to be superior to using a pipette to transfer a portion of the specimen onto a slide when preparing a wet mount. When there are few organisms and a pipette is used, some of the eggs or parasites may be trapped in the pipette, especially when a glass or plastic pipette with a bulb is used. Also, organisms present may remain stuck to the sides in the bottom of the centrifuge tube whereas when the plug is transferred to the slide immediately, less of the material containing eggs and parasites gets trapped in the tube.

When the procedure has been carried out properly, eggs and/or parasites are rarely found in the portion to be discarded and, when found, those under the cover glass are in greater numbers. Only when the number of parasites in the plug is very great are any numbers of organisms found in the portion to be discarded.

Although the author recommends using preserved specimens, it is appropriate to mention briefly the handling of unpreserved specimens.

Concentration of Unpreserved Specimens
Special consideration should be given to the procedure to be used when an unpreserved specimen is submitted for examination. First, chemical changes have taken place in the specimen. The degree of change depends on how long it has been since the specimen was passed, on the bacterial enzymes and chemicals in the specimen, and on the temperature to which the specimen was exposed. Of course, a part of the fecal specimen should be fixed in a proper collecting/preserving solution as soon as practicable to prevent further deterioration of any parasites and eggs that might be present. If there has been a delay, results may be improved by washing the portion to be fixed in 1% saline to remove some of the chemicals that may adversely affect the organisms before fixation. Perhaps the best course of action is to preserve one portion directly and another after washing. The following procedure may be used for washing the specimen.

PROCEDURE

1 Using applicator sticks, thoroughly mix, without straining, 3 to 4 g of the unpreserved fecal specimen with 10 to 12 mL of 1% saline in a conical, polypropylene centrifuge tube.

2 Centrifuge at from 500 to 550 xg for 1 minute and allow the centrifuge to come to a complete stop.

3 Pour off the supernatant fluid without losing any of the plug.

4 Repeat once or until the supernatant fluid is relatively clear. The plug and saline should be thoroughly mixed before each centrifugation.

At this point several alternatives are available:

5a From 0.3 to 0.8 mL of the plug may be concentrated by the CONSED procedure below. Trophozoites will not be recovered by this procedure.

5b The plug may be placed in an appropriate collecting/preserving solution for fixation. If trophozoites are still present and in good condition, they may be recovered from the preserved specimen. The specimen in the vial should stand for at least 1 hour before proceeding further.

5c If the time delay is unacceptable, the washing step may be omitted and 3 to 4 g of the specimen may be thoroughly mixed in 10 to 14 mL of CONSED sedimenting solution to fix the specimen. The mixed specimen is then poured into a vial and dealt with as a preserved specimen. If trophozoites are present and this procedure is followed, they may not be recognizable after concentration since they were not fixed properly before the procedure was carried out.

In order to prepare permanently stained slides from the specimen, a portion of the fecal specimen (or the plug after washing with saline) should be fixed in one of the appropriate fixatives as soon as the specimen is available.

As an alternative, fecal films may be prepared on slides from the unfixed specimen, either before or after washing, and fixed in Schaudinn's fixative and stained by an appropriate staining method (see Stains and Staining Methods).

Formalin-Ether Sedimentation

The Original Method

The formalin-ether sedimentation method was introduced by Richie in 1948 (also called 406th MGL Method after the U.S. Army Medical General Laboratory in Japan where it was developed). Although other methods were better for eggs of a specific group of parasites such as the schistosomes, of all concentration methods tested, formalin-ether sedimentation gave the best overall results for concentrating all of the different kinds of eggs and cysts, especially when different kinds of eggs and cysts were present in the same specimen. Juveniles (larvae) were fairly well concentrated but the procedure was ineffective for trophozoites.

Modifications to the Original Method

Over the years of use, a method may go through a number of modifications. In the original procedure, straining a portion of the fecal specimen through several layers of wet cotton surgical gauze was called for to screen out coarse material present in the fecal sample. Price, in 1977, modified the procedure for MIF-fixed specimens and omitted the straining step using gauze because the parasites preserved in the MIF solution tended to adhere to and be retained in the gauze. Special funnels with plastic or metal screens have been developed that can be substituted for the cotton gauze.

Because of the danger of using ethyl ether in the clinical laboratory environment where parasitology is usually performed, an effort was made to find a substitute for it. Young et al. (1979) published a procedure substituting ethyl acetate for ethyl ether and achieved essentially equal results. Ethyl acetate is now considered the solvent of choice.

Formalin-Ethyl Acetate Procedure for Formalin-Fixed Specimens

The original formalin-ether sedimentation procedure and modifications of it for preserved specimens are being used in many laboratories. The procedure described is applicable for specimens preserved in formalin or in another appropriate collecting/preserving solution (e.g., SAF, MIF, PIF) and it assumes that the specimen is well mixed and in a collection vial with a tight-sealing cap. Ethyl ether may be substituted for ethyl acetate.

Materials and equipment for the procedure are

- Formalin, usually 10%, preferably buffered (buffered formol-saline may be substituted)
- Ethyl acetate (or ethyl ether)
- Conical, 15-mL polypropylene centrifuge tubes with caps
- Cotton-tipped, wooden applicator sticks
- Wooden applicator sticks

- A straining funnel
- Centrifuge with a swing-free head, calibrated

Similar materials and equipment are used in all sedimentation procedures except the stock sedimenting solution may be changed and, in some, special apparatus may be substituted.

PROCEDURE

1 Shake the specimen vial vigorously to make certain that the specimen is evenly mixed in the fixing solution. It may be necessary to use applicator sticks to break up any large particles of fecal materials present in the vial before proceeding further.

2 Pour 2 to 4 mL of the suspended specimen through an appropriate screen into a 15-mL, conical, polypropylene centrifuge tube.(The author recommends straining 10 to 12 mL of the fecal suspension into the centrifuge tube, capping the tube, shaking the tube vigorously, and pouring back into the original vial all but 1.5 to 2.0 mL of the suspension. The portion in the centrifuge tube is used in the procedure.)

3 Add formalin to make a 12 to 14 mL suspension and mix thoroughly.

4 Centrifuge for 2 minutes at 500 to 550 xg and allow the centrifuge to come to a full stop. Carefully pour off the supernatant fluid by quickly turning the tube to a vertical position and then to a horizontal position to retain all of the plug (see Figure 5). If the supernatant fluid is cloudy, repeat this step.

5 Add 8 mL of the formalin solution to the plug, mix thoroughly, add 4 mL of ethylacetate (or ethyl ether), cap, shake vigorously, release gas by carefully loosening the cap, then tightly recap the tube.

6 Centrifuge 2 minutes at the same centrifuge speed as above and allow the centrifuge to come to a full stop. A tube with four layers will result (see Figure 4): a base plug, a fluid layer, a solids interface layer (containing mostly fatty material), and an upper fluid layer (mostly ethyl acetate or ether).

7 Using the clear end of a cotton-tipped applicator stick, free the interface layer from the sides of the tube and quickly turn the tube to a vertical position to pour off the upper three layers while retaining the plug (as in Figure 5). Always hold the tube in a vertical position so that the plug is visible to make certain that a portion of the plug is not poured off with other material. If the plug begins to break up, immediately turn the tube to a horizontal position to retain all of the plug.

8 Remove the debris adhering to the sides of the tube using the cotton-tipped end of the applicator stick, add four to six drops (usually an amount equal to the size of the plug) of formalin (or Dobell and O'Connor's iodine) to the plug, and mix well. If iodine is used, it may be added to either the sample in the tube or to the portion placed on a slide (see Figures 6 and 7).

9 Use a pipette to transfer a portion of the sediment to a glass slide and add a cover glass to prepare a wet mount for examination. Most workers recommend an unstained and an iodine-stained wet mount. Both may be prepared on the same slide if 22 × 22 mm cover glasses are used or if slides wider than 1 inch are used. The author prefers making one slide with saline and one with iodine and using 22 × 40 mm or 22 × 50 mm cover glasses.

Cysts of protozoa and/or eggs and juveniles of helminths present in the specimen vial should be present on the slide. If the plug is large, it may be necessary to prepare several slides to find all organisms present. Wet mounts may be prepared by another method which may reduce the number of slides needed and improve the quality of the slides for examination (see CONSED Sedimenting System for Preserved Specimens, Figures 5 to 13). This is the method described in the original formalin-ether procedure by Richie (1948). The author recommends that all wet mounts be ringed with vaspar to seal the specimen under the cover glass (Figure 12).

Occasionally, specimen vials received by the laboratory do not have enough feces to follow the usual procedure. Levine and Estevez (1983) proposed miniaturizing the procedure to accommodate small amounts of feces. The author feels that using a larger portion of what is in the specimen vial, sometimes the entire specimen, may be a preferable approach. When the specimen is especially small, the straining and washing steps should be omitted and, after concentration, the plug should be transferred to a slide and the cover glass mounted as in Figures 8 to 12.

Acid-Ether Methods

General

The effectiveness of these and other sedimentation methods depends on the separation of coarse, fatty, and oily materials from the eggs and parasites and the clarification of the fecal detritus remaining. Many workers have tested different combinations of acid and ether to concentrate eggs and cysts. The author believes that ethyl acetate can be substituted for ethyl ether in the procedures. The methods do not concentrate trophozoites of protozoa and usually leave them unrecognizable. Two of the methods are given here because of their value in concentrating the heavier trematode eggs, especially those of schistosomes.

Sodium Sulfate-Acid-Triton-Ether Concentration (AMS III Method)

The method was developed at the Army Medical School (AMS) (Hunter et al., 1960) to concentrate helminth eggs and juveniles of nematodes, and it is probably the best one available for concentrating schistosome eggs. It is usually performed on unpreserved specimens but will work on formalin-fixed specimens, although the yield is not as great.

The solution used for the AMS III method is prepared as follows:

- Add together 45 mL of concentrated HCl (37%) and 55 mL of water to make 100 mL (sp. gr. 1.089).
- Dissolve 9.6 g of anhydrous sodium sulfate in 100 mL of water (sp. gr. 1.080).
- Adjust the specific gravity of each solution to 1.080 before mixing.
- Add approximately equal amounts of the two solutions to prepare the working solution.

PROCEDURE

1 Thoroughly mix about 2 g of feces with 14 mL of sodium sulfate-HCl sedimenting solution in a wax-free container and strain through 2 layers of wet gauze into a 15 mL conical, centrifuge tube. (An appropriate plastic funnel and screen may be substituted for the surgical gauze.)

2 Centrifuge at 500 to 550 xg for 1 minute and carefully pour off the supernatant fluid so as not to lose any of the plug.

3 Again, add the sedimenting solution, mix it with the plug, and spin. Repeat the washing step until the supernatant fluid is clear.

4 After the last wash, add 8 mL of the sedimenting solution, 3 drops of a 20% solution of Triton X-100, and 5 mL of ethyl ether to the fecal plug. (The original procedure called for 5 mL of HCl-Na_2SO_4 and 3 mL of ether. Larger volumes of these solutions are easier to use and may give better results. The author found no report on substituting ethyl acetate but believes it could be used instead of ethyl ether.)

5 Cap and invert the tube, shake vigorously, turn the tube upright, and release the cap to allow any gas to escape.

6 Tighten the cap and spin the tube in a centrifuge at 500 to 550 xg for 2 minutes. A tube with four layers, as in the formalin-ether procedure, will result: an upper fluid layer, an interface ring, a lower fluid layer, and a base plug.

7 Free the interface ring with an applicator stick, invert the tube to a vertical position to pour off the upper layers making certain not to lose any of the plug. (The plug is usually not as well packed as in the formalin procedures so special care must be taken to save the plug.)

8 Clean the inside of the tube with a cotton-tipped applicator stick and add a few drops of normal saline (or 10% formalin) and mix with the plug.

9 Transfer all of the material of the plug to a slide by pouring and dragging it with an applicator stick onto the slide (see CONSED Sedimenting System, Figures 5–8).

10 Add the cover glass as shown in Figures 9–12. The amount of specimen remaining in the plug following the procedure is greatly reduced and eggs present are relatively easy to find.

Mathieson and Stoll Technique

The Mathreson and Stoll technique uses a 15% solution of HCl (HCl, concentrated 40 mL and 60 mL distilled water, sp. gr. 1.080) and ether (Faust, 1949; Melvin and Brooke, 1982).

PROCEDURE

1 Mix 1 to 2 g of the specimen (a portion the size of a large marble) with 8 to 10 mL of HCl solution in a test tube or small beaker.

2 Strain approximately 2 mL of the suspension through 2 layers of wet surgical gauze into a 15-mL conical centrifuge tube and adjust level to 5 mL with HCl solution. (A plastic funnel and screen can be substituted for the gauze.)

3 Add an equal amount of ethyl ether (or ethyl acetate) to the tube and mix thoroughly with the fecal suspension.

4 Spin the tube at 500 to 550 xg for 2 minutes in a centrifuge. As in other centrifugation procedures, the interface ring is separated from the tube and is poured off with the supernatant fluid and the remaining plug is examined.

5 Follow the procedure in Figures 7 through 13 for preparing the slide.

Concentration of PAF-Fixed Specimens
Separate concentration methods are used for PAF-fixed specimens to recover either trophozoites of protozoa or cysts, eggs, and juveniles. The usual concentration methods, such as formalin-ethyl acetate, employed in the laboratory are said to be ineffective (Burrows, 1967b).

Concentration of Trophozoites Trophozoites can be somewhat concentrated by washing the specimen with saline.

PROCEDURE

1 Strain the specimen into a centrifuge tube through three layers of 10-cm wide gauze bandage placed in a small funnel (an appropriate funnel with a fixed screen can be substituted).

2 Add saline to about 12 mL level, mix well, and centrifuge at about 500 xg for 2 minutes.

3 Quickly turn the tube to a vertical position to pour off the supernatant fluid, being careful to save all of the plug.

4 With the tube horizontal, wipe inside wall, of the tube with a cotton-tipped applicator stick.

5 Repeat the washing step until the supernatant fluid is clear.

6 Add several drops of saline and mix with the plug.

7 Place a drop of stain on the slide and add a drop of the specimen plug to it, mix well, cover with a cover glass, seal with vaspar, and examine (Thionin or Azure A are used as stains; see Appendix, PAF Fixative).

PAF fixative with the recommended stains is excellent for trophozoites which are often found with pseudopodia still extended in the washed portion of the specimen.

Routine Concentration of PAF-Fixed Specimens
For finding cysts, eggs, and juveniles, the following procedure should be used for PAF-fixed specimens. In this procedure, a surfactant and ethyl acetate are included. This routine method for concentration apparently does not recover trophozoites that are identifiable.

PROCEDURE

1 Add saline (0.9%) to the 10 mL level and then add 2 drops of a 10% solution of Triton X-100 and mix well.

2 Add 2 mL of ethyl acetate, cap, shake, and spin in a centrifuge at about 600 xg for 3 minutes allowing the centrifuge to come to a complete stop.

3 Loosen the fatty interface ring with an applicator stick, pour off the top three layers, and wipe the inside of the tube as in other procedures.

4 Add about 2 mL of saline to the plug, mix, and examine as described above.

Eggs, cysts, and juveniles are found in the final concentrate. The method requires more time and attention to detail than some of the other methods since two procedures are needed. Protozoa fixed in PAF do not stain well on permanently stained slides stained by the Trichrome and Hematoxylin methods. Since satisfactory examinations are only possible on wet preparations, many investigators consider the method as being unsuitable for routine work and it has found limited use in the clinical diagnostic laboratory.

The MIFC (Merthiolate, Iodine, Formalin, Concentration) Technique

The concentration method was developed specifically for specimens fixed in MIF but works as effectively for specimens fixed in PIF. It is less effective than the formalin-ethyl acetate concentration method for specimens fixed in formalin (see Application of Concentration Procedures and Tables 1, 2, and 3). The method is essentially the same as the formalin-ethyl acetate method but substitutes the stock solution of the MIF technique (without iodine) for 10% formalin (see Appendix for MF stock solution formula).

PROCEDURE

1 Thoroughly mix the MIF-fixed specimen in the vial and strain approximately 9 mL through 2 layers of wet surgical gauze into a centrifuge tube. (A suitable plastic funnel with a fixed screen can be substituted for the gauze.)

2 Add 4 mL of ethyl ether (or ethyl acetate) to the suspension in the tube, cap the tube, and shake vigorously.

3 Release the cap to allow any gas to escape and allow the tube to stand for 2 minutes.

4 Tighten the cap, spin in a centrifuge at 450 to 500 xg for 1 minute, and allow the centrifuge to come to a complete stop.

5 As in other procedures, four layers will appear. Free the interface layer from the tube and pour out the supernatant fluid, retaining the plug at the base of the tube.

6 Keeping the tube upside down (or horizontal), remove the debris adhering to the inside of the tube with a cotton-tipped applicator stick.

7 Add a few drops of MF stock solution to the plug and mix.

8 As in Figures 8 to 12, transfer a portion of the plug to a glass slide, add a cover glass, and ring the cover glass with vaspar.

Trophozoites as well as cysts, eggs, and juveniles are found in the concentrate. The method provides a wet preparation with parasites and eggs stained so that they can be more readily identified.

The author prefers to strain the entire specimen through a funnel with a large mesh screen into the centrifuge tube and, after mixing, pour back into the original vial all but about 1 mL of the specimen. Add 8 mL of MF stock solution and 4 mL of ethyl acetate to the specimen in the centrifuge tube and continue the procedure as described.

Concentrating PIF-Fixed and MIF-Fixed Specimens

Several steps are omitted or changed when the formalin-ethyl acetate procedure is performed on specimens collected in PIF or MIF. The washing steps are omitted and a pipette is not used to transfer the specimen to a slide (see Figures 8 to 12). Iodine is not needed for preparing wet mounts because the fixing solution contains iodine in addition to eosin dyes. The procedures used are either the formalin-ethyl acetate method or the CONSED Fecal Concentration System, which usually gives a better yield. In the CONSED System, a special sedimenting solution is substituted for formalin.

Application of Concentration Procedures

In a series of studies, the effectiveness of four different sedimentation methods were compared with direct examination. Results are shown in Table 1. Juveniles of nematodes were not concentrated well using 10% formalin (with either ethyl ether or ethyl acetate) and the procedures using formalin were unsatisfactory for trophozoites of protozoa. Both formalin procedures concentrated eggs better than MIFC but the latter concentrated trophozoites and was better for juvenile nematodes. When the CONSED sedimenting procedure was used, on an

Table 1. Results Obtained in Testing Concentration Methods

	Increase in numbers found over direct examination (gravity sedimentation)				
	Helminths			Protozoa	
	Eggs[a]		Juvenile		
Stages of parasites	**Lighter**	**Heavier [b]**	**Worms**	**Cysts**	**Trophozoites**
Method[c]					
Formalin-ethyl acetate, Formalin-ether	3–6 ×	4–8 ×	*	3–7 ×	**
MIFC[d]	2–5 ×	3–7 ×	2–5 ×	3–6 ×	2–5 ×
CONSED[e]	4–10 ×	6–12 ×	2–6 ×	4–10 ×	3–6 ×

[a] Actual increase depends on species of egg as well as method.
[b] Lighter eggs (nematode and cestode), heavier (trematode).
[c] The apparatus and type of screen used affect the results.
[d] Merthiolate/iodine/formalin concentration (Sapero and Lawless, 1953; Price, 1978).
[e] Materials and procedures available from Alpha-Tec Systems, Inc., PO Box 17196, Irvine, CA 92713.
* Juvenile worms were either not found or were found in no greater numbers than in direct examination.
** Trophozoites were not found in concentrates.
Note: Only a centrifuge with a swing-free head should be used.

average all stages and all forms were concentrated better than they were with other methods. These results pertain only to specimens collected and preserved in the PIF, MIF, or TIF solutions.

The CONSED solution and procedure applied to specimens fixed in either formalin or SAF produced results at least equal to results obtained using the formalin-ethyl acetate method. The number of formalin- and SAF-fixed specimens used in these tests were not adequate to substantiate the results as constant for all specimens.

Examples of Results Obtained with the CONSED System

Results obtained before and after the CONSED sedimentation method was applied to specimens fixed in PIF or MIF are demonstrated in Figures 14–16. In a photomicrograph, the size of the field is approximately one third that of the visual field.

Where on direct examination less than 1 egg per visual field was found at 100× magnification, after CONSED sedimentation an average of 6 eggs per visual field were found. Such results were obtained repeatedly (see Figure 14).

When specimens in which trophozoites and cysts of protozoa were present and were concentrated by the CONSED sedimentation procedure, both trophozoites and cysts were found. In Figure 15, results of examination of a specimen with *Entamoeba histolytica* are shown before and after concentration. The opportunity of finding cysts was increased by about six times by concentrating the specimen.

The fact that trophozoites as well as cysts of protozoa can be found following concentration by the CONSED method is demonstrated in Figure 16 which shows *Trichomonas hominis* trophozoites following concentration.

There is no question that concentration of the specimen can improve the opportunity of finding cysts, eggs, and juvenile nematodes. Concentrating trophozoites in a fecal specimen without altering the level of concentration of cysts, eggs, and juvenile worms is another level of technology. A method that accomplishes this improves the ability of the individual to find and identify parasites and eggs and reduces the number of steps necessary. Selection of the ideal screen, the ideal sedimenting solution, and an appropriate method is the key to effective results. Under these circumstances, it is possible to eliminate the examination of the direct wet preparation and use the concentration to find all stages of all parasites that are present in fecal specimens. The protozoa in the wet preparation made from the concentrated sample are stained, and if stained adequately to make an accurate identification, then the standard procedure for the diagnosis of intestinal parasites can be accomplished by the application of one method.

Some laboratory personnel may have difficulty identifying some protozoa in wet preparations. If the aids in this manual are used, identification should be possible. When protozoa are found in the wet preparation but identification is questionable, permanently stained slides can be prepared from the specimen in the collection vial. The organisms found on permanently stained slides should be compared to those seen in a wet preparation to insure confidence and to make certain that the same organisms are present on the film that were seen in the wet preparation. As individuals become more familiar with identifying protozoa in wet preparations, the need for preparing a permanently stained slide for identification becomes less necessary. It is usually a matter of knowing how protozoa appear in wet preparations and knowing how to properly use the microscope for examining wet preparations.

A repository of permanently stained slides is required by some regulatory agencies as a permanent record on positive specimens, and the policy is followed in most laboratories. The permanently stained slide should not be necessary for making a diagnosis but may be used for confirmation and for the permanent record. Positive specimens should also be retained for review, teaching, and for preparing additional permanently stained slides.

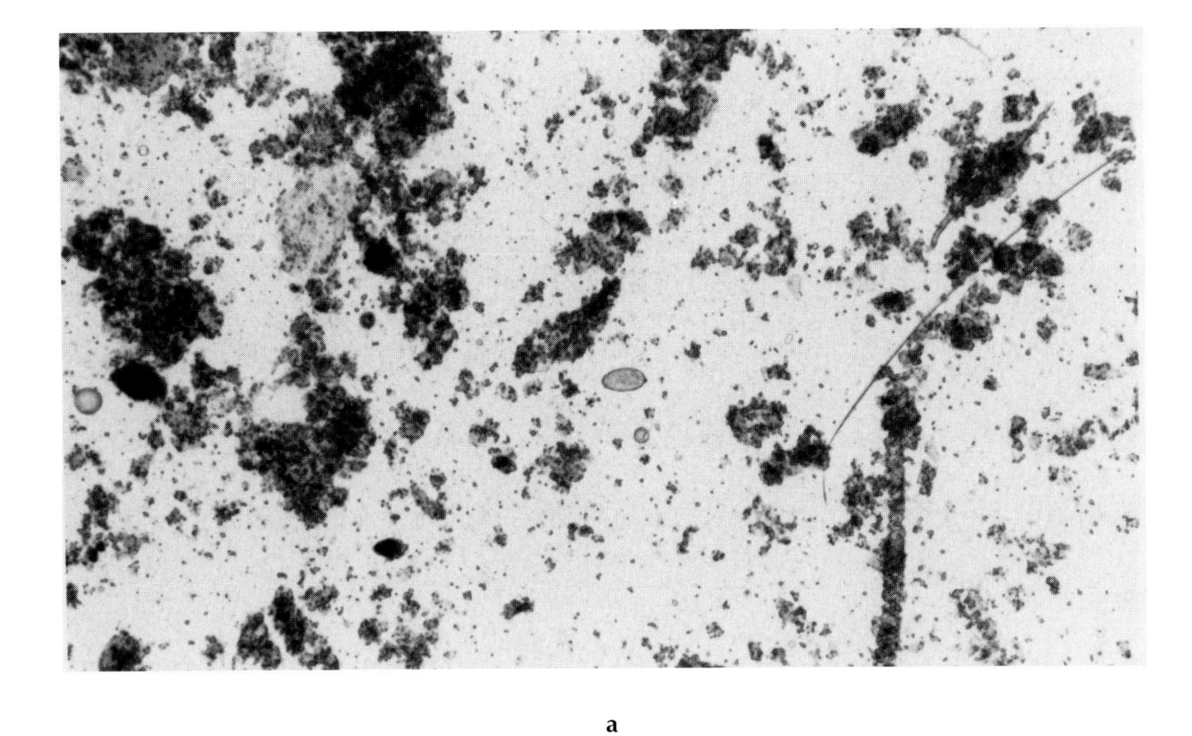

a

b

Figure 14. Number of eggs found before (a) and after (b) CONSED sedimentation. (a) Before sedimentation. An average of less than 1 egg per visual field was found at 100× magnification based on the examination of 40 fields on each of 3 different slides prepared directly from the preserved specimen. One *Clonorchis sinensis* egg is shown in the center of this photomicrograph. (b) After sedimentation. An average of six eggs per visual field were found. Six eggs are seen in this photomicrographic field, one *Trichuris trichiura* and five *Clonorchis sinensis*.

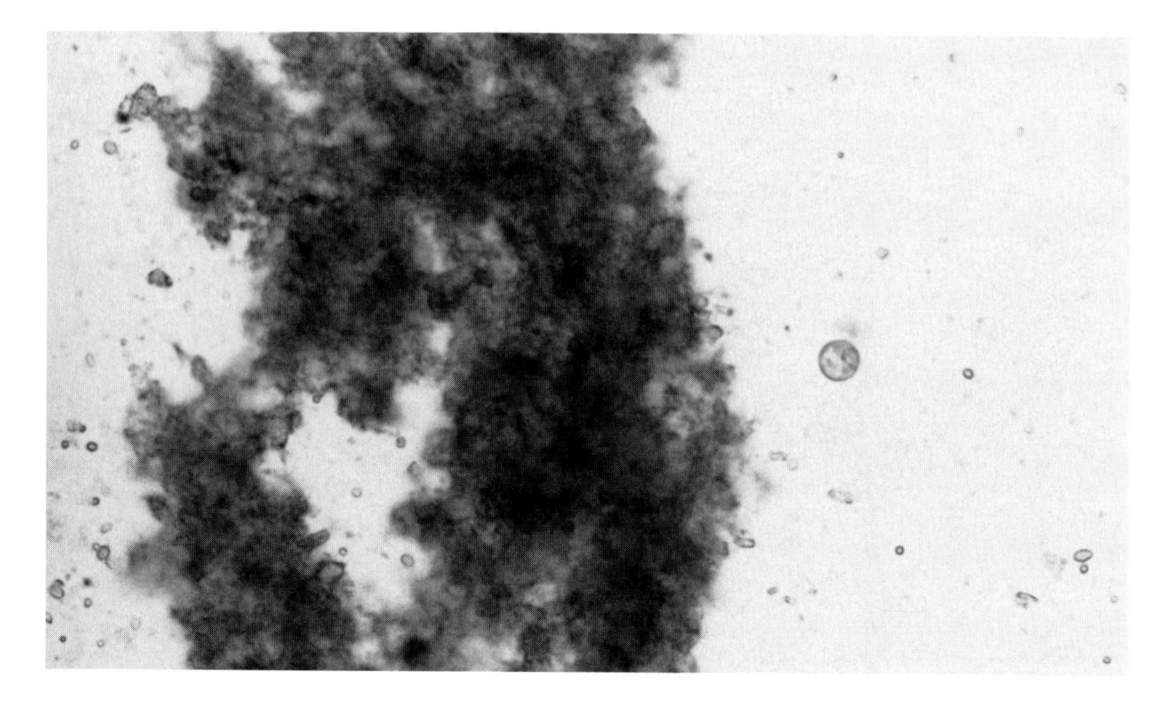

a

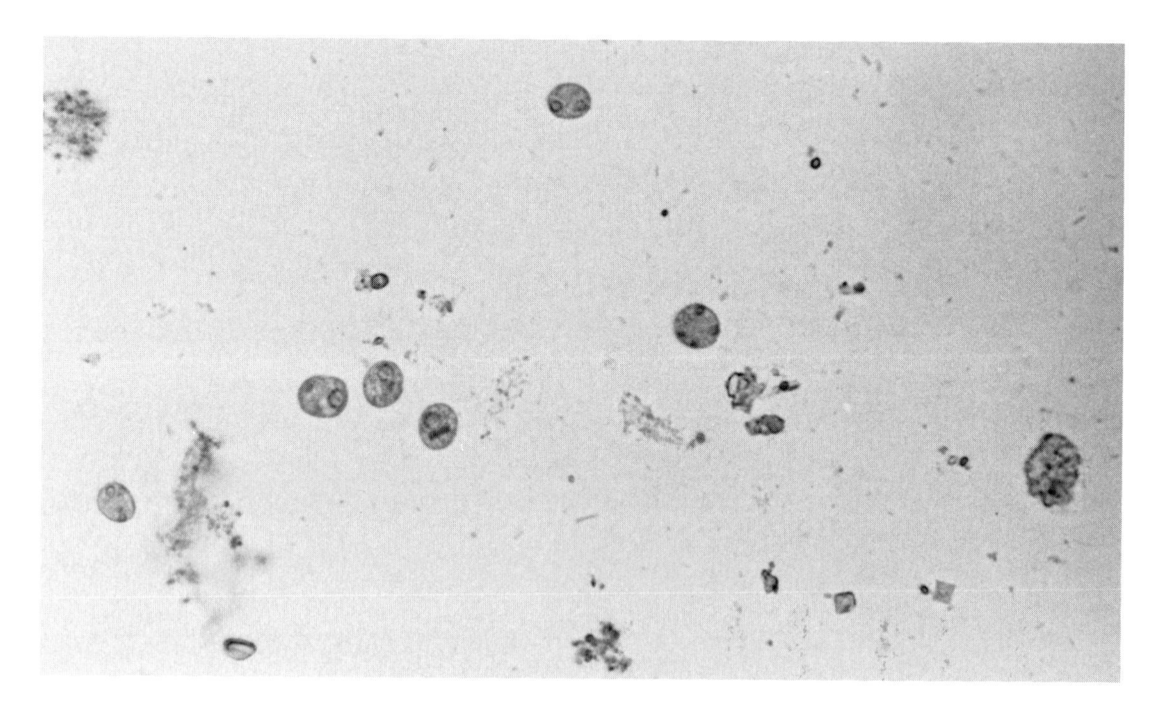

b

Figure 15. The number of protozoan cysts found (a) before and (b) after CONSED sedimentation. (a) Before concentration. On direct examination an average of less than one cyst of *Entamoeba histolytica* in a visual field at 100× magnification was seen. One cyst is shown in the photomicrographic field. (b) After concentration. At least six cysts were seen in every visual field and trophozoites were seen in some fields. Six cysts are seen in this photomicrographic field.

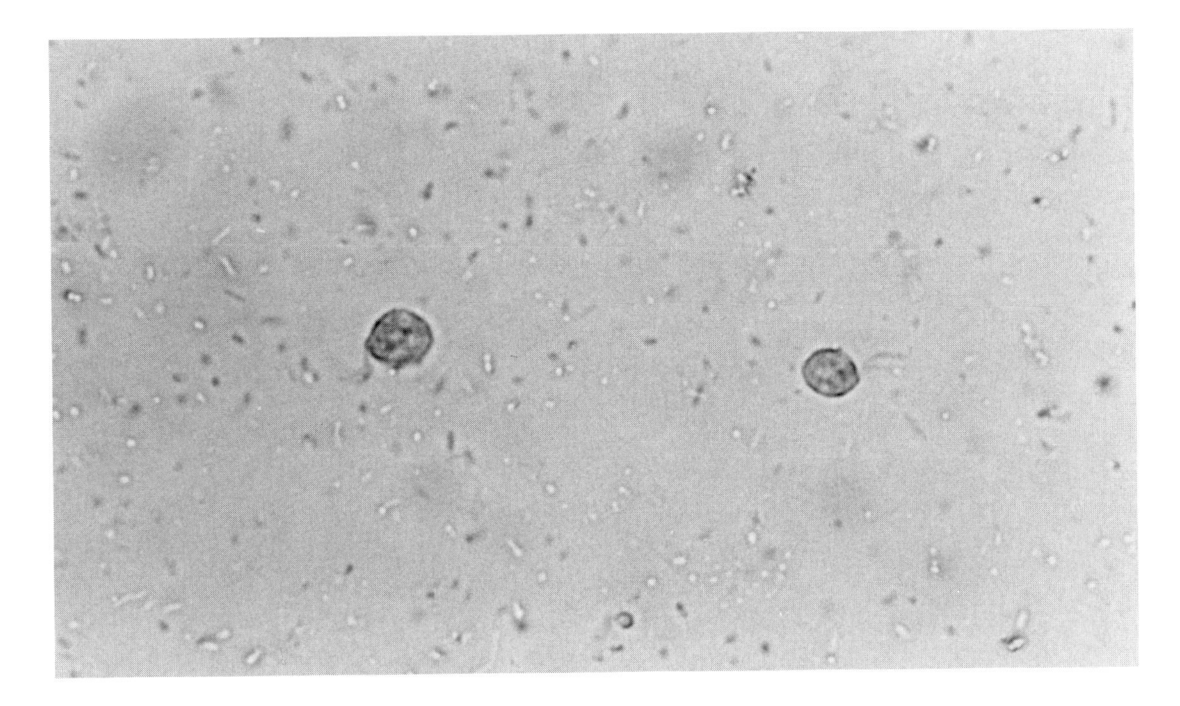

a

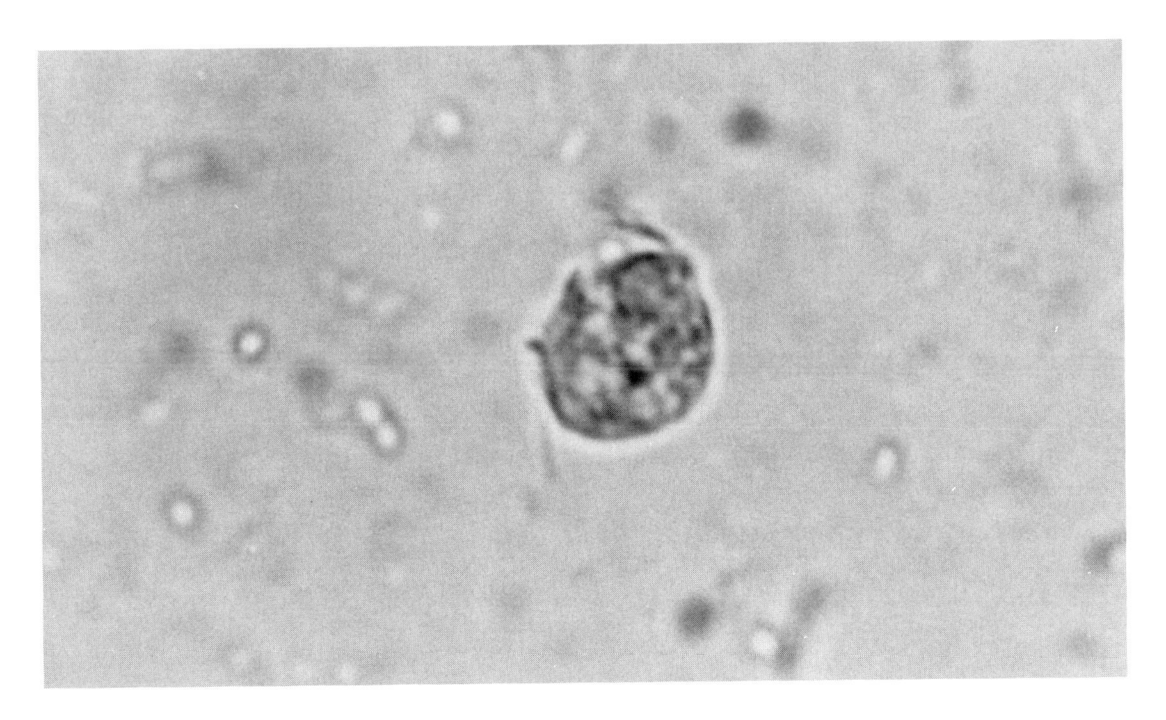

b

Figure 16. Trophozoites of *Trichomonas hominis* found following CONSED sedimentation: (a) one photomicrographic field after sedimentation; (b) an enlargement of one trophozoite. (a) Fewer than one trophozoite was found per ten visual fields on direct examination of the specimen. Following concentration, an average of eight trophozoites were found in ten visual fields. Those in one photomicrographic field are shown here. (b) The photomicrograph of one trophozoite is enlarged to show that they are apparently not affected by the concentration procedure.

Chapter 4

Stains and Staining Methods

STAINS AND STAINING METHODS

Some members of regulating agencies, investigators, and laboratory personnel feel very strongly that protozoa cannot be identified routinely on wet preparations and that a permanently stained fecal film should be made from every soft or unformed fecal specimen received by the laboratory. Protozoa can be recovered from hard, formed fecal specimens as well as from wet, watery ones. If the laboratory director feels that permanently stained slides are necessary for identifying protozoa, then permanently stained slides should be prepared on all fecal specimens.

If in the formalin/PVA system, a wet mount (formalin) and a permanently stained fecal film (PVA) on every fecal specimen arriving at the laboratory are to be examined, technician time will be increased. It takes considerable time for a technician to prepare the specimens for examination and to examine both a wet mount and a permanently stained fecal film. Some activities, such as preparing solutions, organizing materials, etc., may be carried out during the waiting times included in the staining procedure. If permanently stained slides are to be made on every patient's specimen, there are a number of different stains and methods from which to choose. Some of the methods used for staining intestinal protozoa fixed in different collecting/preserving solutions that perform best in various laboratory environments are given in this Section.

PREPARING PERMANENTLY STAINED FECAL FILMS

Several hematoxylin stains are suitable for staining fecal films containing intestinal protozoa. Heidenhain's iron-hematoxylin is perhaps the best method but the long method may take several days. The procedure is complicated because destaining must be carefully carried out using the microscope to observe the process and to determine when destaining has reached the point when films must be moved to the next solution. The method is rarely used for routine parasitology in the diagnostic laboratory because of the time and detailed work involved. For detailed information, refer to Melvin and Brooke (1982).

The phosphotungstic acid iron-hematoxylin method of Tompkins and Miller takes less time and destaining does not require control by microscopic examination. This method is recommended for preparing permanently stained fecal films from SAF-fixed specimens and will work satisfactorily for protozoa fixed in other collecting/preserving solutions (see Table 3 and Appendix).

Trichrome-type stains are most commonly used for intestinal protozoa. Of these, Gomori's One Step Trichrome Stain and the modification by Wheatley (1951) are the most popular. Polychrome IV stain, a closely related stain, was developed specifically for preparing permanently stained fecal films from specimens containing protozoa and fixed in MIF and PIF. It is also suitable for preparing permanently stained slides from specimens fixed in SAF and PVA (see below and Appendix).

In many hospital laboratories, the prepared slides are actually stained in the histopathology section and the stains that are used are those regularly used for staining tissue sections, e.g., Mallory's iron-hematoxylin and Gomori's trichrome. For fecal films prepared from specimens fixed in Schaudinn's solution or PVA, iodine alcohol is required for the removal of mercuric chloride crystals before staining.

PREPARING FECAL FILMS FOR PERMANENT STAINING

General Considerations

In any staining process, there are certain steps that should be considered, regardless of the staining procedure used, to ensure good and consistent results. These steps are often not included as a part of the procedure. Some of these steps are listed here.

Using Clean Slides

Slides used in the staining process must be especially clean in order for the stain to work properly and for the film of the specimen to adhere to the slide during the staining process. This is especially true for fecal films prepared from older specimens. To test if the surface of the slide is clean, a small drop of water is placed on the slide. If the drop rounds up and does not spread out when the slide is slowly tilted to about a 30° angle, the surface is not clean. On the other hand, if the drop spreads out and runs easily, the slide is probably clean enough to use.

In most cases, slides available in the laboratory need to be cleaned further, usually with 70% alcohol. To quickly clean the surface of a slide, dampen a clean 4 × 4 gauze pad with 70% alcohol and vigorously rub the surface of the slide that will be used. Immediately polish the wet surface with a clean, dry 4 × 4 gauze pad. Use only the cleaned surface of the slide for spreading a fecal film. Specially prepared slides can be purchased that supposedly help films adhere to the slide.

Fixing Fecal Films to the Slide

Polyvinyl alcohol is combined with Schaudinn's fixative to make PVA collecting/preserving solution. The PVA helps the fecal film adhere to a slide without appreciably affecting the staining of protozoa. PVA-fixed specimens will adhere to the slide if films are properly made. Samples of specimens fixed in PVA should be spread thinly and evenly to the edges of the slide to help prevent the film from peeling off during the staining process.

Before it was readily available commercially, many laboratories prepared their own PVA fixative. They found the process cumbersome and time consuming and often had difficulties with films peeling off during the staining process. It is usually preferable to purchase vials of PVA from a reliable commercial source where quality control ensures a more consistently reliable product.

Mayer's Albumen Fixative, or another material such as serum, is most often used to help films made from specimens fixed in SAF and the MIF-type solutions adhere to the slide. The adhesive substance is applied to the slide prior to spreading the specimen sample on the slide. Mayer's Albumen Fixative is available commercially and is composed of 50% albumen and 50% glycerin. It should be relatively fresh to be effective.

Scholten (1972) recommended using albumen-glycerin fixative instead of PVA powder. Fecal specimens are fixed directly in Schaudinn's fluid. To help the fecal films adhere to the slide, a portion of a sedimented fecal specimen is mixed with the albumen-glycerin fixative on a slide, smeared over the surface of the slide, and stained by the usual methods.

Fecal films wash off slides no matter what procedures are used, what precautions are taken, or from what collecting/preserving solutions the film is made. If only one slide washes off, another should be prepared and the procedure repeated. If several slides are being run at the same time and most of the films wash off, make certain the slides are clean before additional films are prepared. It may be advisable to recheck the detailed steps in the procedure and the solutions being used before proceeding further.

Carryover of Chemicals During Staining

During the staining process, the carryover of chemicals should be kept to a minimum when slides are being moved from one solution to another. As a slide or slide carrier is moved, this is best accomplished by holding the slide or slide carrier over the staining vessel to allow it to drain, then touching it to some absorbent material such as a paper towel or blotting paper, and finally gently shaking the slide or carrier to remove any excess fluid remaining (drain-blot-shake). Such steps should be taken each time a slide or carrier with slides is moved from one solution to another.

Preparing Fecal Films from Specimens Fixed in PVA

Various ways have been described for preparing films from specimens fixed in PVA for staining. Probably the most commonly used method is that described by Melvin and Brooke (1982).

- Using a pipette, draw up a small amount of the fecal sediment from the bottom of the vial.
- Wipe the outside of the pipette, transfer two or three drops of the specimen from the pipette to a clean glass slide, and evenly spread the sample over the surface of the slide.

A rolling or dabbing motion is preferred over smearing the specimen to spread it on the slide. To help prevent peeling, the fecal film should extend to the edges of the slide (Figure 1). The author has found the method described below for use with PIF- and MIF-fixed specimens to be satisfactory (see Figure 3).

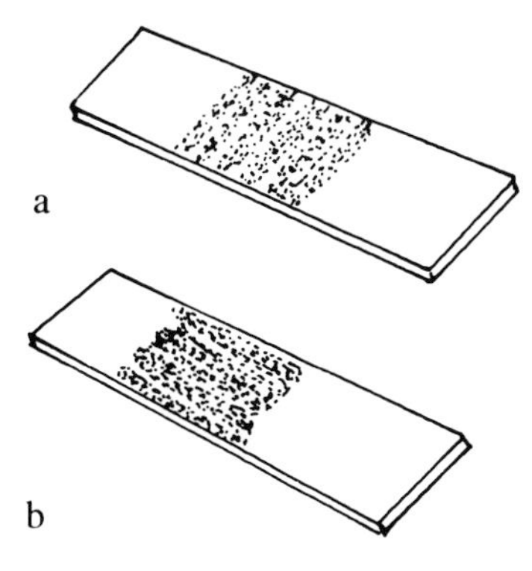

Figure 1. Applying fecal films from specimens fixed in PVA. (1) A few drops of the specimen are placed on a slide. (2) An applicator stick is moved back and forth through the specimen to spread it evenly over the slide to produce a relatively thin film extending to the edges of the slide.

A procedure recommended by Garcia and Bruckner (1988) consists of thoroughly mixing the fixed specimen with the PVA in the vial, pouring a portion of the mixture onto a paper towel, allowing the towel to absorb most of the PVA solution, and preparing slides from the partially dried sample on the towel.

If the specimen in the vial is lumpy and appears to be inadequately mixed with the PVA solution when it reaches the laboratory, Garcia recommends remixing the specimen, spreading the film on a slide, and immediately postfixing the fecal film on the slide in Schaudinn's fixative for 30 minutes or more.

The films prepared from PVA-fixed specimens should be thoroughly dry before staining. Films may be dried overnight at room temperature or in a 37°C incubator for an hour or more. Some laboratory personnel have been able to successfully stain slides after they have dried for 1 hour at room temperature. The slide itself must be exceptionally clean and the fecal film of the proper thickness to prevent the material from peeling off the slide during staining. Protozoa on dried films retain their staining quality for several months at room temperature.

Preparing Fecal Films From SAF or MIF-Type Fixatives

A number of different ways to prepare slides for staining have been described. The usual method for specimens fixed in SAF, MIF, or PIF involves using a slide on which the area over which the fecal film is to be spread has a film of Mayer's Albumen Fixative. The fixative is made of equal parts of albumen and glycerin, is available commercially, and should be relatively fresh. To test its quality, a slide with a smear of the fixative can be stained to make certain that the fixative does not form ridges or stain in a blotchy manner. If it does not stain evenly, either the fixative is old or the slide is not clean.

A drop of the albumen fixative is placed on a slide and immediately spread over the slide using either a cotton-tipped applicator stick or a finger (Figure 2). The adhesive solution used should not be excessive and should be evenly spread over the surface of the slide after which the slides should be laid flat to allow the material to come to an even thickness. Fecal films prepared between 5 and 15 minutes after the adhesive is added to the slides appear to give good results.

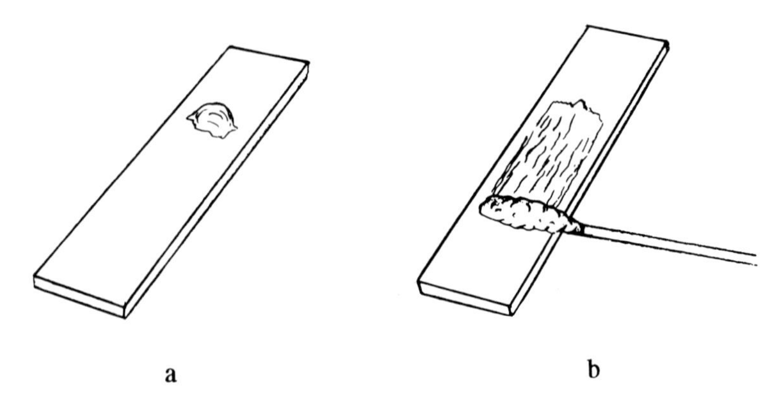

Figure 2. Coating a slide with Mayer's albumen fixative (a) fixative added to the slide; (b) spreading the fixative over the slide surface.

Using a straight-sided glass (or plastic) pipette of proper diameter, a small amount of the fecal specimen is pipetted by capillary action from the appropriate level in the specimen vial (see Chapter 2, Figures 3 and 4). This is especially true for fecal smears prepared from specimens fixed in MIF-type solutions. The outside of the pipette is wiped free of material. While holding the pipette vertically so that its lower opening with the specimen can rest on the slide, it is passed carefully over the area coated with Mayer's albumen fixative applying small amounts of the specimen as it moves (Figure 3). The pipette should not be passed over the same area of the slide more than once so as not to dislodge the adhesive coating.

Films of watery fecal specimens sometimes adhere to the slide better if a drop of albumen fixative and a drop of the specimen twice the size are added together and the specimen and the fixative are spread over the slide. Start in the center of the mixture and use an ever increasing circular motion to evenly spread the specimen over the slide much as one would do in preparing a thick blood film (Figure 4). This method is similar to that of Scholten (1972) for specimens fixed in Schaudinn's solution, described earlier.

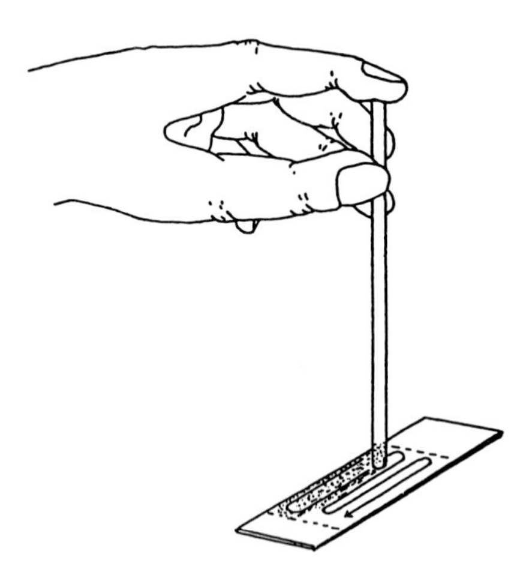

Figure 3. Applying a fecal film over the area of a slide coated with fixative (MIF, PIF, SAF). A portion of the specimen is drawn up by the pipette by capillary action, the pipette is held vertically with the tip resting on the slide, and is then moved back and forth over the slide depositing the specimen evenly over the area coated with the fixative. The film is allowed to dry partially before placing it in alcohol to set the smear on the slide. The method may be used for uncoated slides also.

Protecting Protozoa in Fecal Films from Drying

It is especially important that the area surrounding the protozoa in the sample of specimen spread on the slide stay moist to prevent shrinkage of the organisms. Many workers spray cytology fixative, such as those used to protect PAP smears, over the fecal film to protect it from drying. This has several advantages in that, if the process is interrupted, the fixative will assure that the fecal film remains moist. Some films dry much more rapidly than others and the time between preparing the film and applying the fixative must be judged carefully. If the film is especially wet and the spray is added too soon, rounded droplets will form, making a poor film. Usually a freshly prepared fecal film has dried adequately in 10 to 15 minutes and may be stained immediately. If the film has dried adequately but there will be further delay before staining, the films should be sprayed with cytology fixative to protect them from excessive drying. Fecal films properly protected with cytology fixative may be held for up to a week before staining with minimal reduction in the staining quality of the protozoa present.

When fecal films are prepared without the cytology fixative, attention to drying becomes more critical. If the film dries excessively, the protozoa may be distorted; however, staining is usually a little better

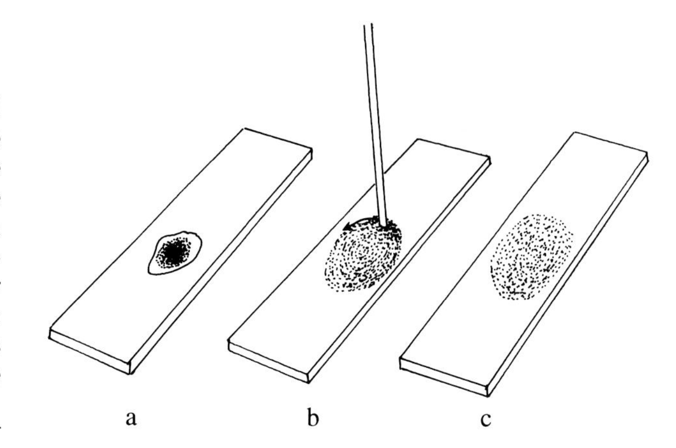

Figure 4. An alternative method for preparing fecal films. (a) Add a drop of Mayer's albumen fixative to the slide, then place a drop of the specimen directly on the fixative. (b) Place the tip of an applicator stick in the center of the material on the slide and move it in a circular motion to evenly spread the fixative and specimen together over the slide. (c) Allow the film to dry for a few minutes (usually 5 to 10 minutes) then place it in the alcohol solution to fix it to the slide. The film is similar to a thick blood film. After mixing on the slide, the film may be spread as shown in Figure 1.

when the cytology fixative is not used. Using a cytology fixative becomes a matter of convenience and/or personal preference.

In some laboratories, the cytology fixative is substituted for the albumen/glycerin coating, i.e., no coating is put on the slide before the fecal film is prepared. The background on fecal films prepared in this manner stains differently but the protozoa stain well.

In order for the albumen fixative to adhere to the slide and hold the fecal film in place during the staining process, it must be set (coagulated). This is usually done by placing the slide with the film in alcohol (95 or 70%) for 3 minutes or longer. When the film appears to be dry enough to adhere to the slide, it is carefully placed in the alcohol while observing to make certain the film does not wash off the slide. If only a small amount of material washes off the slide, leave the slide in the alcohol for at least 3 minutes to allow what has remained on the slide to become set. Then it may be checked to see if enough material remains on the slide to warrant staining. If not, prepare another fecal film from the same specimen.

The albumen may pick up one of the dyes in the stain and that color will be the background for most of the slide. This fact should be considered when preparing the film. If the albumen adhesive is too thick, it may obscure the protozoa. The albumen appears to stain more heavily than the protozoa with certain dyes and hold the stain better than do some of the organisms, i.e., organisms may destain faster than the background. If the albumen does stain heavily, protozoa in areas free of background material may have a dark ring around them. Very thick areas of a fecal film prepared from some specimens may stain red.

If fecal films do not stain well, the problem should be identified and films should be freshly prepared and stained. A single control slide can be stained to check the solutions. How a film is prepared, the stain selected, and the staining procedure used will all affect the results and any method for preparing permanently stained slides should be checked out thoroughly before it is put into general use in the laboratory.

If films are not coated or will not be stained for some time, they may be placed in 95% alcohol for 3 minutes then transferred to 70% alcohol for temporary storage. Films so stored for several days appear to stain well. The author did not find any reference on testing to determine if prepared slides can be held longer than several days in 70% alcohol before staining.

Preparing Fecal Films from Unpreserved Specimens

Fecal films from unpreserved specimens may be prepared in much the same manner as they are from PVA-fixed specimens. With an applicator stick or a flat, wooden toothpick, a small amount of feces is spread thinly and evenly over the surface of a microscope slide (the area should be the approximate size of a cover glass or a cover glass may be used instead of a slide). Mayer's albumen fixative can be coated on the slide to help the fecal material adhere to the slide during staining. The albumen fixative does change the color of the protozoa using some stains.

The slide with the fecal film is immediately placed in Schaudinn's fixative where it is allowed to stand for 30 minutes. After fixation, the fecal film is transferred directly to the first solution of an appropriate staining method (usually 70% alcohol) and the procedure continued.

STAINING METHODS FOR PREPARING PERMANENTLY STAINED FECAL FILMS FROM SPECIMENS FIXED IN PVA

Trichrome-Type Stains

Several methods for staining fecal films with trichrome-type stains are reported in the literature. Wheatley's modified trichrome stain is probably the most popular available for staining fecal films and the method described by Melvin and Brooke is given here as the standard (Table 1). The procedure as given has a minimum time of about 45 minutes. Some workers have reduced times for various steps in the procedure. The step-by-step procedure given by Ash and Orihel (1987) reduces the minimum time to about 24 minutes. Beck and Barrett-Connor (1971) give a procedure with a time range of 8 to 14 minutes (Table 2). A time to carry out the staining procedure between 20 and 25 minutes appears to be most realistic.

In staining fecal films prepared from specimens fixed using Schaudinn's fixative or in PVA collecting/preserving solution, the mercuric chloride must be eliminated from the fecal film or crystals will form that clutter the film which often makes finding and identifying protozoa impossible. A moderately dilute working solution of iodine in 70% alcohol is used which should be an orange, strong-tea color. How the iodine is prepared varies (see Appendix, Iodine Solutions).

Table 1. Procedure for Modified Wheatley's Trichrome Method (Melvin and Brooke, 1982) for Fecal Films Prepared From Specimens Fixed in PVA

Solution	Time
70% Ethanol plus iodine[a]	10–20 minutes
70% Ethanol[b]	3–5 minutes
70% Ethanol	3–5 minutes
Trichrome stain	8 minutes
90% Acid alcohol[c]	5–10 seconds
95% Ethanol (remove destain)	Dip 5 times
95% Ethanol	5 minutes
Carbol-xylene[d]	5–10 minutes
Xylene	10 minutes

[a] Iodine alcohol is prepared by adding stock iodine to 70% alcohol to make a strong-tea-colored solution (see Appendix).
[b] Reagent alcohol, absolute (Formula 3A) may be substituted for pure ethanol.
[c] Acid alcohol is a 0.5% solution of glacial acetic acid in 90% alcohol.
[d] One part of liquefied phenol to 3 parts xylene (see Appendix).
Note: Polychrome IV can be substituted for trichrome stain and staining time shortened to 5 minutes.

Table 2. Procedure for Modified Wheatley's Trichrome Method (Beck and Barrett-Connor, 1971) for Fecal Films Prepared From Specimens Fixed in PVA

Solution	Time
70% Alcohol plus iodine[a]	1 minute
70% Alcohol[b]	1 minute
70% Alcohol	1 minute
Trichrome stain	2–8 minutes
95% Acid alcohol[c]	10–20 seconds
100% Alcohol	Dip 5 times
100% Alcohol	Dip 5 times
100% Alcohol	1 minute
Xylol	1 minute or until refraction at the smear–xylol interface ends

[a] See Table 1, footnote a.
[b] See Table 1, footnote b.
[c] Add one drop of glacial acetic acid to each 10 mL of 95% alcohol.

Following the acid-alcohol destaining step in the procedure for Wheatley's trichrome stain, Melvin and Brooke (1982) call for 95% alcohols, which do not completely stop destaining. Garcia, Price, and others recommend substituting absolute alcohol (100%) for the 95% alcohols so that destaining is stopped during the dehydration steps. Polychrome IV stain can be substituted for trichrome stain in the procedure. The usual time in trichrome stain for fecal films prepared from PVA-fixed specimens is 5 minutes.

Hematoxylin Stains

Procedures for staining PVA-fixed films with hematoxylin take somewhat longer. The faster hematoxylin method by Tompkins and Miller (1947) takes from 33 to 47 minutes, depending on whose modification is used and whether maximum or minimum times are used. A slightly modified procedure is given in Table 3. Colors of protozoa tend to be blue in hematoxylin-stained fecal films and are usually more consistent from slide to slide than when stained with trichrome stains. Protozoa in films prepared from SAF-fixed specimens usually stain well with hematoxylin stains.

Table 3. Procedure for Phosphotungstic Acid Iron-Hematoxylin Method for Fecal Films Prepared From Specimens Fixed in PVA (Tompkins and Miller, 1947)

Solution	Time
70% Ethanol[a]	3–5 minutes
70% Iodine alcohol[b]	3–5 minutes
50% Ethanol	3 minutes
Tap water	3 minutes
Mordant (4% ferric ammonium sulfate)	3–5 minutes
Distilled water	1 minute
0.5% Hematoxylin	1 minute
Distilled water	Dip 5 times, 1 second each
2% Phosphotungstic acid (destain)	3 minutes
Running tap water[c]	3–5 minutes
70% Alcohol (add 3 drops of saturated lithium carbonate for each 59 mL)	3 minutes
95% Alcohol	Dip 5 times
100% Alcohol	1 minute
100% Alcohol or carbol-xylene[b]	2 minutes
Xylene	Dip 5 times
Xylene	3–5 minutes

[a] Reagent alcohol, absolute (Formula 3A) may be substituted for pure ethanol.
[b] See Appendix for formula; also see Table 1 footnote.
[c] The water input should be arranged so that it begins at the bottom of the container and flows up to wash the slides.

STAINING METHODS FOR PREPARING PERMANENTLY STAINED FECAL FILMS FROM SPECIMENS FIXED IN SAF, PIF, OR MIF

Trichrome-Type Stains

Two trichrome-type methods are given for specimens fixed in SAF, PIF, or MIF: modified Wheatley's trichrome and Polychrome IV (available commercially) and either are applicable. The standard procedure for Wheatley's trichrome takes slightly longer, often gives a less satisfactory stain of the cytoplasm of protozoa than Polychrome IV, and yields colors of the protozoa very close to those of the background. On slides stained with Polychrome IV, protozoa are usually stained differently from the background and are often easier to find (see Plates 74 to 84). The cytoplasm is usually stained more distinctly which often helps in identification (see Appendix for a source of Polychrome IV stain).

If several slides are being stained at the same time, it is better to use a slide carrier to move the slides from solution to solution so that all slides have equal times in each solution. A commercially available staining set-up often used in cytology for staining PAP smears with 12 stations is especially good. If more stations are needed for a procedure, smaller units with three stations are available that can be used to set the films or remove the mercury before the rest of the procedure begins (see catalogs of major distributors).

Two control slides should be used for all staining procedures: (1) a slide with a fecal film prepared from a control specimen to be stained with each run and (2) a slide with a previously stained fecal film, made from the same control specimen and stained by the same method (used as the standard). The quality of staining of protozoa in the newly stained fecal smear is compared with that in the standard (see Appendix, Control Slides).

The step-by-step procedure for Wheatley's stain varies from one investigator to another. A relatively quick method for specimens fixed in SAF, PIF, or MIF by Price is given in Table 4 which takes about 20 minutes. Use the "drain-blot-shake" method after each step to reduce fluid carry over.

The procedure for Polychrome IV varies only slightly and takes about 15 minutes (Table 5). Red and blue colors are more predominant than the yellow and green colors of Wheatley's stain and detail of the cytoplasm is often more distinct. Selection of the

Table 4. Procedure for Modified Wheatley's Trichrome Method (Wheatley, 1951) for Fecal Films Prepared From Specimens Fixed in SAF, MIF, or PIF

Solution	Time
95% Ethanol	3–5 minutes
70% Ethanol[a]	2 minutes
Trichrome stain	5–8 minutes
Distilled (or deionized) water[b]	Dip twice, 1 second each
90% Acid alcohol[c]	Dip 5 times
100% Ethanol	Dip 5 times
100% Ethanol	1 minute
100% Ethanol	2 minutes
Xylene	Dip 5 times
Xylene	3–5 minutes

[a] Reagent alcohol, absolute (Formula 3A) may be substituted for pure ethanol. Do not use regular denatured alcohol; obtain "reagent alcohol" from a responsible supplier. Some procedures for SAF recommend placing films in 70% alcohol for 30 minutes to assure coagulation of the albumen fixative.
[b] Step is omitted if Mayer's albumen fixative is not used to set the fecal smear.
[c] Acid alcohol is a 0.5% solution of glacial acetic acid in 90% alcohol.

stain for color and detail may be made by the laboratory director, section supervisor, or persons who will examine the slides.

In both methods, 100% alcohol is used instead of 95% alcohol after destaining. Absolute alcohol stops the destaining, whereas the films continue to destain slightly in 95% alcohol.

Iron-Hematoxylin Stains

Hematoxylin stains are the oldest and are generally the most useful stains available. As with trichrome stains, the use of a rack is advisable when many slides are being stained and two control slides should be used as described above.

A modification of the iron-hematoxylin-phosphotungstic acid method by Tompkins and Miller (1947) is given for staining specimens fixed in SAF, PIF, and MIF (Table 6). The times have been modified for staining fecal films prepared from fecal specimens preserved in these fixatives. The total time is shorter (24 minutes) than that given for staining fecal films prepared from PVA-fixed specimens (average about 40 minutes) mainly because solutions take longer to penetrate the polyvinyl alcohol coating (Table 3).

Table 5. Procedure for Polychrome IV Staining Method for Fecal Films Prepared from Specimens Fixed in SAF, MIF, or PIF

Solution	Time
95% Ethanol[a]	3–5 minutes
70% Ethanol	1 minute
Polychrome IV stain	Dip 5 times, allow to stain 3–4 minutes
Distilled (or deionized) water[b]	Dip twice, 1 s each
90% Acid alcohol[c]	Dip 2 times
100% Ethanol	Dip 5 times
100% Ethanol	1 minute
100% Ethanol (or ½ 100% ethanol and ½ xylene)	1 minute
Xylene	Dip 5 times
Xylene	3–5 minutes

[a] Reagent alcohol, absolute (Formula 3A) may be substituted for pure ethanol. Do not use regular denatured alcohol; obtain "reagent alcohol, absolute" from a responsible supplier.
[b] Step is omitted if Mayer's albumen fixative is not used to set the fecal smear.
[c] Acid alcohol is a 0.5% solution of glacial acetic acid in 90% alcohol.
Note: For specimens fixed in PVA or Schaudinn's Fixative, iodine alcohol must be used to remove mercuric chloride crystals before staining fecal films. After first step, 95% alcohol, iodine alcohol 5–10 minutes, 70% alcohol 2 minutes, and continue procedure.

REUSING AND CHANGING SOLUTIONS IN THE STAINING SETUP

Each laboratory has its way of dealing with fecal specimens, and how and when permanently stained slides are prepared varies. In some laboratories, permanently stained fecal films are prepared once a week whereas in others they are prepared on the same day that other procedures are performed on the specimens. If staining is done only once each week, probably all solutions, except the stain, should be changed weekly. Otherwise, some of the solutions used in the various staining procedures may be reused for several staining runs, some need to be changed frequently, while others should not be reused. The following comments are made as suggestions but are not meant to be hard-and-fast rules and apply only to solutions that appear in the procedure tables for staining.

When a staining setup is used every day, the solutions may be covered and allowed to remain in their containers on the workbench. When the setup is used less frequently, each solution should be placed either in a separate tightly sealing glass container or

Table 6. Procedure for Phosphotungstic Acid Iron-Hematoxylin Method (Tompkins and Miller, 1947) Modified for Fecal Films Prepared From Specimens Fixed in SAF, MIF, or PIF

Solution	Time
95% Ethanol[a]	3–5 minutes
70% Ethanol	1 minute
Tap water	1 minute
Mordant (4% aqueous ferric ammonium sulfate)	3–5 minutes
Distilled water	1 minute
Hematoxylin, 0.5%	1 minute
Distilled water	Dip 5 times
Phosphotungstic acid solution, 2%	3 minutes
Running tap water[b]	2–5 minutes
70% Alcohol (add 3 drops of saturated lithium carbonate for each 50 mL)	2 minutes
100% Alcohol	Dip 5 times
100% Alcohol (or ½ 100% alcohol and ½ xylene)	1 minute
Xylene	Dip 5 times
Xylene	2–5 minutes

[a] Reagent alcohol, absolute (Formula 3A) may be substituted for pure ethanol.
[b] The water input should be arranged so that it begins at the bottom of the container and flows up to wash the slides.

held in a well sealed staining jar or screw-capped Coplin jar.

For purposes of this discussion, a "run" is considered to be five or more slides stained at the same time. If five or more slides are stained together (one run) or if several runs are made, the solutions used in the process that precedes the stain should be filtered through coarse filter paper at the end of the each day.

Those solutions that follow the stain should be watched carefully and filtered when any contamination is seen. The number of times the solution can be reused depends on the number of slides in each run and the time the solution sits after its first use. Individual judgment is needed to maintain a proper staining setup without unnecessary waste. Usually, except for the stain, solutions should not continue to be reused after 1 month of use.

Solutions Used in Trichrome or Polychrome IV Procedures

- 95% Alcohol (used to set Mayer's albumen fixative): filter and reuse several times before replacement.
- 70% Alcohol with iodine: replace each day.
- 70% Alcohol after iodine alcohol: filter and reuse several times.
- 70% Alcohol before stain: filter and reuse several times.
- Water used after stain: change after each run.
- 90% Acid alcohol: use for 3 to 5 runs, then replace.
- First 95% alcohol: use for 3 or 4 runs, then discard and replace with second 95%.
- Second 95% alcohol: use to replace first 95% alcohol after 3 or 4 runs and replace second 95% with fresh alcohol.
- First 100% alcohol (where used instead of 95%): use for 3 or 4 runs then discard and replace with second 100% alcohol.
- Second 100% alcohol: use to replace first 100% alcohol after 3 or 4 runs and replace second 100% with third 100% alcohol.
- Third 100% alcohol: use to replace second 100% alcohol after 3 or 4 runs and replace third 100% with fresh alcohol.
- 100% alcohol and xylene (50/50): use for no more than 8 to 10 runs.
- First xylene: use for no more than 8 runs. (Used xylene has many uses in the laboratory such as removing oil from slides and cleaning off labels. It may be filtered and kept in a properly labeled bottle "Used Xylene".)

- Second xylene: use to replace first xylene after 8 runs, then replace 2nd with fresh xylene.
- Carbol-xylene: use for no more than 8 to 10 runs then discard.
- Xylene (after carbol-xylene): use for no more than 8 to 10 runs, then discard.

Solutions Used in Hematoxylin Staining Procedure

- Solutions that are duplicates of those listed for the trichrome methods above should be treated similarly.
- 50% Alcohol (after iodine alcohol): filter and reuse several times.
- Iron alum mordant: discard daily after use.
- Phosphotungstic acid (destain): discard after three or four runs.
- 70% Alcohol with lithium carbonate: discard after three or four runs.

Most dye mixtures (stains) used for staining intestinal protozoa will improve with age unless some contaminant that affects the stain is introduced. If solutions are changed as necessary, a control slide is stained along with other slides, and the control slide is compared with a previously stained control slide after each run, the quality of the staining setup can be adequately monitored and maintained for proper results. All stains should be filtered through coarse filter paper after three or four runs. As staining volumes in the staining jars are reduced through regular use, more stain can be added to the staining jar to make up for the loss. Different lots of a stain, however, should not be mixed.

Chapter 5

Application of Methods

The methods commonly employed for the diagnosis of intestinal parasites and the procedures used for each method have been discussed in Chapters 3 and 4. As a routine program for the laboratory, the methods are not applied singly but several methods are used in combination to meet the diagnostic needs. Both their application and their effectiveness as systems for routine examination of fecal specimens for parasites and eggs are discussed here.

COMPARISON OF COLLECTING/ PRESERVING SOLUTIONS

Table 1 gives information on some collecting/preserving solutions that may be employed routinely by diagnostic laboratories and how the methods and procedures are applied to the preserved fecal specimens for parasitological examination. The table summarizes information on how effective the solutions are for killing, fixing, and preserving the various stages of parasites and eggs that may be present in the fecal specimen and what one may expect from applying certain methods.

Comparison of Individual Collecting/Preserving Solutions

Formalin-Fixed Specimens

On examination of wet mounts prepared from formalin-fixed specimens, cysts, eggs, and juvenile worms present may be found by direct examination. If a formalin-ethyl acetate or CONSED concentration procedure is performed on a specimen fixed in formalin or formol-saline, the same cysts, eggs, and juveniles should be found and should be in greater numbers than found by direct examination. Because cysts are sometimes difficult to identify in wet preparations, iodine is usually added to facilitate identification. Trophozoites are usually not identifiable in formalin-fixed specimens and the specimens are unsatisfactory for preparing permanently stained fecal films.

PVA-Fixed Specimens

PVA-fixed specimens are not suitable for the preparation of wet mounts for direct examination. The results of the application of concentration methods are far inferior to results obtained using other collection/preserving solutions. Specimens fixed in PVA are especially suitable for preparing permanently stained fecal films for the identification of trophozoites and cysts of protozoa and most of the common staining methods can be used.

SAF-fixed Specimens

Trophozoites, cysts, and eggs may be found on wet mounts prepared from SAF-fixed specimens. They can be subjected to concentration methods as are formalin-fixed specimens with essentially equal results, except that juvenile worms are apparently not preserved as they are in formalin and may not be recovered after concentration. Since protozoa are not stained in wet preparations prepared directly from the specimen or after concentration, iodine is usually used to enhance identification. Protozoa in

Table 1. Effectiveness and Efficiency of Commonly Used Collecting/Preserving Methods

	Formalin or formol-saline	PVA	SAF	MIF or PIF
Fixing time	2–4 hours except eggs of *Ascaris*	1–2 hours	1–2 hours	1–2 hours
Fixes and preserves	Cysts, eggs, and juvenile worms	All parasites and eggs	Trophozoites, cysts, and eggs	All parasites and eggs
Quality of parasites found	Eggs excellent; cysts require stain; juveniles fair	Additional procedure required	Eggs excellent; cysts and trophozoites of protozoa require staining	Trophozoites, cysts, eggs and juveniles excellent
Ineffectual for	Trophozoites; requires additional method	Eggs and juveniles; requires additional method	Juveniles; requires additional method	None
Wet mount, examination	Usually after concentration	No wet mount	Direct examination or after concentration	Direct examination or after concentration
Wet mount, staining	Iodine	No wet mount	Iodine	All parasites stained in bulk in specimen
Concentration methods applicable	Zinc sulfate, formalin-ethyl acetate, CONSED system, acid-ether	Formalin-ethyl acetate	Formalin-ethyl acetate CONSED system	Formalin-ethyl acetate CONSED system
Recovery efficiency; formalin-ethyl acetate	Eggs and cysts, good; juveniles, fair; trophozoites not recovered	Poor recovery from concentration	Eggs and cysts, good; juveniles, poor; trophozoites not recovered	Eggs, cysts, and juveniles, good; trophozoites not recovered
Recovery efficiency; CONSED system	As above	As above	As above, trophozoites are recovered	As above, trophozoites are recovered
Permanent staining	Unsatisfactory	Iron hematoxylin Trichrome Polychrome IV	Iron hematoxylin Trichrome Polychrome IV	Iron hematoxylin Trichrome Polychrome IV
Efficiency				
Iron hematoxylin	N/A	Very good	Excellent	Good
Trichrome	N/A	Excellent	Fair to good	Excellent
Polychrome IV	N/A	Excellent	Fair to good	Excellent
Associated difficulties	Protozoan cysts hard to identify; trophsozoites require an additional method, e.g., PVA permament stain	Eggs and juveniles poorly concentrated; usually requires an additional method, e.g., formalin concentration	Permanent stain often needed to identify protozoa; juveniles not recovered	Collection is slightly complicated by need to add iodine when fecal specimen is added

Note: Statements pertain to specimens well mixed with the fixing solution within 1 hour after passage.

Table 2. Time/Efficiency of Several Procedures Used With Different Collecting/Preserving Systems for the Diagnosis of Intestinal Parasites

	PVA/Formalin	SAF	MIF or PIF[a]
Specimen Collection system	**Two-vial** Collection system	**One-vial** Collection system	**One-vial** Collection system
Procedures performed			
Direct examination, wet mount	Formalin, usually not done; PVA, not applicable	May or may not be done, 15–20 minutes	May or may not be done, 15–20 minutes
Concentration A: formalin-ethyl acetate	Formalin-fixed specimen only; 10–12 minutes cysts, eggs, and juveniles	SAF-fixed specimen; 10–12 minutes; cysts and eggs	MIF- or PIF-fixed specimen; 10–12 minutes; cysts, eggs, and juveniles
Concentration B; CONSED system[b]	Same, results similar, no trophozoites, 8–10 minutes	Same, but trophs are identifiable but not well concentrated, 8–10 minutes	Same, but trophs are stained, concentrated, and identifiable, 8–10 minutes
Examination of wet mount from concentrate	Wet mount preparation and examination, 10–20 minutes	Wet mount preparation and examination, 10–20 minutes	Wet mount preparation and examination, 10–20 minutes
Preparation of fecal film for staining	2–3 minutes; PVA specimen only; film must be dry before staining; 1 hour or more required	2–3 minutes; film mixed with Mayer's albumen fixative, dries in 10–20 minutes	2–3 minutes; film mixed with Mayer's albumen fixative, dries in 10–20 minutes
Staining fecal film	PVA, 20 minutes or more in trichrome or Polychrome IV	35 minutes or more in hematoxylin	15 minutes or more in trichrome or Polychrome IV
Examining stained film	20 minutes or more	20 minutes or more	20 minutes or more
Minimum time expended[c] from starting procedures to completing report	57 minutes + 1 hour drying time = approximately 2 hours to complete report	87 minutes; omit direct examination, 72 minutes; with CONSED, 70 minutes; omit permanent stain with CONSED, 35 minutes[d]	67 minutes; omit direct examination, 52 minutes; with CONSED, 50 minutes; omit permanent stain with CONSED, 20 minutes

[a] The MIF and PIF collection systems require the addition of Lugol's iodine to the specimen vial at the time the feces is added, which some investigators feel complicates the system. It certainly should be as easy as collecting two vials.

[b] Trophozoites present in a fixed fecal specimen will be present and essentially unchanged after the specimen is concentrated using the CONSED concentration system. Trophozoites present in MIF- and PIF-fixed specimens will also be concentrated, i.e., in greater numbers per volume of fecal material than in the unconcentrated specimen.

[c] The times are averaged and based on a single specimen in process. When a number of specimens are processed at the same time, the total time for certain procedures would not be increased by the number of specimens, i.e., it takes the same amount of time to spin six tubes as it does to spin one. In the case of other procedures, e.g., the microscopic examination of stained fecal films, the time is increased directly by the number of slides examined.

[d] Does not include the time for processing and examining an unpreserved specimen for juvenile nematodes.

fecal films prepared from SAF-fixed specimens can be stained as they are in films prepared from specimens fixed in PVA, MIF, or PIF but protozoa in films prepared from SAF-fixed specimens usually stain better when stained with hematoxylin than when stained with trichrome.

MIF-Fixed and PIF-Fixed Specimens

Trophozoites, cysts, eggs, and juvenile worms may be found in wet mounts prepared from specimens fixed in either MIF or PIF. The parasites and eggs are stained in the bulk specimen, so iodine is not used. Parasites, eggs, and juveniles can be concentrated by either the formalin-ethyl acetate or CONSED methods. Trophozoites are recovered only by the CONSED method. In either single vial system, MIF or PIF, the direct examination can be omitted and a permanently stained slide can be used to identify protozoa.

Efficiency of Collecting/Preserving Solutions as Systems

Table 2 summarizes the applicability and time required to perform various procedures on the specimens preserved by the different collecting/preserving solutions and the efficiency of using the different methods and procedures as systems. The time summary at the bottom of the table gives minimum times for completing the procedures on a single specimen.

Formalin/PVA System

Formalin is not a satisfactory fixative for trophozoites of protozoa so it is listed as being ineffectual (Table 1). Therefore, if the laboratory director wants to be able to also find and identify trophozoites of protozoa as well as cysts, eggs, and juvenile worms, an additional collecting/preserving solution or an unpreserved specimen must be used. In practice, formalin and PVA are employed together as a two-part diagnostic system in order to find and identify all stages of the parasites that may be present: (1) a concentration method for cysts, eggs, and juvenile worms (applied to the formalin-fixed specimen) and (2) a permanently stained fecal film for trophozoites and for confirmation of the identification of cysts (prepared from the PVA-fixed specimen).

In Table 2, the PVA/formalin system is seen to have the poorest time/efficiency, i.e., more time is required to complete the examination of a specimen and render a report. Many workers advocate skipping the direct examination of the formalin-fixed specimen since what is found on direct examination is usually present after concentration which does save time. The greatest amount of time expended is in waiting for the fecal film prepared from the PVA-fixed specimen to dry for staining. With good planning, the waiting time need not be wasted.

The SAF System

The SAF system as described by Yang and Scholten (1977) includes a direct examination, a concentration by the formalin-ethyl ether method or a modified concentration method using saline and ethyl ether. Well-preserved trophozoites may be recovered from SAF-fixed specimens and are identifiable after concentration when the CONSED method is used. If a hematoxylin staining method is used to prepare the permanently stained fecal film, the SAF system requires the greatest amount of a technologist's time.

The authors also mention a separate method for the collection and identification of juvenile worms which requires an unpreserved specimen. Although the preserved specimen can be concentrated and permanently stained fecal films can be prepared, the need for an unpreserved specimen for recovering juvenile worms makes the SAF method a two vial system and increases the total time required (processing and examining the unpreserved specimen are not included in Table 2).

The MIF or PIF System

The quality of the protozoa found in specimens properly fixed in MIF or PIF is at least equal to those fixed in PVA or SAF and the time required to process and examine the specimen and render a report is less (Table 2). Trophozoites and cysts that are present in the specimen are stained and can be found and identified in direct wet mounts. The difficulties experienced by individuals result either from improper use of the systems or because the individual using the system is unfamiliar with appearance of parasites fixed in these collecting/preserving solutions (see Part 4).

Cysts, eggs, and juveniles present in either MIF- or PIF-fixed specimens are recovered after concentration with the formalin-ethyl acetate method as they are from formalin-fixed specimens. If the newer CONSED concentration method is applied, trophozoites present are also recovered and are in greater numbers than found by direct examination, a true concentration (see Chapter 3, Table 1, and Figures 14–16). There is less debris in the wet mount when it is prepared by the procedures described in Figures 3–13 in Chapter 3, making it easier to examine microscopically.

Since all stages of parasites that may be present in the fecal specimen are preserved, are stained, and are recognizable in the wet mount prepared after concentration, the direct examination can be eliminated and the permanently stained fecal film becomes a luxury (see Plates 13–35 and 55–73). The total time required to complete the examination of a specimen and render a report on a single specimen is reduced to 20 to 30 minutes. If desired or required by protocol, permanently stained fecal films can be made from specimens in vials fixed in MIF or PIF as they are from specimens fixed in PVA (see Plates 74–84).

Processing Specimens Together

The workload of a laboratory in parasitology can have a very important impact on planning. Smaller laboratories with few calls for eggs and parasites should plan to examine specimens when several can be run together unless there is an immediate demand for results. A batch of specimens on six patients is a convenient number. When a procedure is carried out on several specimens, the time required for each specimen is reduced. Using the formalin-ethyl acetate sedimentation procedure in which there are two centrifuge steps takes approximately 12 minutes. If 6 specimens are run together, the procedure takes approximately 42 minutes, reducing the time for each specimen to 7 minutes. When performing other procedures, e.g., slide examination, the time is increased directly by the number of specimens to be examined and this time is affected by the system used, the ability of the examiner, the type of specimen, and the policy of the laboratory on how much time is spent before a specimen is called negative.

The average technologist can usually handle about 12 patient specimens in an 8 hour period using the PVA/formalin system, assuming that every field in a wet preparation is examined and that trichrome-stained films are prepared and examined on each patient. If all the specimens are on hand at the beginning of the day, the fecal films from the PVA-fixed specimens can be prepared and be set aside to dry while the procedures are run on the formalin-fixed specimen. After slides prepared from the concentrated formalin-fixed specimen are examined, staining of the fecal film can be carried out. Batches of 10 to 12 patient specimens make convenient groups in larger laboratories.

Again, the policy of the laboratory has an impact on the time. Are permanently stained slides prepared on all specimens or only on those specimens that are soft or liquid? What factors determine whether a trichrome or other stain will be done? The number of patient specimens that can be handled in a given period of time is dependent on the policy established. In laboratories that have a much larger workload, a different approach to requests for eggs and parasites may be taken. A larger number of specimens may be batched and different individuals may perform the different procedures on different days.

Regardless of the numbers of specimens involved, communication between the laboratory personnel and physicians and the other personnel directly involved in patient care is essential to establishing an efficient and responsive laboratory program (see Part 1).

SELECTING METHODS FOR THE LABORATORY

In most laboratories, the methods to be performed for the diagnosis of intestinal parasites found in fecal specimens are selected by the laboratory director or someone designated for this task. The solutions available for collecting and preserving the specimen (see Introduction), the method for obtaining the specimen (see Part 1), and the detailed procedures to be used on the specimen (see Part 3) often become a personal decision. In fact, the merits of each collecting/preserving solution, the number of procedures that must be performed to arrive at a diagnosis, how well each different procedure performs, the time involved, the usual workload of the laboratory, the number of people available to carry out the work, and the background and training of the personnel involved are all important considerations. In Tables 1 and 2, some of the information that may have an impact on the selection of a diagnostic system for the laboratory is summarized.

The approach of laboratories doing diagnostic parasitology may be quite different. Requests for "stat" examinations for intestinal parasites are extremely rare, so a laboratory can program its work. Some laboratory protocols call for daily reports while others report every other day. In some laboratories where the MIF-type collecting/preserving solutions are used, two reports are given. The first report is made after examination of a wet preparation (most often prepared from the concentrated specimen), usually on the day the specimen arrives at the laboratory. The second report is made after permanently stained films are prepared and examined. Slides for permanent staining are batched and stained together with controls and reports made once or twice each week.

Technician time required is often a determining factor in deciding which system may be most applicable to the laboratory's program and objectives. As discussed earlier (see Chapter 1, Multiple Specimens), the decision to collect only one specimen (Montessori and Bischoff, 1987; Senay and MacPherson, 1989) or to collect and pool several specimens so as to carry out procedures only once (Peters et al., 1988) rather than collect and examine multiple specimens, appears to be related to technician time. If results using various systems are essentially equal, choosing the one requiring the least technician time seems logical.

As mentioned above, many other factors should be considered before arriving at a decision. No system that appears to be effective for the diagnosis of intestinal parasites should be accepted or rejected before it is evaluated thoroughly and efficiently. Just because a system was reported to be effective when used elsewhere does not mean it is the best system for all diagnostic laboratories. Also, just because a system has been used in a laboratory for many years does not mean that it should not be replaced by a newer and more efficient system when one becomes available. With new innovations, changes are made regularly in the chemistry, hematology, and bacteriology sections; why not in parasitology? Not being willing to change to a better, more time-efficient system is a sign of poor management. Before selecting a system or changing to a new system, know the procedures involved, compare quality of the results obtained, perform time-efficiency studies, and make a decision based on valid information, not on the opinion of someone else.

Chapter 6

Special Methods

There are many special methods that may be applied to the preserved fecal specimen. Some of these pertain to the status of the patient while others relate to teaching or research. Only a few are included here.

ACID-FAST STAINS

In addition to the usual staining methods for intestinal protozoa, some special methods may be necessary, especially acid-fast staining for cryptosporidia or other coccidia. Such stains are usually not a part of the routine diagnostic protocol but are performed on request by the physician ordering the tests.

When elements or organisms are stained with certain dyes and are then subjected to dilute acid solutions and the dye is not removed while other elements on the slide are decolorized, the materials that retain the stain are said to have acid-fastness (being acid-fast). Those individuals who want a more complete discussion of acid-fastness and the techniques that are used, are referred to Lillie (1965).

This property of being acid-fast is shown by *Mycobacterium* spp. and certain other organisms. For example, the shell of an egg of *Schistosoma mansoni* is acid-fast (it holds the dye when subjected to acid) whereas the shell of an egg of *S. haematobium* is not acid-fast. This allows differentiation of eggs that may appear similar in tissues.

Acid-fast staining is used to locate, and sometimes identify, organisms in tissues and to differentiate between organisms that appear similar, in some cases those found in fecal films. For example, acid-fast staining is used to differentiate oocysts of cryptosporidia from yeast cells of about the same size and with about the same refractive index. Because some yeast cells are acid-fast, it is important to compare the morphology of the objects that are acid-fast to make certain of the identification.

Several acid-fast stains are used on fecal specimens, especially for confirming the presence of cryptosporidia. One of the earliest investigators to recognize the acid-fast character of certain bacteria was Ziehl who introduced his procedure in 1882 in Germany. Neelsen modified the procedure in 1883. They used a phenol and fuchsin dye mixture (carbol fuchsin). Kinyoun published another version of a carbol fuchsin stain in 1915 and Bronsdon published a new method specifically for cryptosporidia that used dimethyl sulfoxide in 1984. Price developed another acid-fast stain in 1988 that has been used in several laboratories (see Appendix). There are a number of other methods for acid-fast staining that appear in the literature; these can be substituted for those mentioned. Although how these different methods and stains have been employed has varied somewhat depending on the type of material to be stained and the investigator, one of the fuchsin dye formulas has been employed most often for staining acid-fast organisms.

Table 1. Three Carbol Fuchsin-Type Stains, Components, and Amounts for Each Stain

	Ziehl–Neelsen	Kinyoun	Bronsdon DMSO
Basic fuchsin	0.3 g	4.0 g	2.0 g
Phenol (crystals)	5.0 mL (melted + water = 90%)	8.0 g (melted)	6.0 g (liquid)
Reagent alcohol (95%)	10.0 mL	20.0 mL	12.5 mL
Dimethyl sulfoxide	—	—	12.5 mL
Glycerol	—	—	12.5 mL
Distilled water	95.0 mL	100.0 mL	37.5 mL
Volume	110.0 mL	128.0 mL	81.0 mL

Staining Cryptosporidia in Fecal Films

A number of methods are used to find and identify the oocysts of cryptosporidia in fecal specimens. These include iodine-stained wet mounts, several carbol fuchsin acid-fast stains (see Table 1), auramine O acid-fast stain, and monoclonal antibody methods.

The careful microscopist may be able to distinguish oocysts of *Cryptosporidium* in formalin-fixed, iodine-stained, or PIF- and MIF-fixed wet preparations. Such findings should be confirmed by using one of the more specific identification methods. Of those methods available, acid-fast staining of dried films is most frequently used since the auramine O and monoclonal antibody procedures require the use of a fluorescent microscope.

The oil immersion lens (1000×) and very high contrast are usually needed to distinguish the oocysts of cryptosporidia from yeast cells in wet mounts. Iodine usually stains yeast cells but not the oocysts. In PIF- and MIF-fixed specimens, some of the eosin stain is taken up by the oocyst, but again, very high contrast is needed to differentiate the oocysts from other structures in the preparation (see Plate 72).

Acid-fast stains used to stain oocysts in fecal films include the Ziehl–Neelsen (modified) (Jenrikhen and Pohlenz, 1981), Kinyoun's, Bronsdon's, and Price's (see Tables 1, 2, and Appendix). Good stains should show the outer membrane or oocyst wall and the inner body, or sporocysts when the latter are present, because there are species of yeast cells that are acid-fast and these may appear similar to oocysts of cryptosporidia. Price's stain appears to work exceptionally well for demonstrating the oocyst wall.

Preparing Films for Acid-fast Staining

A small sample of the fecal specimen is spread evenly over an area of the slide the size of the cover glass to be used and allowed to air dry. Films may be prepared directly from the fixed sediment of specimens fixed in PIF, MIF, buffered formalin, or from the plug after a suitable sedimentation method has been performed on the specimen. Films may also be prepared after sugar flotation. When fresh specimens are used and there is danger of worker contamination, slides may be heat fixed by placing them on a heat-controlled warming plate at 70°C for 10 minutes. The oocyst does not appear to be greatly affected by drying or low heat.

Acid-Fast Staining Procedure for Fecal Films

When films are stained, the Ziehl–Neelsen procedure calls for heating the film while staining whereas the Kinyoun, Bronsdon, and Price procedures do not. The microwave oven has been used effectively with the Kinyoun method. Many acid-fast staining modifications have been employed successfully. The procedure given is for either the Kinyoun's or Price's cold method (Bullock-Iacullo, 1988) (see Table 2). Slides may be placed directly in a staining jar or on a staining rack as used for Wright's stain.

A number of different counterstains have been used. The author prefers methylene blue chloride but another counterstain may be substituted according to the preference of the investigator (see Appendix for formulas of some counterstains and solutions).

PREPARING TEMPORARY WET MOUNTS

Permanently stained slides are excellent for many situations and are often required to document positive cases. They may be examined over and over

Table 2. Acid-Fast Staining Procedure for Fecal Films (Kinyoun's or Price's Stain)

Procedure	Time
Post-fix dried fecal films in absolute methanol and allow to dry	10–30 seconds
Place films in acid-fast stain	5 minutes
Rinse films well in running tap water and touch edges to a paper towel to remove excess water	30 seconds
50% alcohol to remove excess stain	Dip 4–5 times
Decolorize in acid alcohol by dipping several times, then allow them to stand until pale pink, 2 minutes maximum	1–2 minutes
Place films in a Coplin jar and wash in running water	1–3 minutes
Note: A piece of tubing should be used to bring the stream of water to the bottom of the Coplin jar so that slides are flushed from the bottom up.	
Remove excess water and counterstain in methylene blue cloride	1–2 minutes
Rinse off excess stain in running tap water	1 minute
Note: Uncovered slides may be examined at 400× magnification. If covered slides are preferred, dehydrate as described below.	
Dehydrate in 95% alcohol	30 seconds
100% alcohol	1 minute
First xylene	Dip 3–4 times
Second xylene	1–2 minutes
Drain excess xylene and mount cover glass (Permount is the preferred mounting medium)	

* See Appendix for counterstains other than methylene blue chloride.

again, but there are situations where only wet mounts of the specimen are appropriate. In such situations, temporary wet mounts that will hold up under repeated use are needed.

Temporary Wet Mounts Sealed with Vaspar, Nail Polish, or Permount

In Part 2, preparing wet mounts sealed with vaspar was discussed. Such wet mounts are not only suitable for routine use in the laboratory but, when sealed properly, are suitable for teaching. Vaspar-sealed mounts may last for weeks or months or they may last only a day depending on how well they are made.

Instead of using vaspar, a wet mount can be sealed in the same manner with nail polish using the brush that comes in the bottle. Permount or other mounting media may be applied using a pipette rather than an applicator stick used with vaspar. Depending on the tightness of the seal, wet mounts sealed with nail polish or Permount should last longer.

Double-Cover Glass Wet Mounts

Double-cover glass wet mounts are especially good for teaching. Such mounts are prepared by using cover glasses of different sizes. A drop of the specimen is placed on the larger cover glass and the specimen carefully covered with the smaller one using forceps. The suspension should reach the edge of the smaller cover glass with no air bubbles formed. If the fluid extends beyond the edge of the smaller cover glass, it must be removed so that the exposed edges of both cover glasses are free of fluid. A drop of the permount is placed in the center of a slide. Using forceps, the two cover glasses with the specimen between are carefully turned over so that the smaller one is down and placed on the permount. The permount should run to the edge of the larger cover glass and completely seal the smaller one below.

The cover glasses with the specimen between can be inverted and placed on a dry slide and a slightly thinned permount medium can be placed at the edge of the larger cover glass and drawn under by capillary action to seal the cover glasses on the slide. This is a much slower method but some excellent slides have been prepared in this manner.

Another method for preparing double-cover glass wet mounts is accomplished by sealing a cover glass to a slide with the permount and setting the slide aside until the permount dries; usually, 24 hours is sufficient. A drop of the specimen is placed on the mounted cover glass and carefully covered with another cover glass of the same size. As soon as the edges of the two cover glasses are free of liquid, the edges of the upper one are ringed with the permount so as to completely seal it to the slide while sealing the edge of the lower one also. This method is often used for eggs or larger elements because it is thicker than the others described (see also Ash and Orihel, 1987; Garcia and Brucknew, 1988; Isenberg, 1992).

With a cover glass sealed to a slide, a drop of specimen can be placed on the cover glass and covered with a larger cover glass. When the specimen appears to be contained within the area of the smaller cover glass underneath, diluted permount can diffuse under the edges of the larger cover glass by capillary action.

As the permount dries, bubbles may form if the seal is not adequate. Well-sealed wet mounts prepared by these methods may last several years.

Using a Slide Ringer to Prepare Temporary Wet Mounts

Preparing sealed wet mounts is time consuming, so if many slides are being prepared for teaching, it may be appropriate to purchase a "Slide Ringing Turntable" and use round cover glasses. Turntables are becoming rare and one may be hard to find. Either a ringed well can be prepared on the slide with a drop of the specimen placed in the well then covered and sealed, or the specimen and cover glass can be mounted directly on the slide and ringed. To use the turntable, a slide is fastened to it with microscope stage clips so that the cover glass is centered and the turntable is turned while applying a mounting medium to the edge of the cover glass. The result is a wet mount with a clean, even sealing ring.

Use of Temporary Wet Mounts

All temporary wet mounts should remain flat (horizontal) at least until they are thoroughly dry and should be stored in that position if possible.

There are disadvantages to temporary wet mounts: parasites and eggs that are stained may lose their color, eggs and cysts may collapse, or parasites and/or eggs in some way may lose their integrity. Temporary wet mounts stored between teaching sessions should be checked periodically for quality. For teaching purposes, if good preserved specimens are available, it is always preferable to prepare slides at the time they are needed as they would be prepared in the clinical laboratory environment.

EGG COUNT PROCEDURES

There are times when it is important to be able to estimate the number of eggs in a specimen and several methods for making estimates have been reported (Stoll and Hausheer, 1926; Dunn, 1968; Ash and Orihel, 1987; Garcia and Bruckner, 1988; Isenberg, 1992). For specimens collected in formalin, SAF, PIF, or MIF, the method described here has been found to be relatively consistent and therefore relatively accurate.

The specimen in the collecting/preserving solution is allowed to stand for 1 hour or more, except formalin-fixed specimens which should stand overnight or longer. Either the formalin-ethyl acetate or the CONSED sedimentation method may be used. In practice, the CONSED method has yielded more consistent results and numbers have been higher.

The specimen vial is shaken vigorously and 12 to 14 mL of the specimen poured into a preweighed, 15-mL polypropylene centrifuge tube; remixed with wooden applicator sticks, and all but about 1 to 1.5 mL poured back into the vial. The specimen is not strained, but any large particles or strands of vegetable material are removed with an applicator stick at this time. The portion of fecal suspension remaining in the centrifuge tube is used for the washing procedure.

Rarely, a specimen will have too much fiber to pour easily into the centrifuge tube. In such cases, the specimen is shaken and the fluid portion of the specimen with its suspended materials is poured into a centrifuge tube. Then, 10 mL of sedimenting solution is poured into the original vial still containing most of the fibrous material, the vial is capped, shaken, and the fluid portion in the vial is poured into a second tube.

Next, the first tube is shaken and all but 1.0 mL of the specimen is returned to the original specimen vial. The fluid in the second tube with its suspended material is added to the 1.0 mL in the first centrifuge tube. The tube is shaken vigorously and all but 1.0 mL is poured back into the vial and the 1.0 mL remaining in the tube is used for the washing procedure. The fecal suspension is washed by adding 10 mL of the CONSED solution (or 10% formalin) to the centrifuge tube, the fecal suspension is mixed with the solution, and the tube is spun in a centrifuge with a free-swinging head at 500 to 550 xg for 2 minutes.

After centrifugation, the fluid layer is poured off, making certain that the base plug does not break up. The inside of the tube above the plug is wiped free of debris with a cotton-tipped applicator stick, and the tube placed in a rack upside down on a paper towel for 5 minutes to allow any excess fluid to drain off. The inside of the tube is wiped again and reweighed to determine the actual amount of fecal material that constitutes the plug. The CONSED (or formalin-ethyl acetate) method is run on the specimen using a centrifuge force of 600 xg and the specimen plug transferred to a glass slide (as described in the CONSED procedure, see Figures 7 and 8 in Part 3, Chapter 3).

When estimating the number of eggs, it is important to examine as much of the material making up the plug as possible, except for the very coarse material. This may require preparing more than one slide. After transferring the plug to a slide, some of the material can be transferred to a second slide. Touch the long edge of the first slide to another slide and tilt it so that some material is transferred. A cover glass is then mounted on each slide, sealed with vaspar, and the specimen examined by microscopy (as in Figures 9 through 13 in Part 3, Chapter 3).

All eggs in the specimen should be counted systematically covering the entire specimen under the cover glass. When the eggs are too numerous to count, all of the eggs in several microscopic rows (at 100× magnification) are counted and the total estimated based on the number of rows under the cover glass (usually 10 longitudinal per cover glass).

The weight of the specimen examined is adjusted to 1 g and the equivalent number of eggs in a gram of feces is recorded. In the example below, there were approximately 57 eggs in 1 g of the specimen. Where there is a mass of fibrous material as discussed above, its bulk (weight or volume) is not considered in the weight or volume of the specimen for the purpose of egg counts.

$$\frac{\text{number of eggs counted}}{\text{weight of specimen sample}} = \frac{x\ (\text{number in } 1\text{ g})}{1\text{ g}}$$

$$\frac{16}{.28\text{ g}} = \frac{x}{1\text{ g}}$$

$$.28x = 16$$

$$x = 57$$

The value of egg counts for most helminths is questionable. Counts of eggs of parasites such as *Trichuris*, *Ascaris*, and hookworms that pass eggs more or less continually may be used in making estimates of worm burdens and in evaluating the results of specific treatment regimens, but caution should be used in making interpretations. This is especially true of counts of trematode eggs where many factors affect the number and frequency of eggs passed. The procedure can be used in the protocol for evaluating the effectiveness of different concentration procedures. For such evaluations, the straining step is also included.

POSTTREATMENT EXAMINATIONS

Posttreatment examinations are often requested by the physician or are performed to check the effectiveness of some therapeutic regimens. These examinations are usually more important in cases where the clinical episode has been severe. If parasites are not completely eliminated by the treatment regimen, "How soon after treatment do parasites recover sufficiently to be found by the usual diagnosic procedures?" is a commonly asked question. Although individuals may respond differently to each type of treatment, a general estimate is needed to establish a reasonable protocol. Posttreatment fecal examinations should be done even though the patient has shown no return of symptoms.

For giardiasis, a series of 3 to 5 specimens taken 2 or 3 days apart should be examined 4 to 5 weeks after treatment has been completed. The patient should be free of symptoms.

For amoebiasis, examination should be repeated after 2 weeks if symptoms persist. In patients free of symptoms, a procedure similar to that for giardiasis should be followed about 2 months after treatment; if no parasites are found, an additional series of at least 6 specimens taken 3 to 4 days apart should be examined.

Most helminths usually recover faster so a posttreatment examination may be made after 2 to 3 weeks and, if no eggs are found, repeated after 6 weeks.

If tapeworms are not eliminated, especially *Taenia* sp., they may take longer to produce gravid proglottids (depending on how much, if any, of the worm was lost during treatment). The patient should be asked to watch for proglottids evacuated with the feces. The posttreatment examinations should be made 2 to 3 months after treatment when no proglottids are seen earlier. If the examination is negative, it should be repeated after 6 months. Finding the scolex of a tapeworm immediately after treatment may prevent the long delay in evaluating the treatment regimen (see below).

Recovery of Helminths or Eggs After Treatment

As part of his evaluation following treatment, the physician may wish to know if the scolex of a tapeworm or if the adults of *Ascaris, Trichuris*, or hookworms were passed.

There are several ways to recover adult helminths from feces. Perhaps the most effective method begins by giving the patient a purge before treatment and then collecting all feces passed for several days following the initiation of treatment. The feces is finely blended with water (or 1% saline if examinations for eggs are to be made), usually by shaking and passing through a series of 5 or 6 stacking laboratory sieves with screens that gradually decrease in mesh size from a relatively large mesh screen to the finest mesh screen with about 30 openings per linear inch. The initial material that passes through the screens may be discarded or held for examination for cysts and eggs. After the initial screening, the material remaining in the series of screens are further washed until nothing more appears to wash away. The materials remaining in each of the sieves is then carefully examined for helminths or helminth parts.

If there is a reason to find eggs and/or cysts present in the fecal mass, the portion of feces initially passed through the series of screens should be retained. It can be concentrated by gravity sedimentation in several conical flasks or in large pharmaceutical graduates (see Gravity Sedimentation). Formaldehyde may be added to the fluid to make a 10% formalin solution. This fixes the organisms present and prevents possible infection through handling. Specimens with large numbers of cysts and/or eggs can be preserved in any of the appropriate collecting/preserving solutions for teaching or other uses.

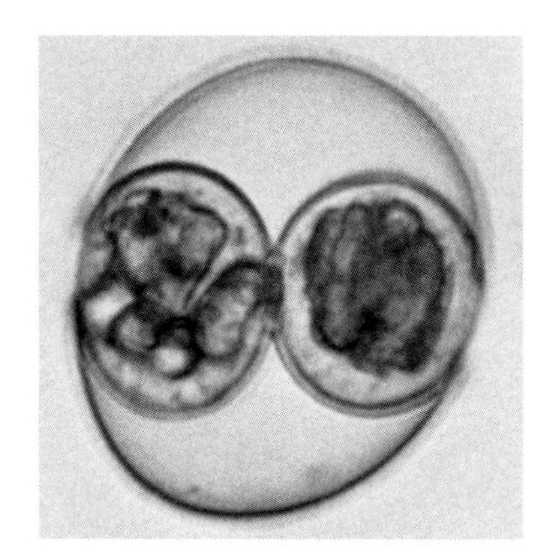

Part 4

Chapter 7

Introduction to the Parasites

PARASITES IDENTIFIED BY EXAMINING FECAL PREPARATIONS

As pointed out in Part 1, a variety of things may be present in a fecal specimen. The parasites and eggs found on microscopic examination of human feces usually represent one or more of eight different groups of organisms — four helminth and four protozoa. One parasite, *Blastocystis hominis*, which for many years was considered to be a yeast, has been reclassified as a protozoan (Zierdt et al., 1967) but it does not fit into any of the current taxonomic groups.

The diagnostic criteria for the identification of the various intestinal parasites has been addressed in different ways by a variety of authors (Markel et al., 1942; Chitwood and Chitwood, 1950; Hunter et al., 1960; Beek and Barnett-Connor, 1971; Sun, 1982; Ash and Orihel, 1987; Garcia and Bruckner, 1988; Bullock-Iacullo, 1988; Isenberg, 1992). Some have placed the emphasis on descriptions, some on photomicrographs, and others on a combination of different aids. Specific techniques have been suggested as ideal for identifying certain parasites; however, these may be inadequate for other parasites that may be present in the same specimen. In general, the principle needs of the clinical laboratory for the diagnosis of intestinal parasites are several products and procedures that be can used for collecting and preparing the specimens for examination, and aids that are adequate for making accurate identifications of what is found on microscopic examination.

The products and procedures for collecting and preparing the specimens have been addressed in the previous Parts. The specific criteria needed for identification of what is found by microscopic examination is covered in Part 4 with descriptions, diagrams, photomicrographs, and keys as the necessary aids.

HELMINTH PARASITES IN GENERAL

The helminths include a diverse group of metazoan parasitic animals. The term "helminth" comes from the Greek, *helmins*, *-inthos*, meaning worm, and originally referred principally to intestinal worms.

Although members of four phyla of helminths — Platyhelminthes, flatworms; Acanthocephala, thorny-headed worms; Nematoda, round worms; and Nematomorpha, hairworms — have been reported in the intestinal tract of man, only members of the phyla Platyhelminthes and Nematoda are considered to be of major importance.

The organization of these groups into phylogenetic categories is not fixed but varies somewhat with the classifier. For the purposes of this presentation, Nematoda is considered to be a phylum. (Chitwood and Chitwood, 1950). Although the nematodes are divided into two classes (see Table 1), they will be considered simply as nematodes or roundworms for the purposes of this manual.

Table 1. Helminths Infecting the Intestinal Tract of Man

Phylum	Class	Parasitic
Platyhelminthes	Trematoda (flukes)	All
	Cestoda (tapeworms)	All
Nematoda	Aphasmidia (roundworms lacking phasmids)[a]	Few
	Phasmidia (roundworms having phasmids)	Many

[a] Phasmids are caudal chemoreceptors (Chitwood and Chitwood, 1950).

Generally, the trematodes (flukes), cestodes (tapeworms), and nematodes (roundworms) are accepted as major groups of helminths infecting man and the number of species involving man is greatly limited. A number of species of trematodes are parasitic in man but many more infect mammals and lower vertebrates. In man, trematodes of different species infect the liver, lung, intestine, or blood vessels.

Among the cestodes, only a few species in the adult stage are parasitic in humans, primarily infecting the intestine, but man may become an accidental, aberrant, intermediate host (harboring the larval stage) of some others.

A number of the nematodes are parasitic in man, and man may become the accidental host of others. Nematodes may infect the blood vascular system, lymphatics, muscle, intermuscular fascia, other organs, or the gastrointestinal tract. A larger number of roundworms are parasites of mammals, lower animals, and plants, and many more are free-living.

Platyhelminthes

The phylum Platyhelminthes includes the Turbellaria, which are free-living, the Trematoda (flukes) which are all parasitic, and the Cestoda (tapeworms) which are all parasitic.

The adult members of the phylum Platyhelminthes are bilaterally symmetrical, (mostly) flattened dorsoventrally, and are commonly referred to as flatworms. They have no body cavity (coelom); the internal organs are embedded in a loosely connected parenchyma. They have no circulatory or respiratory systems, no exoskeleton, and no definitive anus. The excretory system is comprised of flame cells and ducts (protonephridia).

The parasitic flukes and tapeworms of man produce eggs which usually reach the intestinal tract (in some cases the respiratory or urogenital tract) and are evacuated with the feces. The entire proglottid of some cestodes may be evacuated before it disintegrates and releases the eggs.

The Trematodes, All Parasitic

The trematodes of man are a very diverse group that includes adult flukes that range in size from the very small *Metagonimus yokogawai*, which may be only 0.5 mm in length, to the larger *Fasciolopsis buski*, which may reach 7.5 cm in length. Most flukes are monecious (hermaphroditic), but the schistosomes are diecious, having male and female organs in separate individuals.

The life cycles of trematodes involve one or more intermediate hosts and/or helping hosts. The eggs are passed from the definitive host (the host in which the sexually mature stages occur) and in most cases must reach water for the life cycle to continue. A miracidium (a ciliated larva) emerges from the egg either in the water or after the egg is eaten by the first intermediate host, a mollusk, usually a snail, and penetrates the tissues of this host.

▶

Plate 2. The comparative size of various helminth eggs. The average length of the egg of each species is given in microns. 1. *Metagonimus yokogawai*, 26; 2. *Heterophyes heterophyes*, 28; 3. *Clonorchis sinensis*, 31; 4. *Opisthorchis viverrini*, 28; 5. *Opisthorchis felineus*, 30; 6. *Taenia* spp., 35; 7. *Dicrocoelium dendriticum*, 40; 8. *Capillaria philippinensis*, 40; 9. *Hymenolepis nana*, 45; 10. *Enterobius vermicularis*, 57; 11. *Trichuris trichiura*, 58; 12. Hookworm, either *Ancylostoma duodenale* or *Necator americanus*, 63; 13. *Ascaris lumbricoides*, fertilized, 57; 14. *Diphyllobothrium latum*, 65; 15. *Hymenolepis diminuta*, 75; 16. *Trichostrongylus* sp., 85; 17. *Ascaris lumbricoides*, unfertilized, 91; 18. *Paragonimus westermani*, 97; 19. *Schistosoma japonicum*, 89; 20. *Schistosoma mansoni*, 145; 21. *Fasciola hepatica*, 140.

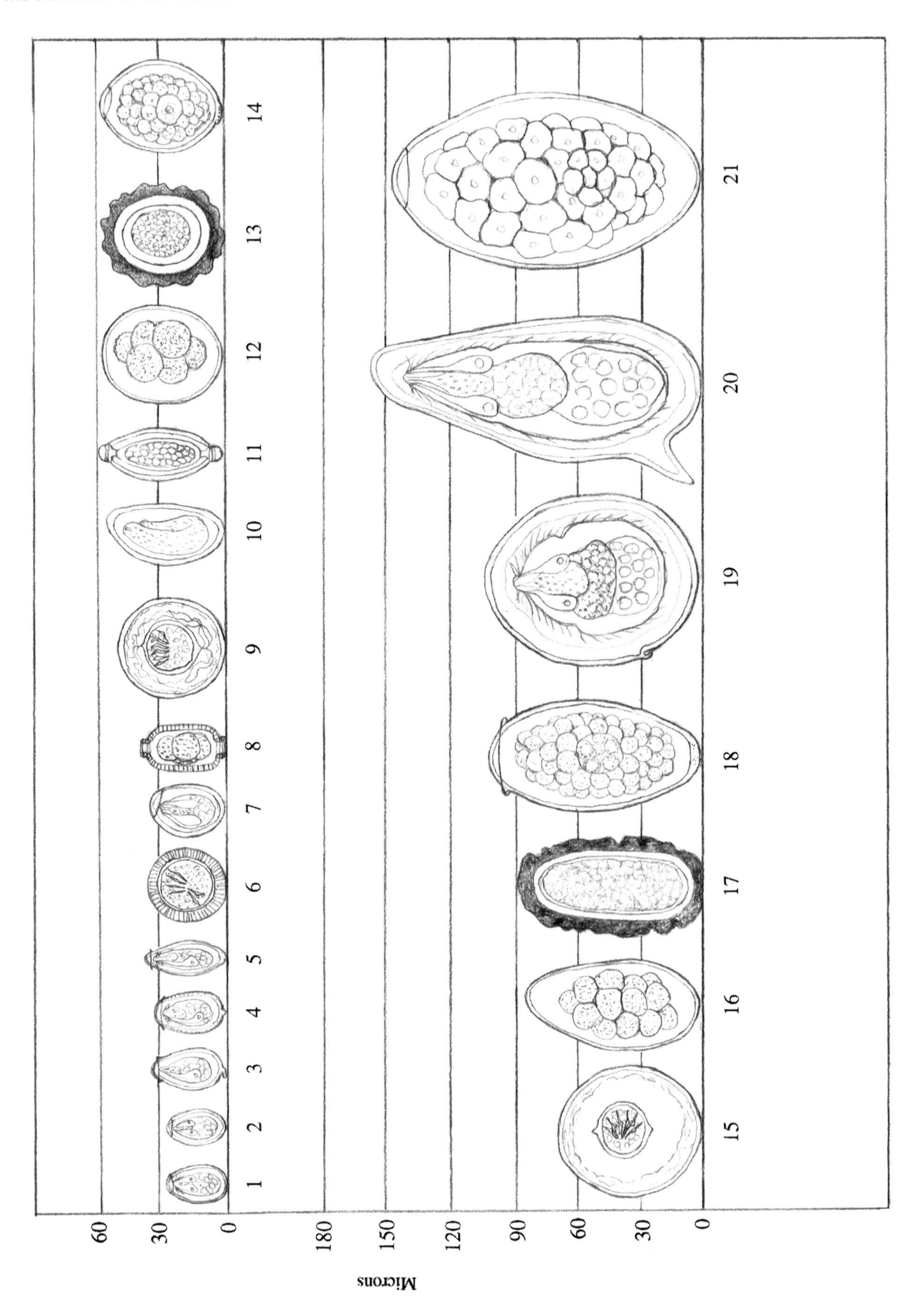
1
2
3
4
5
6
7
8
9
10
11
12
13
14
15
16
17
18
19
20
21
60
30
0
180
150
120
90
60
30
0
Microns

Several stages of larval development take place in the molluscan intermediate host, the stages varying with the species of trematode, after which free-swimming cercariae emerge (except the Dicrocoeliidae). The first intermediate hosts of this latter family of flukes are land snails that become infected when snails ingest viable eggs from feces. The cercariae are passed from the snail in slime balls that are deposited on fresh vegetation, especially grass, as the snail moves about. Certain species of ants eat the slime balls and become infected with the metacercariae. Infection of the definitive host occurs when the infected ants are accidentally ingested.

Cercariae of most species of monecious trematodes either seek an aquatic invertebrate or vertebrate intermediate host, or attach to a helping host, such as an aquatic plant, where they encyst as metacercariae in or on this second intermediate host. Infection of the definitive host with a monecious fluke requires the ingestion of a metacercaria, usually when the second intermediate or helping host is eaten as food. The cercariae of the diecious schistosomes (blood flukes) are infective and can directly penetrate the body surface of the definitive host. The definitive host usually becomes infected when it immerses a part of its body in infested water.

The Cestodes, All Parasitic

The cestodes are also a diversified group of monecious helminths. The mature adults of species for which man is the definitive host (all found in the intestinal tract) vary in size from the small *Hymenolepis nana,* measuring from 25 to 40 mm in length, to the large *Diphyllobothrium latum,* which may attain a length of 20 m. The anterior end (scolex, head) is either radially or bilaterally symmetrical and bears the organs of attachment: four cup-shaped suckers or two bothria (grooves) and in some species a rostellum with hooks. The remainder of the body (the strobila) is a series of segments, called proglottids, which begin just below the scolex at the neck (the germinal center). New proglottids are continually formed at the neck. As each new proglottid is formed, older ones as they mature, are moved down the chain. The mature proglottids contain both male and female reproductive organs and eggs are produced that eventually fill the uterus.

The life cycles vary greatly. The adults of *Diphyllobothrium latum*, a pseudophyllidian tapeworm, inhabit the small intestine of man. The mature proglottids are pushed farther and farther from the neck as new proglottids are formed and operculate eggs are passed singly from the genital pore of each gravid proglottid. Eggs are extruded continually from the gravid proglottids as long as they are being produced. The most terminal proglottids disintegrate and are sloughed off when egg production within the proglottid ceases.

If an egg reaches water at a suitable temperature, a ciliated larva (coracidium) emerges. If the coracidium is eaten by a suitable aquatic crustacean, it will lose its cilia, penetrate the gut wall, and enter the hemocoele (body cavity) of the crustacean where it develops into a procercoid (first larval stage). When the infected crustacean is ingested by a vertebrate, usually a fish, the procercoid is released, penetrates the wall of the digestive tract, and enters the muscle where it develops into a plerocercoid (sparganum). Man usually becomes infected when he ingests the plerocercoid when eating infected fish.

In the cyclophyllidian tapeworms, egg production continues and eggs are not released but accumulate within the uterus. When the proglottid is fully gravid and no more eggs are being produced, the segment either breaks from the chain or ruptures releasing the eggs, and either proglottids or eggs are passed with the feces.

The larvae of cyclophyllidian tapeworms are parasitic in a variety of animals, both invertebrate and vertebrate. Most species require an intermediate host but eggs of *H. nana* are infective for man when passed in the feces either in the proglottid or free. If the eggs of *H. nana* are freed from the proglottid while still in the small intestine, the hexacanth embryo may emerge from the egg and attach to the mucosa initiating a new infection in the same individual (autoinfection). An intermediate host of a cyclophyllidian tapeworm becomes infected when it ingests proglottids or eggs passed in the feces. The larvae take different forms depending on the species of tapeworm. Man becomes the definitive host when he ingests the intermediate host (or part of it) as food or accidentally.

The larval stages of some species, such as the pseudophyllidian tapeworm *Spirometra mansonoides*, and the cyclophyllidian tapeworms *Taenia solium, Multiceps multiceps, Echinococcus granulosa, E. multi-locularis* can develop in man. Infection with the larval stages of cyclophyllidian tapeworms occurs when eggs are accidentally ingested. Man becomes an intermediate host of *S. mansonoides* when he accidentally ingests an infected crustacean. In such cases, man becomes an aberrant, intermediate host. With the exception of *T. solium*, the adults of these tapeworms do not live in the intestine of man and therefore eggs are not present in human feces.

The Nematodes, Parasitic and Free-Living

The nematodes (both classes in the phylum Nematoda will be referred to simply as nematodes or roundworms) are extremely diverse but as a rule they include helminths that are unsegmented and cylindrical, with four main longitudinal chords, a triradiate esophagus, a definite anterior-posterior axis, a true digestive tract, three body layers, a body cavity unlined with mesothelium and without flame cells or cilia, and usually distinct separation of sexes.

Most of the nematodes are free-living or parasites of plants and animals. Those nematodes parasitic in the intestine of man vary in size and form. Adult females of *Strongyloides stercoralis* are microscopic while those of *Ascaris lumbricoides* may reach 8 in. in length.

Nematodes of the Gastrointestinal Tract

Some species of nematodes inhabit the small intestine (hookworms and *Ascaris*) while other inhabit the ceacum and colon (*Trichuris trichiura* and *Enterobius vermicularis*). Those nematodes diagnosed by fecal examination, i.e., those inhabiting the human gastrointestinal tract, produce eggs that either are passed with the feces or hatch in the intestine releasing first stage (rhabditoid) juveniles that are passed. These species do not require an intermediate host. Man becomes infected by ingesting eggs containing infective but immature stages (usually rhabditoid juveniles), or when infective (filariform) juveniles penetrate the body surface (hookworms and *Strongyloides stercoralis*).

Diagnostic Stages of Intestinal Helminths

The diagnosis of a helminth infection related to the gastrointestinal tract and associated organs is usually made by finding eggs or, less frequently, juvenile nematodes or proglottids of tapeworms in the feces (Table 2). In preceding Parts, how to collect, preserve, process, and examine fecal specimens were discussed and many of the materials and procedures employed were described. In this section, we are concerned with identification of what is found in the portion of feces examined.

Eggs of Helminths

The eggs that are passed in the feces vary greatly in size, as was stated in Part 2. When an egg is found on microscopic examination of a fecal specimen, perhaps the first question to be answered is, what is its size? The relative size of the various eggs found in human feces is presented in Plate 2. The drawing of each egg is numbered, it appears on a scale that shows its average length, and its general appearance is shown. For the size range of each egg, refer to the "Key" in Part 4, Chapter 8, The Helminths. In the legend, the helminth producing the egg is named and the average length of the egg is given in microns. Reference to this plate is a starting point in identifying eggs of helminths found in feces. Not all eggs that might be found are included on the Plate. In Part 4, Chapter 8, drawings of helminth eggs are presented in several Plates. These include additional species and give greater morphologic detail.

PROTOZOAN PARASITES IN GENERAL

The phylum Protozoa (unicellular animals) is represented by four major groups: Sarcodina (amoebae), Ciliophora (ciliates), Mastigophora (flagellates), and Apicomplexa (formerly Sporozoa). All members of the subphylum Apicomplexa that infect the intestine of man belong to the coccidia.

Blastocystis hominis, formerly considered to be a yeast, is now grouped with the protozoa (Table 3) (Zierdt et al., 1967). Because of the apparent variation in the forms seen in feces, what is identified as *B. hominis* may represent more than a single species. At present, the parasite does not fit into any of the currently recognized taxonomic groups.

Table 2. Helminths: What to Look for in Fecal Specimens

Parasite group	Found in feces
Trematodes	Eggs
Cestodes	Eggs or segments (proglottids)
Nematodes	Eggs or juveniles (larvae)

Table 3. Protozoa Infecting the Intestinal Tract of Man

Phylum	Group	Parasitic	Infecting man
Protozoa	Sarcodina (amoebae)	Many	Yes, some species
	Ciliophora (ciliates)	Many	Yes, one species
	Mastigophora (flagellates)	Many	Yes, few species
	Apicomplexa (coccidia)	Many	Yes, some species
	Blastocystis (unclassified)	One	Only one species recognized

Protozoa Infecting the Gastrointestinal Tract of Man

Considering the vast number of protozoa that have been described, only a relatively small number infect man and only a portion of these are present in the gastrointestinal tract. The largest of these is a ciliate, *Balantidium coli*, which is said to reach over 150 µm in length by 75 µm in width. The usual size range of the trophozoite is 50 to 80 by 30 to 60 µm and that of the nearly round cyst is 50 to 65 µm. Probably, the smallest protozoan parasites of the intestinal tract of man are the coccidia, *Cryptosporidium parvum*. Its oocyst, which may be present in human fecal specimens, usually measures no more than 5 µm in diameter.

A few species of amoebae are parasitic in man, usually in the intestine. Sizes range from the largest, *Entamoeba coli* (15 to 50 µm) to *Endolimax nana* (6 to 12 µm). *Entamoeba histolytica*, which is potentially pathogenic, usually infects the lower intestine, but under special circumstances it can invade other tissues.

The only ciliate that infects the intestinal tract of man is *Balantidium coli*.

A few species of flagellates are found in the intestine while others are found in the blood stream and in various tissues. The only intestinal flagellate that is considered to be pathogenic is *Giardia lamblia*. The size of the trophozoite is 10 to 20 by 6 to 19 µm and the cyst is 8 to 14 by 7 to 10 µm.

Stages of the five groups of protozoa found in fecal samples include trophozoites and cysts (oocysts or sporocysts in the case of coccidia) and cyst-like structures in the case of *Blastocystis hominis* (see Table 4).

The groups of protozoa most commonly seen in fecal examinations are the intestinal amoebae and flagellates. Certain structures (morphological characteristics) are common to all intestinal protozoa. Those structures that are unique to a particular group are identified in several Plates in Part 4, Chapter 9 that follows.

Intestinal Amoebae

Based on the presence or absence of peripheral chromatin in the nucleus, the amoebae can be placed into two groups. This grouping usually helps in making an identification of an amoeba seen on microscopic examination. In Plate 3, the four species of amoebae having a nucleus with peripheral chromatin are shown. The trophozoite and mature cyst stage of *Entamoeba histolytica*, *E. coli*, and *E. hartmanni* and the trophozoite and immature cyst stage of *E. polecki* are depicted.

Plate 4 shows the three species of amoebae having nuclei without peripheral chromatin. The trophozoite and mature cyst stage of *Endolimax nana* and *Iodamoeba butschlii* are shown, while only the trophozoite of *Dientamoeba fragilis* is shown since the cyst stage is unknown. Although *D. fragilis* is classified as a flagellate, it is included with the amoebae since it is seen only in the amoeboid form in fecal specimens.

In Part 4, Chapter 9, there are several Plates of drawings and of photomicrographs of intestinal amoebae

Table 4. Protozoa: What to Look for in Fecal Specimens

Parasite group	Found in feces
Sarcodina (amoebae)	Trophozoites and cysts
Mastigophora (flagellates)	Flagellated trophozoites or cysts (in some species)
Ciliophora (ciliates)	Ciliated trophozoites or cysts
Apicomplexa (coccidia)	Oocysts or sporocysts
Blastocystis (unclassified)	Cyst-like structures

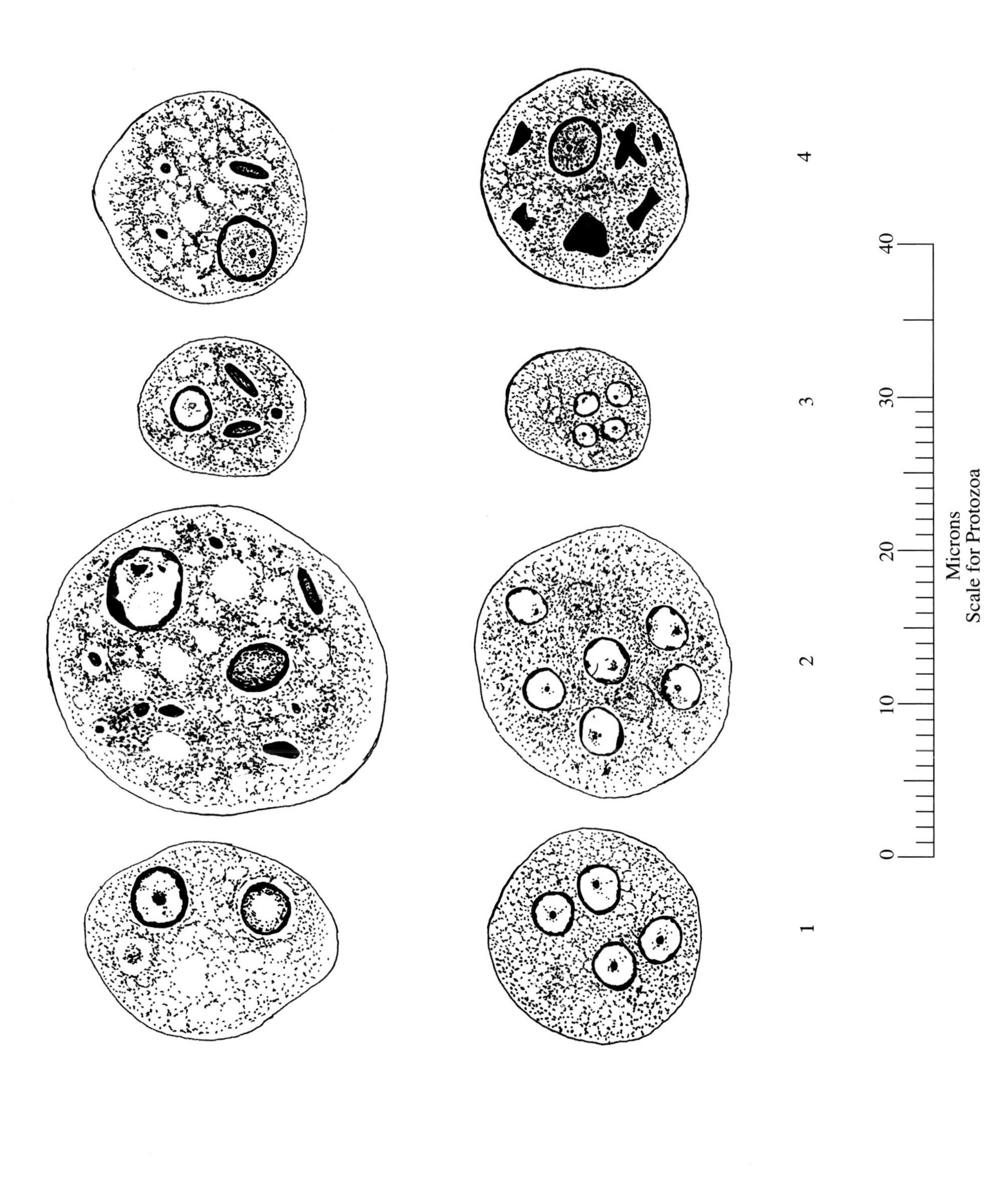

Plate 3. Species of intestinal amoebae having a nucleus with peripheral chromatin. Upper row, trophozoites and lower row, cysts. 1. *Entamoeba histolytica* trophozoite and mature cyst. 2. *Entamoeba coli* trophozoite and mature cyst. 3. *Entamoeba hartmanni* trophozoite and mature cyst. 4. *Entamoeba polecki* trophozoite and cyst. The scale used here is the standard for all drawings and photomicrographs of protozoa unless otherwise indicated.

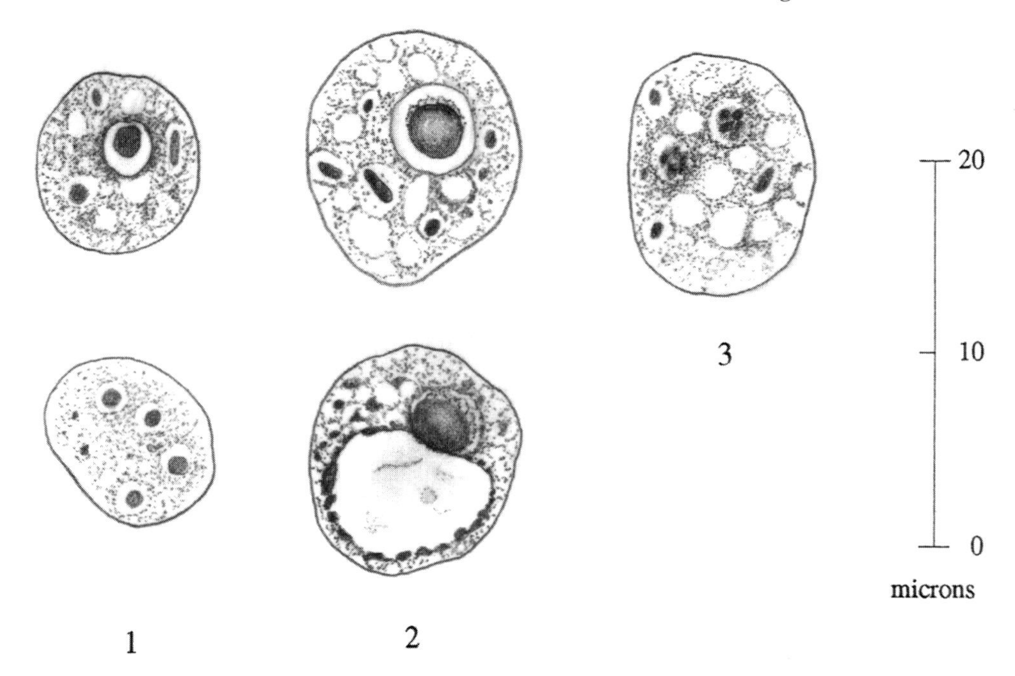

Plate 4. Species of amoebae having a nucleus without peripheral chromatin. Upper row, trophozoites and lower row, cysts. 1. *Endolimax nana* trophozoite and cyst. 2. *Iodamoeba butschlii* trophozoite and cyst. 3. *Dientamoeba fragilis* trophozoite; cyst unknown. Shown with amoebae since it is seen only in the amoeboid form in fecal specimens.

that depict the various characteristics observed in each species. There is also a key for differentiating the species of amoebae found in fecal specimens. A size scale is given in each Plate of drawings and the photomicrographs comprising each Plate were printed at the same scale so that relative sizes can be determined.

Intestinal Flagellates

Of all species of intestinal protozoa, *Giardia lamblia* is most commonly found and is the only intestinal flagellate that is pathogenic (excluding *Dientamoeba fragilis*, which is included herein with the amoeba). Two other species are present occasionally in fecal specimens, *Chilomastix mesnili* and *Trichomonas hominis*, while others, *Enteromonas hominis* and *Retortomonas intestinalis*, are rarely found. *G. lamblia* and *C. mesnili* have both trophozoite and cyst stages while *T. hominis* appears in fecal specimens only in the trophozoite stage. The general appearance of the three more commonly seen flagellates is depicted in Plate 5.

Intestinal Ciliates

The ciliate, *Balantidium coli*, is the only ciliate and the largest of the protozoa found in the intestinal tract of man (see schematic drawing, Plate 39). It has been reported from many different animals, especially pigs and some primates. Human cases of balantidiasis are relatively rare and are often found in individuals in low economic groups associated with pigs.

Intestinal Coccidia

The parasites belonging to the taxonomic order, coccidia, have been reclassified and placed in the subphylum Apicomplexa. Those who wish to review the taxonomy of this complicated group of protozoa are referred to *Protozoan Parasites of Domestic Animals and of Man* by Levine (1973).

For the purpose of this manual, three genera, *Isospora, Sarcocystis*, and *Cryptosporidium*, having species diagnosed by fecal examination, will be referred to as coccidia (see Plate 6).

Members of the genus *Isospora* parasitize dogs, cats, swine, humans, and some songbirds. *Isospora belli*, probably the only valid species in this genus infecting man, has a worldwide distribution but apparently has a low prevalence. Diagnosis is based on finding oocysts (or occasionally mature sporocysts) in fecal specimens.

Man is recognized as being the definitive host of two species of the genus *Sarcocystis, S. hominis* and *S. suihominis*. The intermediate hosts are cattle and swine, respectively. Diagnosis is made by finding either the oocysts or more frequently sporocysts in

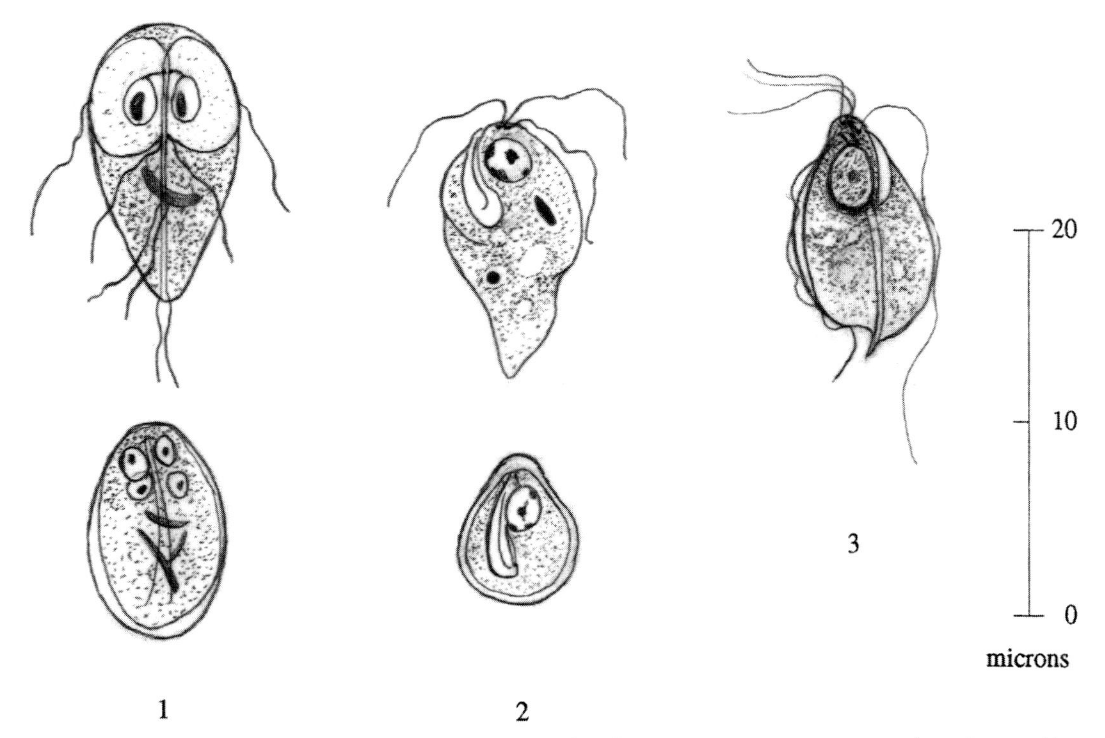

Plate 5. The species of flagellates more commonly seen in fecal specimens. Upper row, trophozoites and lower row, cysts. 1. *Giardia lamblia* trophozoite and cyst. 2. *Chilomastix mesnili* trophozoite and cyst. 3. *Trichomonas hominis* trophozoite. Cyst form unknown.

fecal specimens. There are very few reports of intestinal infection of humans with *Sarcocystis*. There have been a number of reports of man as an intermediate host where sarcocysts have been found in the tissues (Beaver et al., 1979).

Some oocysts of coccidia from animals may be passed in human feces but these are usually spurious infections. Most spurious infections are associated with eating game animals and birds, but man may become a transient, accidental host of some animal species.

Cryptosporidia, which inhabit the enterocytes of the villus border of the small intestine, are extracellular and produce sporozoites without a sporocyst wall, free in the oocyst, thereby differing from other coccidia. *Cryptosporidium parvum*, the only species in the genus recognized as infecting man, is now considered to be an important parasitic coccidia.

Protozoa of Undetermined Classification

Blastocystis hominis, which has been classified as a protozoa, does not appear to fit into any of the general groups of protozoa. It too appears to be responsible for diarrhea in patients having a compromised immune system (see Plate 6).

SOURCES OF INFECTION AND MODES OF TRANSMISSION

Up to this point, information pertinent to the diagnosis of intestinal parasites has been the only theme of this manual. It seems appropriate to deal briefly with the sources from which infection is acquired and the way (mode) in which transmission can occur before proceeding with more detailed information on parasite identification.

Sources of Infection

There are two distinctly different general types of sources from which infection by intestinal parasites can be initiated: contaminated sources and infected (or infested) sources.

Contaminated Sources

These sources are represented by media, such as soil and water, or objects including dirty utensils, dishes, and vegetables. In each case, the source has in some way become contaminated with feces and the infective stage of the parasite is present.

The source may be immediately infective. Eggs of the tapeworms *Hymenolepis nana* and *Taenia solium*

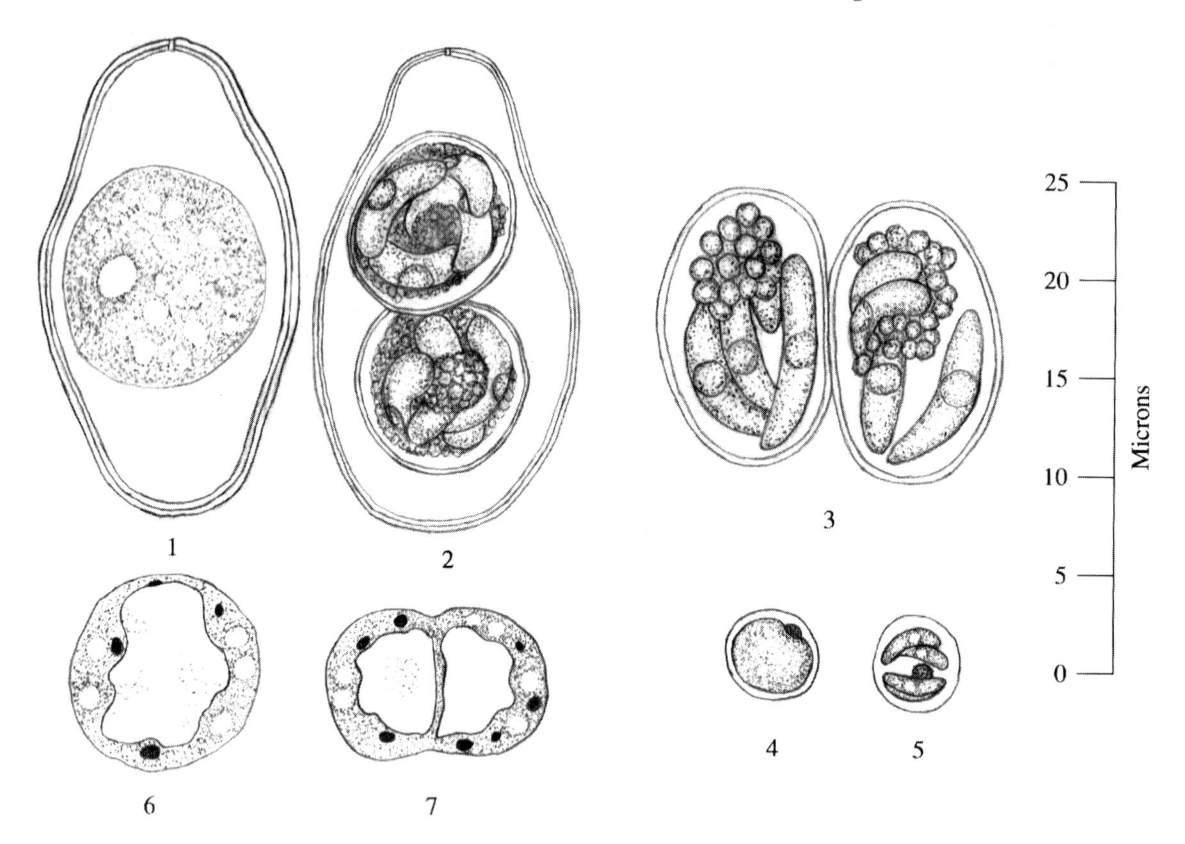

Plate 6. The species of coccidia more commonly seen in fecal specimens, and *Blastocystis hominis*. 1 and 2. *Isospora belli*, 1. Immature oocyst. 2. Mature oocyst with two sporocysts, each containing four sporozoites. 3. Paired sporocysts of *Sarcocystis* sp., each with four sporozoites as frequently seen in fecal specimens. 4 and 5. *Cryptosporidium parvum*, 4. An immature oocyst. 5. A mature oocyst with four sporozoites (note that there is no sporocyst wall surrounding the sporozoites. 6 and 7. *Blastocystis hominis*, 6. Typical cyst-like structure seen in fecal specimens. 7. A form in division.

are infective when passed with the feces. Eggs of the nematode *Enterobius vermicularis* are infective shortly after being laid in the folds of the anus. The mature cysts (or oocysts) of protozoa (*Entamoeba histolytica*, *Giardia lamblia*, and *Cryptosporidium parvum*) are also infective when passed with the feces. The oocysts of coccidia, if not developed to the infective stage when passed, can continue development outside of the host, whereas immature cysts of the amoebae do not continue to develop.

Eggs of some nematodes develop to the infective stage outside of the host and must remain in the soil for a period of time before becoming infective. A delayed contaminated source evolves when eggs of some nematodes (or oocysts of coccidia) must have time outside of the host for the parasite in the oocyst or egg to reach the infective stage, e.g., eggs of *Ascaris* and *Trichuris*. Although changes may have taken place within the egg, the form (oocyst or egg) that initiates a new infection is the same as the form passed in the feces. Regardless of the type of contaminated source, infection is initiated only by ingestion of the infective stage of the parasite. The organism may be infective immediately after the source is contaminated or after the parasites have had time to develop to the infective stage in the oocyst or egg.

Infected or Infested Sources

An infected source is one in which or on which the infective stage of a parasite is living. It may be an insect (beetles, *Hymenolepis diminuta*), a mammal (cattle, *Taenia saginata*), or a plant (watercress, *Fasciola hepatica*). An infested source is a medium or object that harbors the infective stage as a transient such as water (cercariae of *Schistosoma mansoni*) or soil (infective juveniles of *Strongyloides stercoralis*).

The eggs of many helminth intestinal parasites are not the infective stage. Some intestinal parasites require development within an intermediate host, especially trematodes and cestodes, therefore the eggs are not infective (or are not the infective stage for man) when passed in the feces but must develop in another host (to a different stage or form) before becoming infective. Many of the trematode parasites have two intermediate hosts. In such cases, the egg or the stage of the parasite in the egg is infective for an intermediate host, not for man. Some nematode parasites must change form, e.g., egg to juvenile worm, and develop outside of the host to become infective. In these cases, the eggs are not infective for man, although there is no intermediate host involved. Soil or ground vegetation becomes the infested source of infection.

Examples of parasites that are transmitted to man from an infested source are *Strongyloides* and hookworms, nematode parasites, and *Schistosoma mansoni* and *S. haematobium*, trematode parasites.

Eggs of hookworms must reach soil where the rhabditoid juvenile worm that develops in the egg emerges. This stage is not infective but develops to a second stage, and then to a third stage, filariform juvenile worm, which is the infective stage for man.

In the case of *Strongyloides*, the rhabditoid juvenile worm is passed in the feces and goes through a development similar to that of the hookworms. It also has an alternative developmental cycle in which it may have one to several free-living generations of male and female worms before the filariform juvenile that is infective for man develops. What triggers this change is unknown. The soil, grass, and other ground vegetation become an infested source of these parasites for man.

The miracidia of schistosome parasites of man that emerge from the eggs must find a suitable snail and penetrate the body surface to establish an infection in the intermediate host. The parasites go through several larval stages to develop to the cercarial stage. The cercariae leave the snail and remain free in the water awaiting a suitable host. The water becomes an infested source of infection for man.

Examples of parasites that are transmitted to man from an infected source are *Clonorchis sinensis* and *Taenia saginata*. The egg of *C. sinensis*, and other monecious trematodes infecting man, must reach fresh water in which an appropriate snail intermediate host is present in order for the life cycle to continue. The miracidia of some species of monecious trematodes emerge in the water and find and infect the first intermediate host, a snail (e.g. *Fasciola hepatica*). In the case of *C. sinensis*, the egg containing a miracidium is ingested by the snail intermediate host and the miracidium emerges in the snail's intestine and penetrates its intestinal wall to initiate an infection. The parasite goes through several larval stages in the snail host and eventually cercariae emerge from the snail. These cercariae must find a suitable fish within a limited time or die (a transient free-living stage in water). Those that find an appropriate fish, attach to it and invade its tissues where they develop into encysted metacercariae. The fish becomes the infected source for man.

In the case of *T. saginata*, cattle ingest eggs or proglottids of the tapeworm that have been passed in the feces of man to contaminate the soil or ground vegetation where the cattle are feeding. The hexacanth embryo emerges from the egg in the gastrointestinal tract of the animal, penetrates the gut wall, and finds its way, usually via the blood stream, to the fascia between muscles. There it becomes encysted and develops into a cysticercus, the infective stage for man. The meat of the animal becomes an infected source of infection for man.

Modes of Transmission

Transmission may be either active or passive. In cases of schistosomiasis, strongyloidiasis, or hookworm infection, the parasite enters the body by direct penetration of the body surface by the infective stage. When either the parasite or the intermediate host plays an active role in transmission, it is considered to be active transmission. When neither a host nor a parasite plays any role, transmission is considered to be passive transmission. Passive transmission of the infective stage from either a contaminated source or an infected source is by ingestion.

Active Transmission

Active transmission can occur when stages of parasites such as the infective, filariform juveniles of hookworms or *Strongyloides stercoralis*, penetrate the skin of the host. Sitting on the ground at a picnic or walking barefoot through the grass can expose individuals to infection by bringing them into direct contact with the infective stage of the parasite. Although humans are usually responsible for contaminating the source (soil), in the case of *Strongyloides*, there are a few instances when pet dogs have been incriminated (Georgi and Sprinkle, 1974). When the juveniles reach the filariform, infective stage, they can actively initiate the infection by finding a suitable area of body and penetrating the body surface. These cases represent active transmission from an infested source.

Active transmission is exemplified by the schistosomes. When cercariae of schistosomes that have emerged from the snail intermediate host infest water in which individuals are swimming or bathing, they are capable of directly penetrating the unbroken skin of the definitive host. In fact, simply rinsing ones hands in infested water can bring an individual into contact with the infective stage of the parasite. By attaching to the host with its suckers and using fluid extruded from its penetrating glands, a cercaria can penetrate the body surface to gain entrance into the tissues and eventually the blood vessels of the host thereby initiating an infection. Again, active transmission from an infested source.

Autoinfection, i.e., reinfection with a parasite derived from the same host, may occur in patients with strongyloidiasis and is another form of active transmission. When juvenile worms are retained in the intestine until they become infective, a new infection can be initiated when these filariform juveniles invade the mucosa without having left the host.

Autoinfection is possible in individuals infected with *Hymenolepis nana*. Eggs are infective when the proglottid becomes fully gravid. As pointed out earlier, if a fully gravid proglottid begins disintegration before leaving the small intestine and the eggs are released, the delicate egg shell may be readily broken, the hexacanth larva released from the egg where it can attach to the mucosa of the intestine, and establish a new infection without the eggs leaving the original host. Autoinfection represents active transmission from an infected source.

Passive Transmission

Passive transmission is the more frequent mode of infection with intestinal parasites and neither the parasite nor its intermediate host takes an active role in the transmission.

Passive transmission from contaminated sources commonly occurs with the accidental ingestion of the infective stage of an intestinal parasite that produces resistant eggs or cysts, e.g., *Ascaris lumbricoides*, *Trichuris trichiura*, *Entamoeba histolytica*, *Giardia lamblia*, and *Cryptosporidium parvum*. The infective stage may be acquired from any of a variety of contaminated sources: soil, water, food, or even fomites (door knob, drinking glass). Salads prepared from vegetables harvested from contaminated soil may harbor infective parasites or eggs. Unfiltered water, which is usually safe, may from time to time become contaminated. The potability of water supplies in many cities as well as in many rural areas is considered questionable. Plates and utensils in restaurants that have been washed in cold water or have been dried with dish towels that are repeatedly used may be contaminated sources of infection.

Passive transmission from infected sources is less common but does occur. Parasites and eggs that are infective when evacuated from the definitive host can be passed directly from one infected individual to another (direct transmission) or even back to the same individual (hand to mouth transmission). The individual is infected and the parasite is unintentionally and accidentally transmitted from the infected to an uninfected individual. Examples of direct and hand-to-mouth types of transmission are seen in institutions and day care centers where behavioral patterns of the individuals and sanitary control are more conducive to such modes of transmission (Yoeli et al., 1972).

Many animals or plants may function as intermediate or helping hosts of parasites. Infection with the adult stage of *Taenia solium* or *T. saginata* is acquired by deliberate ingestion of infected, inadequately cooked pork or beef, respectively (pigs and cattle are the intermediate hosts). Infection with *Clonorchis sinensis* can be acquired by eating portions of infected fish that may be used to prepare certain uncooked dishes, such as Japanese "sushi" or Chinese "yue-shan chuk" (raw carp with rice soup). Small infected arthropods may be accidentally ingested when eating grains or dried fruit (*Hymenolepis diminuta*). These are all examples of passive transmission from an infected source.

Infection with most intestinal parasites is initiated following the ingestion of the infective stage of the parasite either from a contaminated or an infected source.

Ectopic and Aberrant Infections

Rarely, extraintestinal infection with *Entamoeba histolytica* may occur. If a wound or cut becomes contaminated with feces containing trophozoites, they may invade the surrounding tissue to initiate an infection. Some animal parasites can invade the tissues of man to initiate a transient, aberrant infection. The juveniles of animal hookworms may penetrate the body surface causing creeping eruption, the cercariae of bird schistosomes may penetrate the skin, causing cercarial dermatitis, but such infections are beyond the scope of this manual which deals exclusively with parasitic infections of the gastrointestinal tract and adjacent organs that are diagnosed by microscopic examination of fecal specimens.

Chapter 8

The Helminths

DIAGNOSIS OF HELMINTH INFECTIONS

Helminths are cosmopolitan and play a significant role in the morbidity and mortality of people in many parts of the world, especially among the poorer classes in third world countries. Most prevalent are the intestinal helminths and infection with these helminths is most often diagnosed by finding and identifying eggs and/or juvenile nematodes when examining fecal specimens. Either fresh or preserved specimens may be used for diagnosing an infection. Although the utilization of certain specific products, procedures, and techniques for collecting, preserving, and examining fecal specimens may be given emphasis in this manual, most of the included material that relates to parasites applies regardless of the collection method, specific product, or procedure employed.

Identifying an egg found but not readily recognized often presents a problem for the laboratory worker and may require looking in several reference sources before an identification can be made. If an appropriate reference source is not readily available, identification may take several days or sometimes longer. This manual is presented as a single reference source that provides methods, procedures, techniques, and information to enable the individual to properly collect and prepare the specimen, and rapidly find and accurately identify the parasites and eggs present and found on microscopic examination of fecal specimens.

The Structure of Helminth Eggs

In Chapter 7, the relative size of various eggs was shown in Plate 2. In addition to size, there are certain structures or combination of structures (morphologic characteristics) that distinguish one egg from another. In Plate 7, drawings of representative eggs from each group, trematodes, cestodes, and nematodes are shown and some of their structures are identified. These structures are used also in the "Key to Helminth Eggs Found in Feces" and most are defined in the Glossary. The eggs drawn in the Plate are representative of those with more simple structure (e.g., hookworm) and others with more complicated structure (e.g., *Hymenolepis diminuta*), and some just beginning development (e.g., *Ascaris lumbricoides*) and others with complete embryos (e.g., *Schistosoma mansoni*). All of the structures identified in the Plate may not be evident in eggs found in a fecal preparation. Usually, in specimens examined or preserved immediately after passage there are sufficient features for the observer to make an identification. In some cases, closely related species have eggs that appear identical (e.g., *Taenia solium* and *T. saginata*). Since the eggs of these two species cannot be readily differentiated, characteristics of the gravid proglottid or scolex are used to identify the species present. Other members of the family Taeniidae produce similar eggs (members of the genera *Echinococcus* and *Multiceps*) but man is not infected by the adult stage and therefore eggs are not found in human feces.

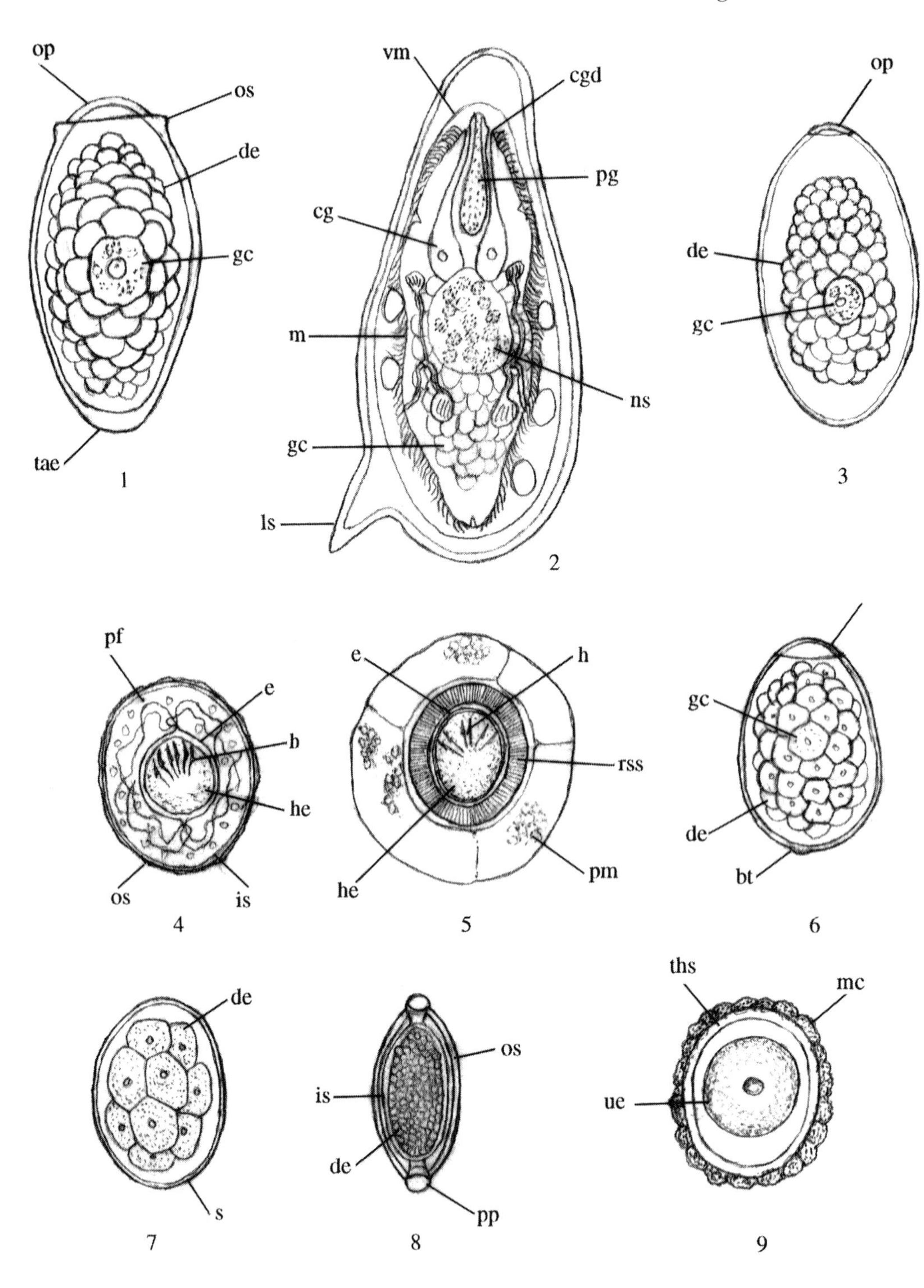

Plate 7. The structures seen in eggs representing each helminth group. **Trematodes:** 1. *Paragonimus westermani*, 2. *Schistosoma mansoni*, 3. *Echinostoma ilocanum*. **Cestodes:** 4. *Hymenolepis nana*, 5. *Taenia* sp., 6. *Diphyllobothrium latum*. **Nematodes:** 7. Hookworm (*Ancylostoma duodenale* or *Necator americanus*), 8. *Trichuris trichiura*, 9. *Ascaris lumbricoides*. **Key:** bt, button-like thickening; cg, cephalic gland (lytic, penetrating gland); cgd, cephalic gland duct; de, developing embryo; e, embryophore; gc, germ cell; h, hooklet; he, hexacanth embryo; is, inner shell; ls, lateral spine; m, miracidium; mc, mammillated coat (cortex); ns, nervous system; os, outer shell; op, operculum; ops, opercular shoulders; pf, polar filaments; pg, primitive gut; pm, primary membrane; pp, polar plugs; rss, radially striated shell; s, shell; tae, thickened abopercular end; ths, thick hyaline shell; ue, unsegmented embryo; vm, vitelline membrane. Not to scale.

The smaller trematode eggs, *Clonorchis sinensis, Opisthorchis viverrini, Heterophyes heterophyes,* and *Metagonimus yokogawai* are very similar and often differentiating between them is difficult. (Refer to Plates 8, 9, 13–17.) If eggs can be put into one of the two infection groups, i.e., species inhabiting the intestinal tract or species inhabiting the bile ducts of the liver, further identification is generally not necessary. When there are many eggs in the specimen, an identification usually can be made; however, in patients infected with several species of trematodes producing small eggs, which is not uncommon, identification of the eggs becomes more complicated. Plate 8, shows the structural characteristics that may be used to separate the small trematode eggs into the two infection groups. The eggs of the intestinal trematodes, *M. yokogawai* and *H. heterophyes* have paired organs, especially cephalic glands, whereas the species inhabiting the bile ducts do not, as shown in the drawings and photomicrographs.

To assist those who have the responsibility for identifying eggs of helminths, three aids are used: (1) photomicrographs of the eggs of helminths more commonly found on microscopic examination of fecal specimens, all taken at the same magnification with enlargements of some eggs to help with detail; (2) drawings (or diagrams) of eggs that may offer some detail not readily seen in a two-dimensional photomicrographs; and (3) a "Key to Helminth Eggs Found in Feces". These three, along with some discussion and suggestions on technique, should provide considerable support to the individuals responsible for finding and accurately identifying eggs of helminths.

TREMATODE EGGS

The Monecious Trematodes of the Liver, Intestine, and Lungs

Because of the difficulty in differentiating between the small trematode eggs, more attention is given to them in the drawings and photomicrographs. As stated above, the organs of the intestinal flukes, *Metagonimus* and *Heterophyes* are symmetrical and their shoulders at the operculum are not distinct whereas the organs of the liver flukes, *Clonorchis* and *Opisthorchis* are asymmetrical and the shoulders are prominent (see Plates 8 and 13).

Metagonimus eggs (Plate 9:1; Plate 13:1; Plate 14) are about 26 × 16 µm, without shoulders, and usually with a small papilla at the abopercular end. If the details of the internal structures of the miracidium can be discerned, the paired cephalic (cytolytic) glands may be seen. The eggs are usually barrel shaped and broadest below the middle, toward the abopercular end.

The eggs of *H. heterophyes* are about 28 × 16 µm and are similar but usually have a more narrow operculum, are broadest at the middle, and often have no papilla at the abopercular end (Plate 8:3 and 4; Plate 9:2; Plate 13:2; Plate 15).

C. sinensis eggs are usually ovoidal (light-bulb shaped) with distinct shoulders and a small buttonhook projection at the abopercular end. They average 31 × 16 µm in size. The operculum is usually rounded (Plate 8:1; Plate 9:3; Plate 13:4; Plate 17).

The eggs of *Opisthorchis viverrini* have a broader and flatter operculum, and the shell appears to be rough or striated in comparison. There is a small, rounded papilla at the abopercular end. The average size of the eggs are similar, about 28 × 16 µm (Plate 9:4; Plate 13:3; Plate 16).

Another species, *Opisthorchis felineus* (not shown) is found in Eastern and Central Europe. It is about 30 × 12 µm, usually thinner, and tapers from the middle toward both ends. If the species of these small trematode eggs cannot be easily determined, a report of *Clonorchis*-like eggs may be acceptable, depending on the condition of the patient, since treatment is similar for all five species.

Dicrocoelium dendriticum eggs average 38 × 26 µm, have thick shells slightly flatter on one side, a broad operculum on a slight angle, and a fully developed miracidium with symmetrical organs (Plate 9:5; Plate 18). Eggs may appear in the feces after an individual has eaten liver or sausage containing liver from an infected animal (spurious infection). It is mainly a parasite of the liver of sheep and other herbivorous animals but man is rarely an accidental host.

Echinostoma locanum is one of a group of intestinal flukes that have been reported from man. The eggs appear similar to *Fasciola* but are about 2/3 the size

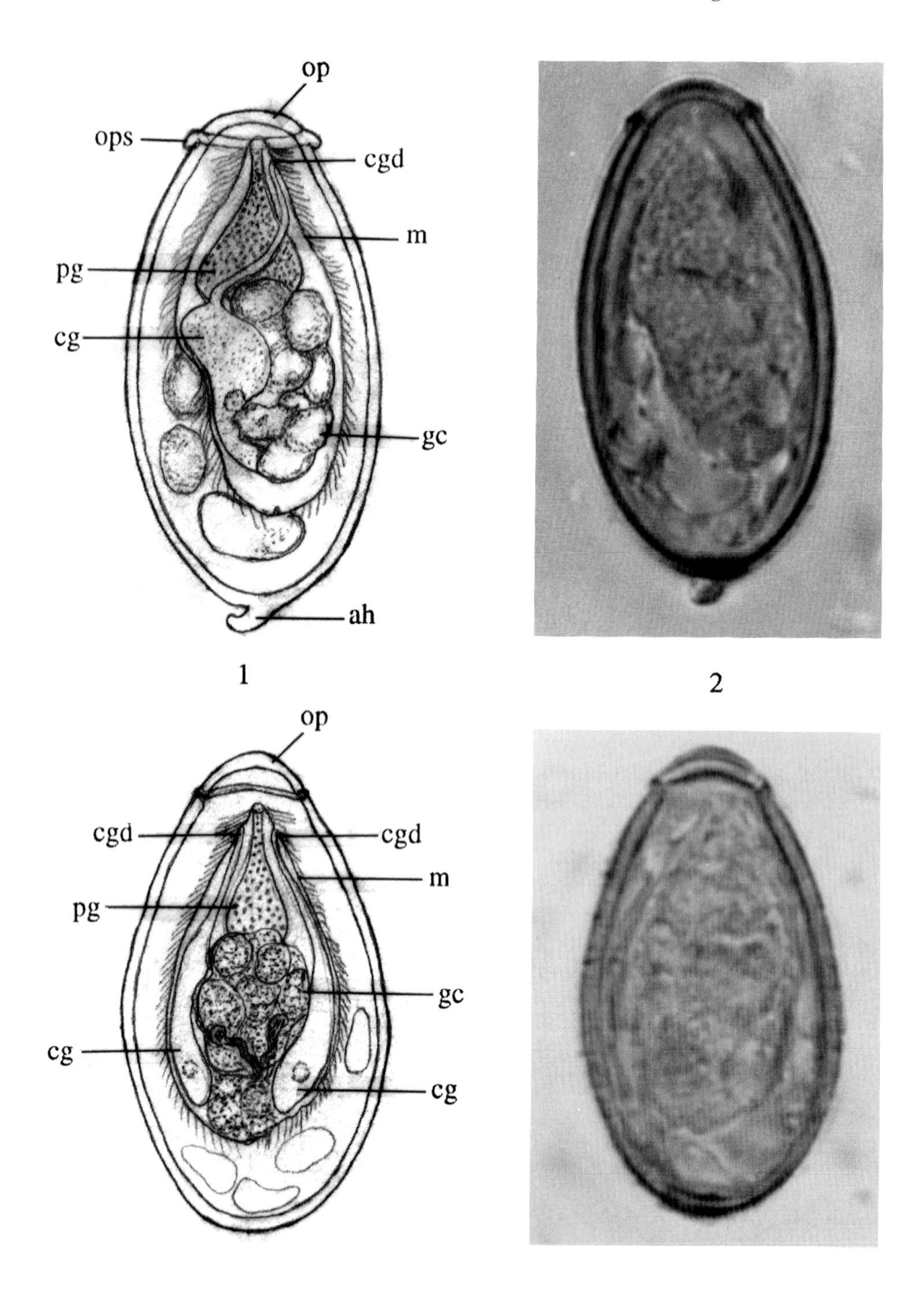

Plate 8. Comparative structures in the eggs of *Clonorchis sinensis* (1. diagram, 2. photomicrograph); and *Heterophyes heterophyes* (3. diagram, 4. photomicrograph). The miracidia in the eggs of the liver flukes, *C. sinensis, Opisthorchis viverrini*, and *O. felineus*, are asymmetrical, having a single cephalic gland on one side, while those of the intestinal flukes *H. heterophyes* and *Metagonimus yokogawai* are symmetrical, having paired glands. This difference in symmetry allows the separation of the eggs into those flukes inhabiting the ducts of the liver (the first three species) from those inhabiting the intestine (the latter two). ah, abopercular hook; cg, cephalic gland; cgd, cephalic gland duct; gc, germinal cells; m, miracidium; op operculum; ops, opercular shoulders; pg, primitive gut. Not to scale.

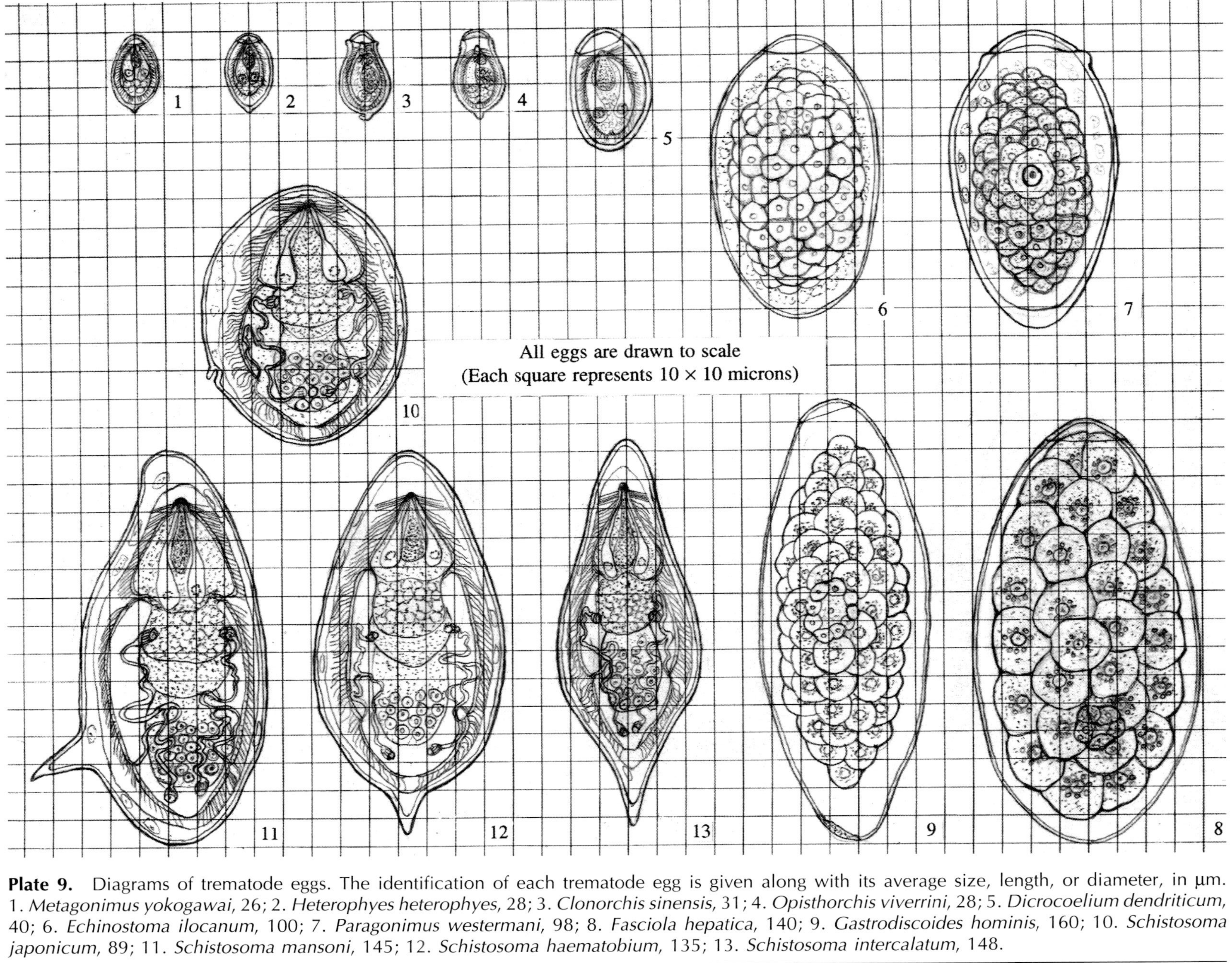

Plate 9. Diagrams of trematode eggs. The identification of each trematode egg is given along with its average size, length, or diameter, in μm. 1. *Metagonimus yokogawai*, 26; 2. *Heterophyes heterophyes*, 28; 3. *Clonorchis sinensis*, 31; 4. *Opisthorchis viverrini*, 28; 5. *Dicrocoelium dendriticum*, 40; 6. *Echinostoma ilocanum*, 100; 7. *Paragonimus westermani*, 98; 8. *Fasciola hepatica*, 140; 9. *Gastrodiscoides hominis*, 160; 10. *Schistosoma japonicum*, 89; 11. *Schistosoma mansoni*, 145; 12. *Schistosoma haematobium*, 135; 13. *Schistosoma intercalatum*, 148.

(about 100 × 55 μm). They are thin-shelled, with a narrow, indistinct operculum and an undeveloped miracidium (Plate 9:6; Plate 21, Figures 1 through 3). Differentiating between those of the rarely found intestinal flukes that produce eggs between the sizes of *Echinostoma ilocanum* and *Fasciola hepatica* can be very difficult. Members of the genus *Echinostoma* inhabit the intestine of a variety of mammals and are usually geographically restricted to countries in the Far East and to islands in the Pacific region. Man is an accidental host.

Fasciola hepatica, F. gigantica, and *Fascioloides magna* are parasites of cattle, sheep, and wild herbivores. The immature and adult flukes invade the parenchyma and bile ducts of the liver where they cause considerable damage. They rarely infect man. The eggs of these flukes vary somewhat in size between species (*F. gigantica* eggs are usually larger), but eggs within the same size range are so similar that differentiating between them in fecal specimens is not usually possible. When eggs are passed in the feces, they are undeveloped (Plate 9:8; Plate 19).

The eggs of *Fasciolopsis buski*, an intestinal fluke, are also similar in size and shape and usually cannot be differentiated from those of the large liver flukes. When determining whether the parasite is an intestinal or liver fluke, it may be necessary to collect bile fluid from the opening of the common bile duct to determine the source of the eggs. When adult flukes are in the bile ducts, eggs are usually present in the bile fluid.

The lung flukes, represented here by *Paragonimus westermani*, have heavier eggs usually with broad, distinct shoulders and a thickened abopercular end. The size of the eggs may vary from short and broad (75 × 65 μm) to long and narrow (120 × 45 μm). The miracidium is undeveloped (Plate 9:7; Plate 20). Eggs are usually coughed up with sputum from the bronchial tubules and are present in the sputum. Some sputum containing eggs is swallowed and eggs pass unaltered through the intestine and are evacuated with the feces. Because of the route the immature fluke takes to arrive at its typical site in the lungs, i.e., migrating through the abdominal cavity and passing through the diaphragm into the pleural cavity, adult flukes may be found in widely distributed locations in the host. Adult parasites may be located in the intestinal wall from which eggs are passed directly into the lumen of the intestine. From other ectopic foci, eggs may have no route of exit from the host and remain trapped in the tissues.

Although *Gastrodiscoides hominis* is a relatively small intestinal fluke, from 5 to 10 × 4 to 6 mm, its eggs are large, about 160 × 65 μm, and are rhomboidal in shape. The operculum is indistinct and on a slight angle (Plate 9:9; Plate 21:4–6). The adult fluke attaches to the mucosa of the caecum and ascending colon where it causes inflammation and attendant symptoms of diarrhea. It has a broad geographic distribution but only rarely does man become an accidental host.

The Diecious Trematodes of Blood Vessels

There are five species of schistosomes that are considered to be human parasites, *Schistosoma japonicum* and *S. mekongi* in the Far East, *S. mansoni* in Africa and the Western Hemisphere, *S. haematobium* in Africa and the Middle East, and *S. intercalatum* in Africa. Members of the genus *Schistosoma* infecting man inhabit blood vessels of the extrahepatic portal and caval venous system, especially the mesenteric venules rich in nutrients coming from the intestine. Eggs laid by the female worms jam the small vessels adjacent to the intestinal and bladder wall. Some eggs are carried by the vessels to the liver and a lesser number to other organs. Other eggs, probably a majority, become trapped in or pass through the wall of the intestine (*S. mansoni, S. japonicum, S. intercalatum*) or bladder (*S. haematobium*). Those that reach the lumen of the intestine are evacuated with the feces and those that reach the lumen of the bladder are passed with the urine.

The eggs of schistosomes are quite distinct. They are large and contain a fully developed larva (miracidium) that usually emerges from the egg as soon after passage as it reaches suitable water. All eggs have some kind of spine, but the spine of the eggs of *Schistosoma japonicum* and *S. mekongi* are very small and rarely visible. The eggs of *Schistosoma japonicum* (Plate 9:10; Plate 22) are about 90 × 68 μm and often have a loose coating covering the shell which makes them more difficult to find. When eggs are free of the coating, the small spine may be visible. The parasite is endemic in the Orient and Pacific islands as far east as the Philippines.

The eggs of *Schistosoma mansoni* (Plate 9:11; Plate 23) are large, averaging about 145 × 60 μm. They have a lateral spine toward the thicker posterior end making them relatively easy to recognize. Sometimes, an egg is lying with the spine up or under the egg and care must be taken to find the spine (Plate 23). The parasite is endemic in Africa, parts of South and

Central America, and on many of the islands of the Caribbean including Puerto Rico.

There are several schistosome eggs with terminal spines. The eggs of *Schistosoma haematobium* average 135 × 65 µm, are broadest at the end toward the spine and taper toward the opposite end (Plate 9:12; Plate 24:1). The species is endemic in many countries in Africa and the Middle East. It usually infects the small venules adjacent to the bladder and eggs are usually passed with the urine, although they are found rarely in fecal specimens. *S. haematobium* is the scourge of the Nile river valley in East Africa and Egypt, especially in the Delta region, probably with 10,000,000 people infected.

S. intercalatum is known to be endemic in seven African countries south of the Sahara (Schwetz, 1951). It inhabits the venules of the mesenteries adjacent to the intestine and, like *S. mansoni*, the eggs are usually present in the intestine and are passed with the feces. Whereas the eggs of *S. haematobium* are broad and somewhat oval, those of *S. intercalatum* taper slightly toward both ends and average 190 × 68 µm (Plate 9:13; Plate 24:2) and appear to be very similar to *S. haematobium*. Although the eggs of *S. intercalatum* are usually larger, those of the two species have been drawn at approximately the same size to show the structural differences. If many eggs with terminal spines are found in the fecal specimen and/or if they are in the larger size range, the species is probably *S. intercalatum*. A urine specimen should be checked also to see if eggs in the same size range are present to eliminate *S. haematobium*. It is possible for an individual to be infected with both species at the same time since they occur in some of the same geographic areas.

Some species of schistosomes that usually infect animals and produce eggs with terminal spines (*S. bovis*, and *S. mattheei*) have been reported from man (Kuntz, 1955; Pitchford, 1961; Pitchford, 1965). Cattle, camels, goats, pigs, and sheep may act as the definitive host

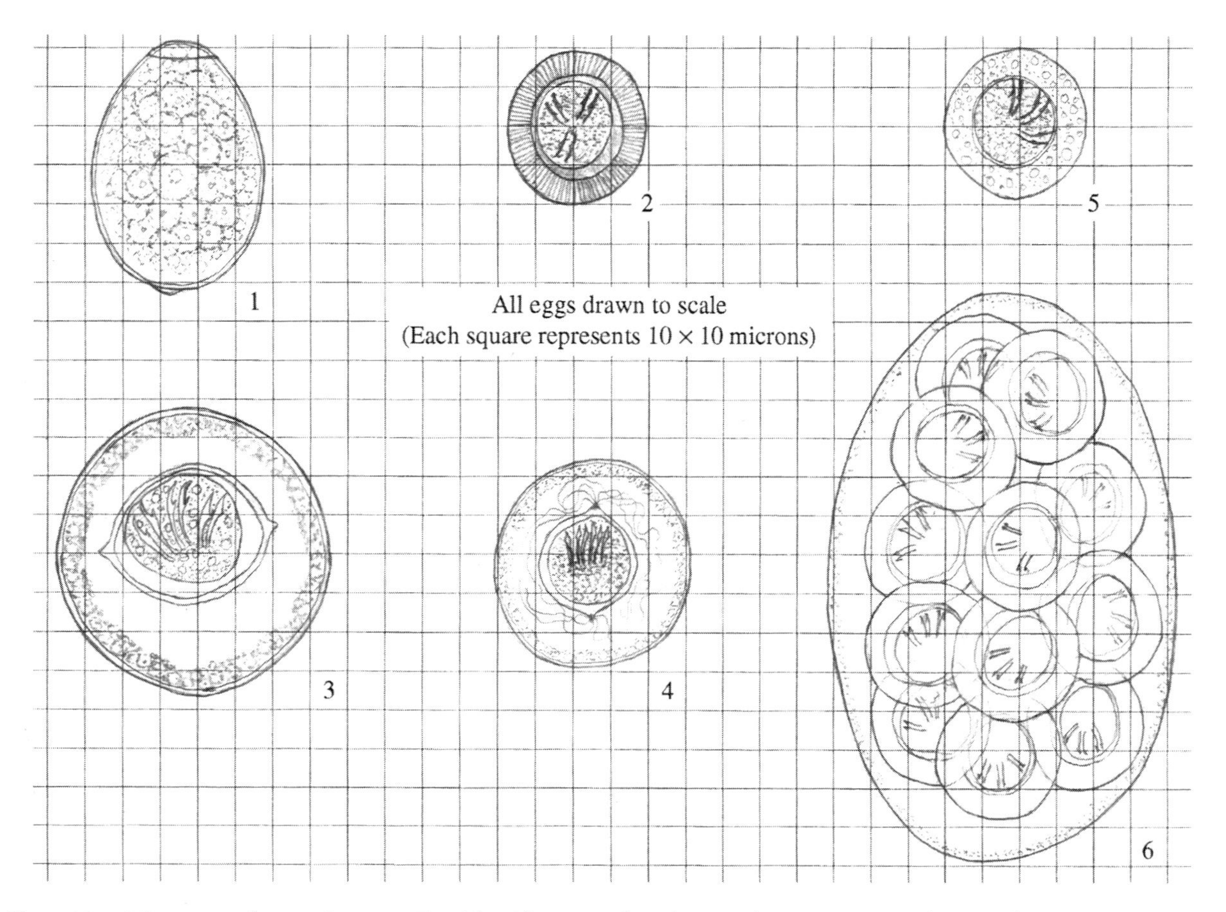

Plate 10. Diagrams of cestode eggs. The identification of each cestode egg is given along with its average size or diameter in µm. 1. *Diphyllobothrium latum*, 65; 2. *Taenia* sp., 35; 3. *Hymenolepis diminuta*, 75; 4. *Hymenolepis nana*, 45; 5 and 6. *Dipylidium caninum* (usually in packets). 5. individual egg, 35; 6. packet of 10 (or more) eggs, 150.

of *S. bovis*. A few human infections have been reported. Spurious infections do occur (see Plate 24).

Cattle, goats, sheep, and a number of wild mammals are naturally infected with *S. mattheei* (Alves, 1949). In southern Africa, *S. mattheei* is a frequent, accidental parasite of man and eggs are found in both the urine and feces of the infected individual. Eggs found in the feces may be confused with those of *S. intercalatum* which are about the same size. Although eggs are generally larger, averaging about 200 × 64 µm, those found in the urine may be confused with *S. haematobium*. Experimentally, the parasite has been shown to hybridize with *S. haematobium* (Pitchford, 1961).

Man is rarely infected with *S. rodhaini* (Schwetz, 1951), a species infecting rodents, canines, and felines. The eggs have a spine that is usually just to one side of terminal (subterminal) and a curved knob at the opposite end. Infection has been reported from man in Zaire (formerly the Belgian Congo).

The cercariae of many species that infect birds and mammals can penetrate human skin and cause schistosome dermatitis.

CESTODE EGGS

The adult stages of several species of pseudophyllidian tapeworms have been reported from man but only one, *Diphyllobothrium latum,* is commonly diagnosed. Its unembryonated eggs are broadly ovoidal (barrel-shaped), averaging about 65 × 47 µm with a moderately thick shell, and with a broadly rounded operculum that is often hard to see. At the abopercular end there is usually a button-like thickening of the shell slightly off center which is characteristic for the species (Plate 10:1; Plate 25).

The cyclophyllidian tapeworms produce eggs that have a hexacanth (six-hooked) embryo (onchosphere). Eggs of species of *Taenia* are all so similar that they cannot be differentiated. The thick egg shell is made up of truncated prisms cemented together so that they appear as striations (Plate 10:2; Plate 26). Eggs are spherical or subspherical, and average from 30 to 40 µm in diameter, with an indistinct embryophore, and rarely surrounded by the primary membrane (Plate 10:2; Plate 26).

In human taeniasis, it is important to know if the adult tapeworm is *T. solium* or *T. saginata*, since the eggs of *T. solium* are infective for man and accidental ingestion of the eggs can lead to infection with the larval stage (cysticercus) and the possibility of serious complications. Specific diagnosis is usually based on the the structure of the scolex or the mature or gravid proglottid (see Part 3, Chapter 6, Recovery of Helminths or Eggs After Treatment).

The adult tapeworms of *Hymenolepis diminuta* are common parasites of the intestines of rats and mice. The intermediate hosts are fleas, beetles, and cockroaches. Man becomes an accidental definitive host after ingesting the intermediate host infected with the larval stage, most probably when eating dried fruit infested by small infected beetles.

The eggs of *H. diminuta* (Plate 10:3; Plate 27) are ovoidal and from 60 to 85 µm in diameter. The onchosphere occupies about one third of the diameter of the egg and is surrounded by a membrane (embryophore) that has a thickening at both poles which is often difficult to see. The egg shell has a thin, inner lining with a coarse outer covering. Often in slide preparations, the outer covering breaks off (Plate 27:5), leaving the very thin inner shell exposed (Plate 27:6).

Hymenolepis nana is a common parasite of the intestine of mice. No intermediate host is necessary, and the eggs are infective in the mature proglottid or after being freed from it. Once the infection is established in man, it can be transmitted from individual to individual through contamination. Autoinfection is also possible when a proglottid disintegrates in the intestine since eggs are directly infective. Because the usual treatment of choice, niclosamide, breaks up the proglottid releasing the eggs, treatment must continue for 5 days or more to ensure that autoinfection does not occur.

The eggs of *H. nana* (Plate 10:4; Plate 28) are ovoidal and from 30 to 60 µm in diameter. Fine filaments extend from the poles of the embryophore into the space between the membrane and the inner shell. As in *H. diminuta,* there is an outer and inner shell and the outer shell may be lost in slide preparations leaving the inner shell exposed (Plate 28, Figure 6).

Dipylidium caninum is a common intestinal parasite of dogs, cats, and certain wild carnivores and the

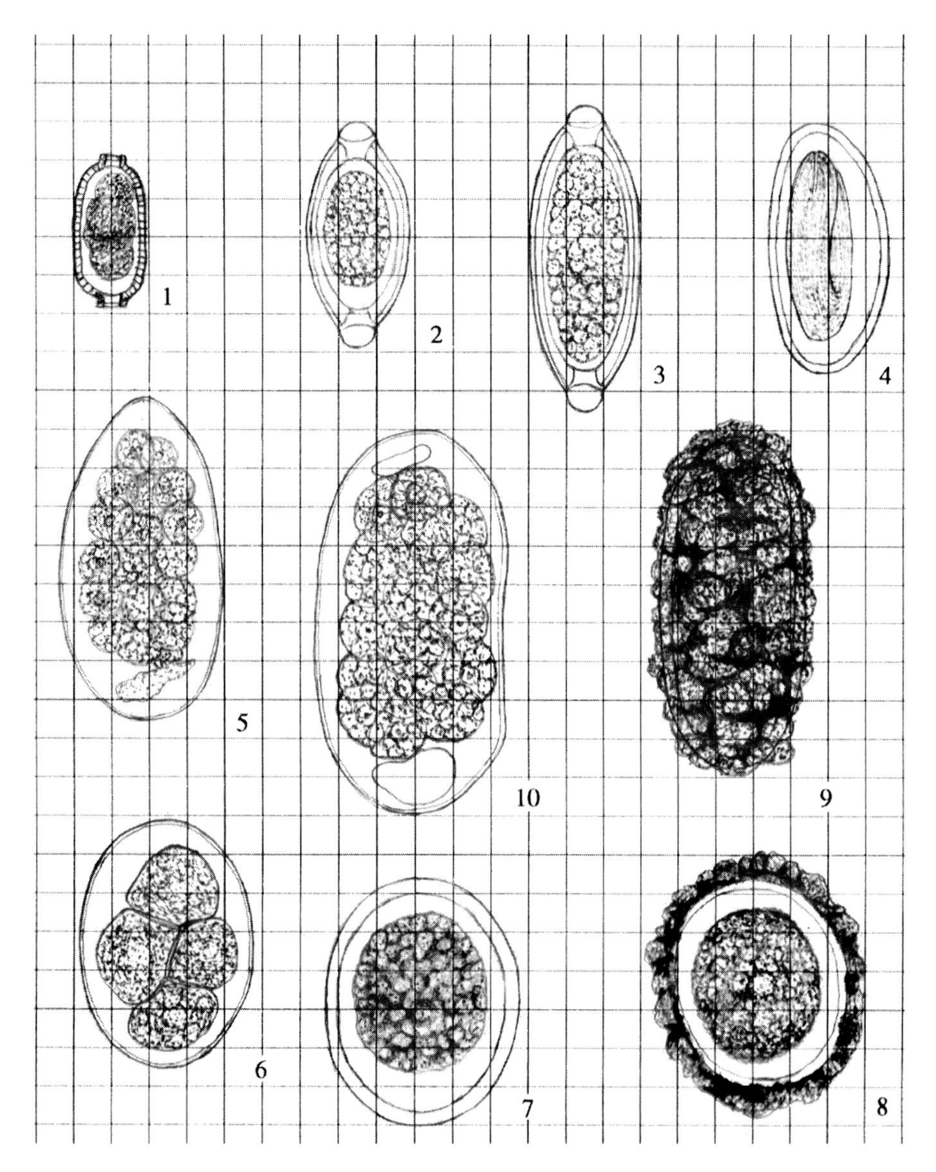

Plate 11. Diagrams of nematode eggs. The identification of the nematode egg is given along with its average size in microns. 1. *Capillaria philippinensis*, 41; 2. *Trichuris trichiura*, 58; 3. *Trichuris vulpis*, 79; 4. *Enterobius vermicularis*, 58; 5. *Trichostrongylus orientalis*, 83; 6. Hookworm (*Ancylostoma duodenale, Necator americanus*), 63; 7 through 9. *Ascaris lumbricoides*, 7. decorticated egg, 58; 8. typical egg, 60; 9. unfertilized egg, 90; 10. *Heterodera* sp., 100.

intermediate hosts are usually fleas. Most human infections have been in children who inadvertently ingest infected fleas and become an accidental definitive host. The infection in dogs and cats is often made by finding the small proglottids stuck to the hairs near the anus.

The eggs of *D. caninum* are usually seen in packets of 10 to 25 (Plate 10:6; Plate 29). The individual eggs (Plate 10:5) appear to be without an outer shell, are spherical, with a thin and nearly transparent shell, and measure from 25 to 50 μm in diameter.

NEMATODE EGGS

The eggs of nematodes vary greatly in size and shape and, with the exception of hookworms, most have characteristics that make the different species readily recognizable.

The adult parasites of *Capillaria philippinensis* inhabit the intestine and eggs are passed in the feces. The life cycle is not confirmed but infection is thought to occur from eating uncooked, fresh-water fish. Although initially thought to be limited to the island of Luzon in the Philippines, a few cases have been reported from Thailand. Autoinfection is believed to occur and is the explanation given for the extremely heavy worm burdens reported.

Capillaria philippinensis eggs (Plate 11:1; Plate 34:1–3) are small, averaging about 41 × 21 µm, with flattened sides and blunt polar plugs. The egg shell has radial striations and the embryo may be either developed or undeveloped when the egg is passed in the feces. If autoinfection is a possibility, then some eggs must contain fully developed embryos.

The adults of *Trichuris trichiura* are often referred to as whipworms because of their whip-like shape. The parasite is cosmopolitan but is more common in rural, agricultural areas. In many parts of the world, it competes with *Ascaris lumbricoides* as being the most common intestinal infection. Eggs must incubate outside of the host 21 or more days before becoming infective.

The eggs of *T. trichiura* average 58 × 26 µm and are football shaped, tapering toward both poles. There are mucoid plugs at each pole. The embryo is undeveloped when the egg is passed in the feces (Plate 11:2; Plate 30).

The eggs of *Trichuris vulpis* (Plate 11:3; Plate 34:3 to 5) are very similar but larger, averaging about 80 × 28 µm and often have partially flattened sides at the center then taper to the poles. Like *T. trichiura*, they also have mucoid plugs at the poles. The adult parasites infect the caecum of dogs and other canines throughout the world. The eggs of *T. vulpis* may reach the infective stage in 9 to 10 days under ideal conditions.

The adult parasites of *Enterobius vermicularis* inhabit the colon but eggs are usually deposited in the folds of the perianal region. The eggs mature *in utero* when the adult migrating female worm reaches the lower level of the large bowel. Those eggs deposited in the upper colon do not become infective. Eggs of *E. vermicularis* (Plate 11:4; Plate 31) are flattened on one side but this is not seen when the egg is lying with the flat side up or down. Eggs passed in the feces measure about 55 × 25 µm and contain a partially developed embryo. Rarely, males or immature females are present in the feces.

Members of the genus *Trichostrongylus* are parasites of the digestive tract of ruminants. *T. orientalis* is a relatively common parasite of man in the orient, especially in agricultural areas of Japan, Taiwan, and Korea. Other species more rarely are accidental parasites of man. The eggs (Plate 11:5; Plate 34:7–9) are thin-shelled, elongated, and may be generally ellipsoidal, slightly tapering toward one end, or sometimes both ends, and measure about 85 × 35 µm. Typically, the embryo is developed to near the morula stage. Infections diagnosed by finding eggs in fecal specimens may be spurious infections resulting from the ingestion of food contaminated with dung from an infected reservoir host, especially when many eggs are present. Rarely, juveniles of *Trichostrongylus* are seen in fecal specimens.

Two species of hookworms are common parasites of the small intestine of man, *Ancylostoma duodenale* and *Necator americanus*. Because of their cosmopolitan distribution and the movement of people throughout the world, hookworm eggs may be found in fecal specimens examined in any locality.

Eggs of the two species of human hookworms are indistinguishable one from the other. They have thin, hyaline, oval shells, and average 63 × 43 µm in size. The embryo is usually 4 to 16 cells (Plate 11:6; Plate 32:1–3), but if passage is delayed they may continue to develop to become fully embryonated (Plate 32:4–8) or even hatch (Plate 32:9; Plate 35:4, 5, 7).

Ascaris lumbricoides has been recognized since ancient times and is one of the most common helminth infections throughout the world. The adult worms are also the largest of the intestinal nematodes of humans, the mature females occasionally reaching 49 cm in length.

A. lumbricoides inhabits the small intestine where it seldom causes overt symptoms except when present in large numbers. Under certain circumstances, it may cause serious complications and even death. Large numbers of adult worms may cause intestinal obstruction and migrating adults may enter the bile or pancreatic ducts where they cause severe blockage which, if it persists, may cause death. Occasionally, they may perforate the intestinal wall causing peritonitis. They may enter the appendix causing acute appendicitis. Whenever possible, in infected individuals, *Ascaris* should be given priority consideration before other conditions are treated. Medications given to a patient that are ineffective for *Ascaris* may stimulate the worm to migrate, which can lead to serious complications.

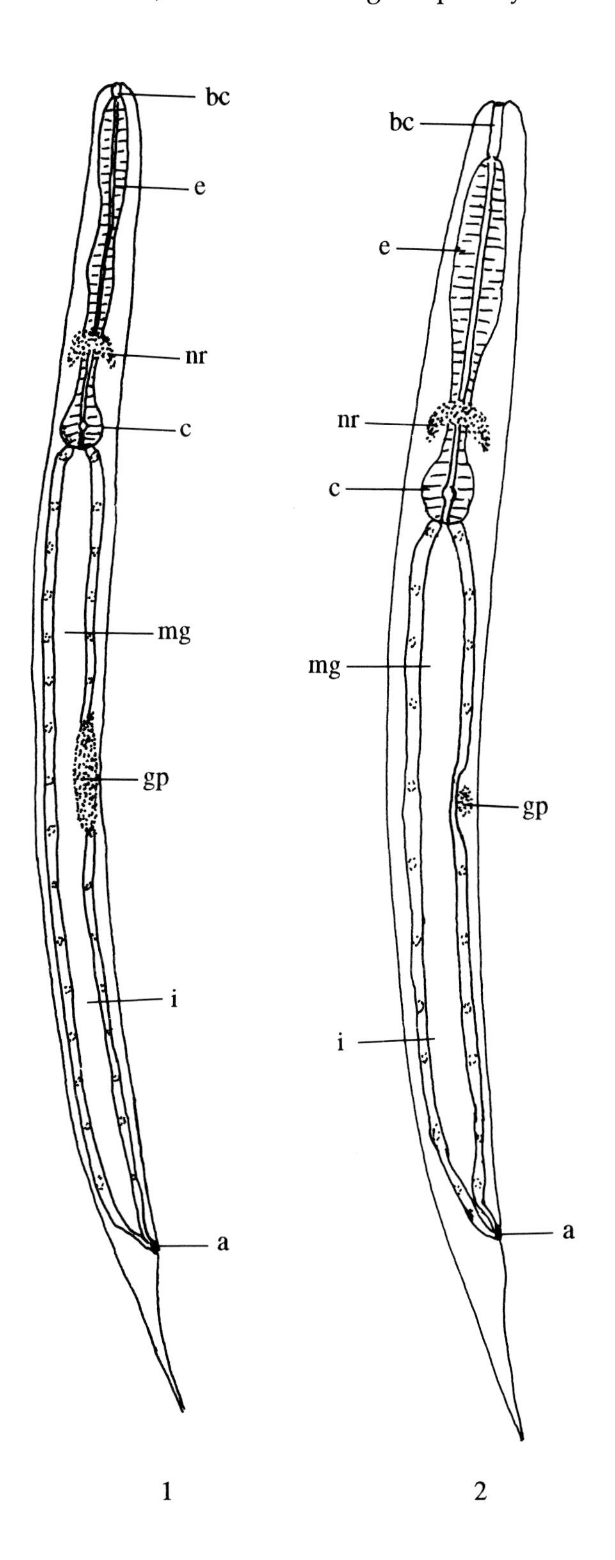

The fertilized eggs of *A. lumbricoides* (Plate 11:8; Plate 33:1–4) measure about 60 × 43 µm, and are broadly ovoidal with a thick shell, which has a coarsely mammillated, albuminous outer coating (the cortex). The outer coating may be missing (Plate 11:7; Plate 33:5, 6) revealing the thick shell. The embryo usually has from one to four cells. If unfertilized, the eggs are elongate oval or rhomboidal (Plate 11:9; Plate 33:7, 8) and average in size 92 × 43 µm. The outer coating is irregular when present, but may be missing as in the decorticated egg (Plate 11:7; Plate 33:9). In the unfertilized egg there is no embryo.

Plant nematode eggs, such as *Heterodera* sp. (Plate 11:10), occasionally may be present in fecal specimens. They are large, bean-shaped, measuring about 100 × 33 µm.

NEMATODE JUVENILES

In addition to eggs of helminths, juvenile worms (larvae) may be found in fecal specimens. Infection

Plate 12. Diagrams of juveniles of the nematodes *Strongyloides stercoralis* and hookworm seen in fecal specimens. 1. *Strongyloides stercoralis*; 2. Hookworm (*Necator americanus* or *Ancylostoma duodenale*). a, anus; bc, buccal cavity; c, cardiac bulb of esophagus; e, esophagus; gp, genital primordium; i, intestine; mg, midgut; nr, nerve ring.

The juvenile nematodes in fecal specimens vary in length depending on the time lapsing between emerging from the egg and fixation. In fresh specimens sent to the laboratory, juveniles will continue to develop and may even molt to the second stage before being fixed, depending on the time fixation has been delayed. Also, the juveniles of *S. stercoralis* will develop to the second and even to the third stage if their passage down the intestinal tract is delayed sufficiently. The diagrams depict juveniles shortly after emerging from the egg.

The diagnostic features are the length of the buccal cavity and the size of the genital primordium. The buccal cavity of the juvenile of *S. stercoralis* is short, about 4 µm, while that of hookworm is 8 to 10 µm in length. The genital primordium of *S. stercoralis* is relatively large and visible while that of hookworm is very small and not generally visible. Not to scale.

with *Strongyloides stercoralis* occurs throughout tropical and temperate areas of the world but is more prevalent in the warm, wet regions of the tropics. Man is considered to be the typical, definitive host of *S. stercoralis,* but dogs and cats may also be infected and act as reservoir hosts for man (Georgi and Sprinkle, 1974). The female worms of *S. stercoralis* inhabit the crypts of the duodenum or the first section of the jejunum where they lay partially embryonated eggs. The juvenile develops rapidly, emerges from the egg, and usually migrates to the intestinal lumen. Juveniles pass down the intestinal tract and are evacuated with the feces.

If movement through the intestine is delayed, the juveniles may reach the infective filariform stage and invade the mucosa or the perianal region to establish autoinfection that can lead to hyperinfection and even death. Infections that do not produce overt symptoms may be overlooked. When infection is diagnosed, it should be taken seriously because of the possible exacerbation of the disease in some individuals, especially those who have a depressed immune system. Treatment with thiabendazole is usually effective, but when dealing with seriously ill or immune deficient patients, follow-up examinations should be carried out for 14 consecutive weeks to insure that the infection is eliminated.

Diagnosis depends on finding the typical juvenile worms in the feces. The buccal cavity is extremely short, measuring no more than 4 µm and a little below midway down the length of the body lies the distinct, oval, genital primordia (Plate 12:1; Plate 35:1,2,6). Since juveniles are not passed on a regular basis, a positive finding may require examination of several specimens. The majority of cases are diagnosed on routine examination of fecal specimens in patients not suspected of having strongyloidiasis.

The juvenile worms are distinctive when seen in fresh or adequately preserved fecal specimens. In the early, first-stage juvenile of *S. stercoralis* that has passed rapidly through the intestinal tract, the short buccal cavity, the esophagus, the nerve ring, the esophageal bulb, the intestine, and the anus are usually visible (Plate 12:1; Plate 35:1–2). After a few hours in an unpreserved fecal specimen, the juvenile may be 12 to 15% longer (Plate 35, compare 1 and 2). They usually vary in size from 200 to 300 µm relative to the time after emerging from the egg and examination or preservation of the fecal specimen. Development may continue if stool passage is delayed or if fresh fecal specimens are not fixed immediately after passage (Plate 35:3).

Juvenile worms of other species of nematodes may be present in fecal specimens. Of these, the most frequently seen are those of hookworms. These juveniles have a longer buccal cavity, about 8 µm long, and the genital primoidia is not distinct (Plate 12:2; Plate 35:4, 5, 7). Other species of juveniles that may be present in fecal specimens also have a long buccal cavity. In addition, eggs of the same species as the juveniles, except *S. stercoralis*, are usually present in the fecal specimen along with the juvenile worms which may help in the identification.

INTRODUCTION TO THE "KEY TO HELMINTH EGGS FOUND IN FECES"

In developing the Key, the author has relied on many reference sources for basic information. Those eggs that are most commonly seen in fecal specimens have been included. Some species, such as *Capillaria philippinensis* and *Ternidens deminutus*, seen only in limited geographic areas, have been included because of the ease of travel from those areas. Other species, such as *Dicrocoelium dendriticum* and the free-living *Heterodera* sp., seen only as spurious infections, have been included also. Others that are rarely seen, for example *Capillaria hepatica* and *Schistosoma mekongi*, have been omitted.

It was necessary to rely on various literature sources to arrive at the outer extremes of the sizes of eggs used in the Key. Although some of these have been confirmed using specimens available, confirming all size ranges was not possible. In a few instances, sizes of eggs consistently exceeded reports in the literature and the size range was changed to reflect these differences. The author would welcome comments based on actual measurements of eggs from properly identified species that vary from those given in the Key.

In addition to those species of helminths infecting man, eggs of mites, insects, helminths of animals, and free-living helminths may be accidentally ingested, move through the gastrointestinal tract, and pass with the feces. The presence of such unusual eggs in fecal specimens adds to difficulty in making accurate and positive identifications. Eggs in spurious infections are usually present only for 1 or at the most 3 days and would not be found in subsequent fecal specimens, which further substantiates the importance of collecting several specimens on alternate days from each patient being examined. Obviously, it is impractical to include in the Key all eggs that might be found in fecal specimens (see Part 4, Chapter 10).

There are some places in the Key where one may have some difficulty in making a decision. For example, in Set 3 one must decide if a small operculate egg, 35 μm or less, has a miracidium with symmetrical or asymmetrical organs, particularly cephalic (cytolytic) glands. Anyone, even an experienced parasitologist, may have some trouble with this decision (refer to Plate 8). It will be necessary to use high magnification (1000×), and even turn the egg around to make the determination.

The most recent treatment is the same for many of the trematodes. Those producing small operculate eggs, e.g., *Clonorchis sinensis, Opisthorchis viverinni*, infecting the liver and *Heterophyes heterophyes* infecting the intestine can be treated similarly. The pathologic effects on the host are different for different species, so a specific diagnosis should be made when possible.

Regardless of some difficulties, the Key does provide a practical, effective approach to the identification of helminth eggs that are found in fecal specimens but not readily recognized, particularly when supported by photomicrographs and drawings. Those who have used the Helminth Egg Key have found it helpful.

PHOTOMICROGRAPHS OF HELMINTH EGGS

The photomicrographs that follow were taken of helminth eggs seen on wet preparations prepared from specimens fixed in either formalin, MIF, or PIF collecting/preserving solution. A magnification of 400× was used as a standard for the photomicrographs of helminth eggs and 1000× was used when higher magnification was needed. To make certain that the printed enlargement of each negative was directly comparable to all other prints, a standard scale was prepared. Where there are two photomicrographs taken at different magnifications, the standard scale for photomicrographs taken at 400× is designated "a" and the scale for those taken at higher magnification is designated "b". The enlarger was set according to the scale and a print made from the negative. Exceptions to the standard scale are indicated where they occur. The scales appear either on the Plate or in the Legend.

Because of the limits of the depth of field (or focus), only what is in focus at a single focal plane is shown in a photomicrograph. In drawings, however, those structures present at each focal plane can be added. The drawing then becomes a composite of several planes of focus.

It is appropriate to review information on the depth of field given in the Appendix before closely examining photomicrographs. The thickness of the portion of the organism or section presented in a photomicrograph varies with the magnification used. A photomicrograph taken at 400× magnification represents a focal plane 1.01 μm thick no matter how much it may be enlarged. For example, if a photomicrograph of an egg taken at 400× magnification is printed at three different sizes, one 2 inches long, a second 3 inches long, and a third 4 inches, all three will represent exactly the same depth of field (thickness) of 1.01 μm. A photomicrograph may be projected on a screen and be enlarged many times, but the thickness of the plane represented by the photomicrograph does not change.

Several photomicrographs appear on the same page as a "Plate" and all are at the same scale. In the legend, the number of the figure, the name of the helminth producing the egg, and the medium in which the egg was preserved is given. For example:

1. *Clonorchis sinensis*, MIF; 2. *Opisthorchis viverrini*, PIF.

Comments that relate to each parasite and/or photomicrograph appear in the legend.

KEY TO HELMINTH EGGS FOUND IN FECES

Plate

1. a. Egg with operculum, sometimes inconspicuous 2
   b. Egg without operculum 13
2. a. Small egg, $\leq$ 35 µm; containing a larva when passed 3
   b. Larger egg, $>$ 35 µm; with or without a developed larva 7
3. a. When passed, organs of larva are not obviously symmetrical but are asymmetrical 4
   b. When passed the organs of the larva, particularly cephalic glands, are obviously symmetrical 6
4. a. Egg ovoidal (light-bulb-shaped) with pronounced shoulders; narrow, raised operculum; abopercular end with a buttonhook projection, sometimes incomplete; 27 to 35 × 12 to 20 µm; in feces or from duodenal drainage. = ***Clonorchis sinensis***
   b. Egg with pronounced or moderate shoulders; abopercular end with raised papillae or knob 5
5. a. Egg ovoidal (light bulb-shaped) with a slightly raised, relatively broad operculum; abopercular end with a prominent papilla; 23 to 33 × 12 to 20 µm; in feces or duodenal drainage. =***Opisthorchis viverrini***
   b. Egg more slender and tapering slightly from center to both ends; papillae on abopercular end small; average size 30 × 12 µm; in feces or from duodenal drainage. = ***Opisthorchis felineus***
6. a. Egg without distinct shoulders; broadest below the middle; slight or no thickening at abopercular end; 24 to 28 × 15 to 17 µm; in feces. = ***Metagonimus yokogawai***
   b. Egg without distinct shoulders; broadest at the middle; slight or no thickening at abopercular end; 26 to 30 × 15 to 17 µm; in feces. = ***Heterophyes heterophyes***
7. a Egg very large, over 130 µm in length 8
   b. Egg less than 125 µm in length 10
8. a. Egg oval in shape 9
   b. Egg rhomboid, widest in middle and tapering toward both ends; 150 to 170 × 60 to 70 µm; in feces. = ***Gastrodiscoides hominis***
9. a. Egg 130 to 145 × 60 to 75 µm; with yolk granules evenly distributed throughout yolk cells; in feces. = ***Fasciolopsis buski***
   b. Egg 130 to 155 × 65 to 90 µm; with yolk granules concentrated around nuclei of yolk cells; in feces or from duodenal drainage. = ***Fasciola hepatica***

It is often impossible to distinguish between the latter two species by the examination of the eggs. Collecting bile fluid from the opening of the common bile duct may be necessary to confirm or eliminate *Fasciola hepatica*. A few cases of human infection with *F. gigantica* have been reported. The eggs are extremely large, 150 to 196 × 90 to 100 µm and should be distinguishable from *F. hepatica* or *Fasciolopsis buski*.

10. a. Egg less than 50 µm and containing a larva when passed; with a thick shell usually more rounded on one side; 35 to 45 × 22 to 30 µm; in feces or from duodenal drainage; usually seen as a spurious infection. = ***Dicrocoelium dendriticum***

    b. Egg over 50 µm and not containing a larva when passed 11

11. a. Egg with flattened operculum and pronounced shoulders; abopercular end of shell thickened; 75 to 120 × 45 to 65 µm; in sputum; or in feces in ⅓ to ½ of the cases. = ***Paragonimus westermani***

    b. Egg without shoulders, operculum rounded and often indistinct 12

12. a. Egg relatively thin-shelled; oval in shape with a narrow operculum, sometimes difficult to see; 85 to 115 × 45 to 65 µm; in feces. =***Echinostoma ilocanum***

    b. Egg relatively thick-shelled; broadly barrel-shaped with a relatively broad operculum often difficult to see; abopercular end of shell with a button-like thickening slightly off center; 55 to 75 × 38 to 55 µm; in feces. =***Diphyllobothrium latum***

13. a. Egg contains a ciliated larva; has a conspicuous spine or minute knob, the knob often difficult to see 14

    b. Egg does not have a ciliated larva; without a spine or minute knob 17

14. a. Egg with a conspicuous spine 15

    b. Egg broadly oval, with a minute knob on one side near base of egg; often with a loose coating covering the shell; 70 to 108 × 55 to 80 µm; in feces. = ***Schistosoma japonicum***

15. a. Egg with a terminal spine 16

    b. Egg large, elongated, with a lateral spine; 110 to 180 × 45 to 75 µm; in feces, rarely in urine. = ***Schistosoma mansoni***

16. a. Egg with terminal spine; 140 to 220 × 50 to 90 µm; tapering slightly toward both ends and thickest in the center; in feces. = ***Schistosoma intercalatum***

    b. Egg with terminal spine; 100 to 170 × 50 to 80 µm; narrowly ovoidal; usually in urine, rarely in feces. = ***Schistosoma haematobium***

S. intercalatum infection in man has been reported from seven countries in Africa. It has a larger size range than *S. haematobium* (140 to 220 µm in length). Since the size ranges overlap, it may be necessary to use means other than size to differentiate between the two species. In South Africa, other species with terminal spines, *S. matthei* (180 to 232 µm in length) and *S. bovis* (180 to 232 µm in length), are rarely seen in human feces. Human infection with these latter species is accidental and some reports may reflect spurious infections. The eggs of *S. rodhaini*, a species infecting rodents, having a subterminal spine, and a rounded knob at the opposite end, have been recovered from the feces and from tissues of humans in Africa.

17. a. Egg fully embryonated; containing an embryo without cilia and with three pair of hooklets, sometimes difficult to see 18

    b. Egg either fully embryonated or not fully embryonated but never with hooklets 21

18. a. Egg with a single, thick, dark, radially pitted shell; spherical, 30 to 60 μm; subspherical, 30 to 40 × 20 to 30; in feces. = ***Taenia* sp.**

    b. Egg shell moderately thin without radial pitting; embryophore encasing hexacanth embryo separated from shell by a relatively large space 19

19. a. Egg single; embryophore with polar thickenings 20

    b. Eggs usually in packets of 10 to 25; single eggs without thickenings or filaments on the embryophore; shell thin and nearly transparent; spherical, 30 to 60 μm; in feces. = ***Dipylidium caninum***

20. a. Egg oval with thin shell, composed of two layers; embryophore with polar thickenings from which filaments extend into space beneath inner shell; diameter 30 to 60 μm; in feces. =***Hymenolepis nana***

    b. Egg round to slightly oval; outer shell thicker and dark; embryophore occupies about one-third of space within shell, with polar thickenings but without polar filaments; diameter 65 to 85 μm; in feces. = ***Hymenolepis diminuta***

The shells of *Hymenolepis* spp. eggs have two parts; an outer dark covering that may be lost in fixed specimens exposing an inner, almost clear shell wall.

21. a. Egg with a thick, dark shell 22

    b. Egg with a clear, transparent shell 26

22. a. Egg barrel-shaped; with mucoid plug at each pole; with smooth shell 23

    b. Egg round to oval, without mucoid plugs at the poles; shell covered with a rough mammillated coating 25

23. a. Egg small with flattened sides and blunt ends; shell with radial striations, 36 to 45 × 19 to 22 μm; in feces. = ***Capillaria philippinensis***

    b. Egg tapering to both poles; shell without striations 24

24. a. Egg tapering from center toward both poles; 50 to 65 × 22 to 30 μm; in feces. = ***Trichuris trichiura***

    b. Egg larger, 70 to 88 × 25 to 30 μm, often slightly flattened at middle and tapering to both poles; in feces. = ***Trichuris vulpis***

25. a. Egg round to broadly oval; 45 to 75 × 35 to 50 μm; in feces. (normal fertile egg) = ***Ascaris lumbricoides***

    b. Egg elongate, oval or rhomboidal; no organized embryo present; 88 to 95 × 40 to 45 μm; in feces. = ***Ascaris*** (unfertilized)

Either fertile or infertile eggs may lose their cortex in the specimen. In such cases, the very thick clear shell helps differentiate *Ascaris* eggs from those of similar size.

26. a. Egg shell very thick; often seen in older preserved specimens or those where fixation has been delayed 27

    b. Egg shell thin 28

27. a. Egg contents not segmented beyond four cell stage; 43 to 68 × 33 to 48 μm; in feces. = ***Ascaris*** (decorticated)

    b. Egg with larval stage inside; 48 to 52 × 32 to 36 μm; in feces. = ***Physaloptera caucasica***

28. a. Egg flattened on one side 29

    b. Egg not flattened on one side 30

29. a. Egg contains a partially developed, rhabditoid juvenile; 50 to 65 × 20 to 30 μm; relatively rare in feces; or with a fully developed juvenile, found in folds of anus. = ***Enterobius vermicularis***

    b. Egg bean-shaped with air spaces at poles; larva developed to morula stage; 80 to 120 × 20 to 45 μm; in feces. = ***Heterodera* sp.**

30. a. Egg with both poles rounded 31

    b. Egg with one pole more pointed than the other; embryo usually developed to the morula stage; 70 to 100 × 24 to 45 μm; in feces. = ***Trichostrongylus* sp.**

31. a. Egg developed to the four to eight cell stage; 56 to 70 × 35 to 50 μm; in feces. = **Hookworm sp.**

    b. Egg with embryo developed beyond eight cell stage 32

32. a. Egg with embryo from 16 cells to rhabditoid juvenile stage; 56 to 70 by 35 to 50 μm; in feces. = **Hookworm sp.**

    b. Egg with embryo developed to eight cells or greater but not to juvenile stage; 75 to 85 × 46 to 55 μm; in feces. = ***Ternidens deminutus***

Hookworm eggs in human feces may represent either *Ancylostoma* sp. or *Necator americanus*. Differentiation must be made from adult or filariform juvenile stage. The filariform stage is obtained for identification by culturing the eggs or juvenile worms (Melvin and Brooke, 1982; Ash and Orihel, 1987). When hookworm eggs are delayed in moving through the intestine, they may develop to the juvenile stage. Some juvenile worms may emerge in feces before it is passed so, in addition to eggs, occasionally juvenile nematodes of hookworms are seen in the fecal specimen. The juvenile stages of *Strongyloides stercoralis* also may be present in fecal specimens where hookworm eggs are present. In such cases, the juveniles must be positively identified (see Plates 12 and 35). Hookworm eggs are differentiated from *Ternidens deminutus* eggs only on the basis of size.

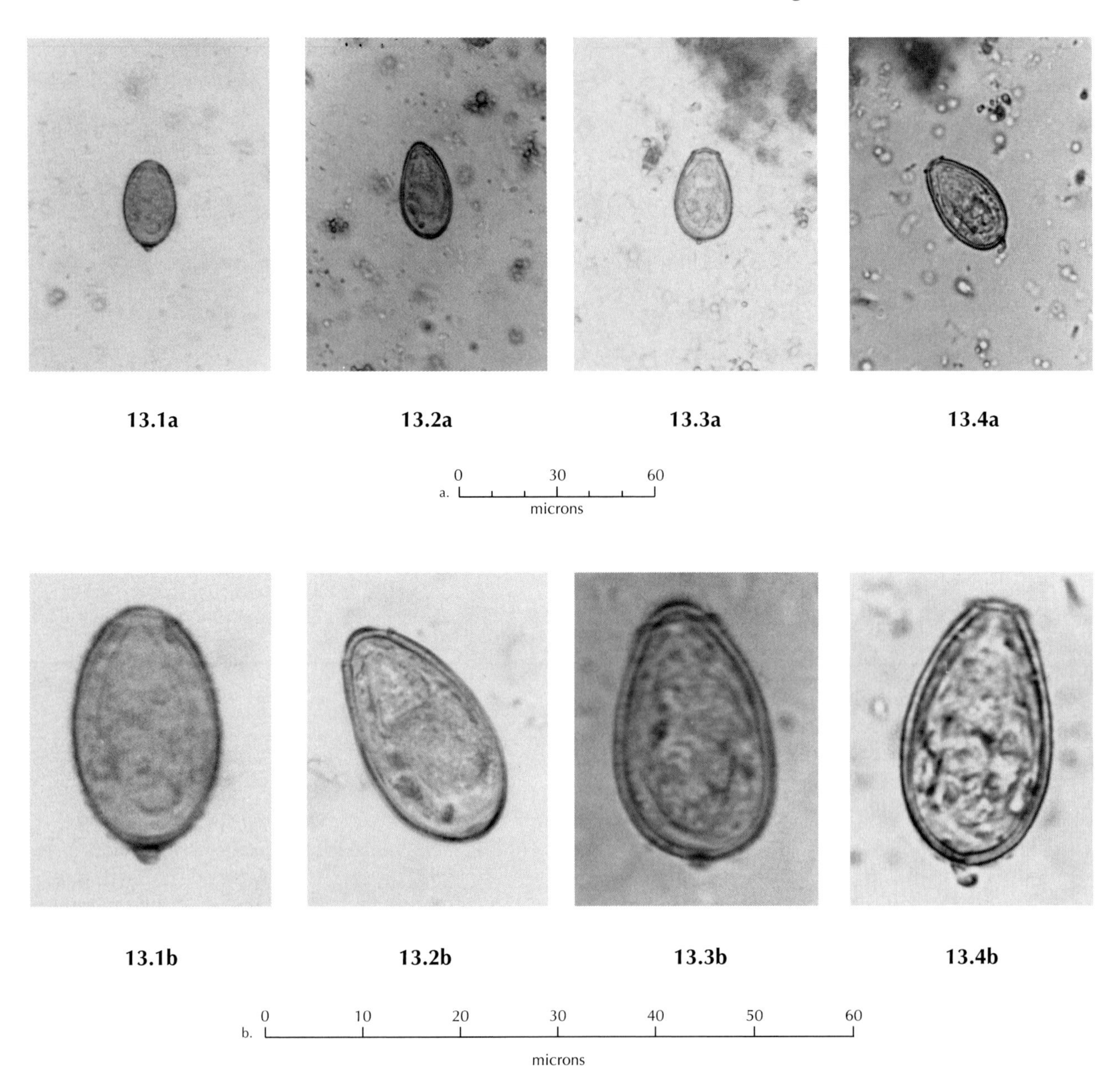

Plate 13. Eggs of four trematodes 35 μm or less in length. 1. *Metagonimus yokogawai*, a and b: MIF. The egg is broadly barrel-shaped, with a wide, rounded operculum without shoulders; with or without a small knob (papillae) at the abopercular end; miracidium with paired (symmetrical) cephalic glands. 2. *Heterophyes heterophyes*, a and b: PIF. The egg is ovoidal, with a narrow, slightly raised operculum without shoulders; with or without a small knob at the abopercular end; miracidium with paired (symmetrical) cephalic glands. 3. *Opisthorchis viverrini*, a and b: MIF. Egg is broadly ovoidal, with a relatively wide, rounded operculum with distinct shoulders; usually with a small knob at the abopercular end; miracidium with a single cephalic gland (asymmetrical). 4. *Clonorchis sinensis*, a and b: MIF. The egg is ovoidal, with a narrow operculum having distinct shoulders, with a buttonhook projection at the abopercular end; with a single (asymmetrical) cephalic gland.

Fully developed miracidia are contained in the eggs of these four species. Dead eggs, without an embryo or typical shell, are sometimes present and their identification is impossible. Even intact eggs may vary in size and shape. Where many eggs of one species are present, identification becomes less difficult. Photomicrographs marked "a" were taken at 400× and enlarged at the standard scale; those marked "b" were taken at 1000× and enlarged equally so appear 2.5–3.0 times larger.

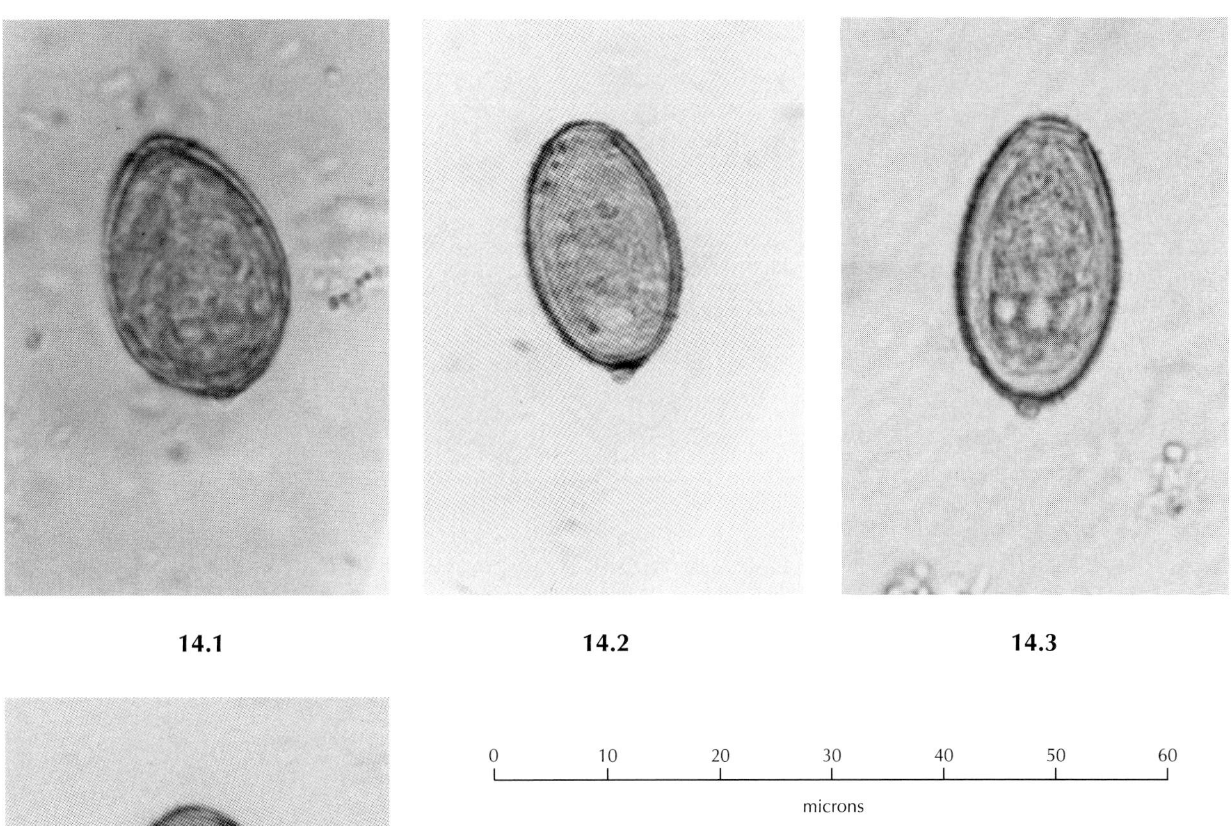

14.1 **14.2** **14.3**

14.4

Plate 14. *Metagonimus yokogawai* eggs. 1–4: MIF. The eggs of *M. yokogawai* and those of *Heterophyes heterophyes* (Plate 15) are very similar and are occasionally found in the same specimen. Eggs of both species are without shoulders and miracidia have paired cephalic glands (Plate 13:1), separating them from the eggs of small flukes inhabiting the bile ducts of the liver. The more broad operculum (2, 4) and general barrel shape (2–4) of the egg does not always allow clear separation of *H. heterophyes* and *M. yokogawai* since the latter is often broadly ovoidal (1). The fluke is endemic in eastern Asia and in small foci in countries bordering the Mediterranean Sea. The fluke is a typical parasite of the intestine of pigs, dogs, and cats and has been reported from pelicans. Infection in man, animals, and birds is initiated by the ingestion of raw fish infected with metacercariae.

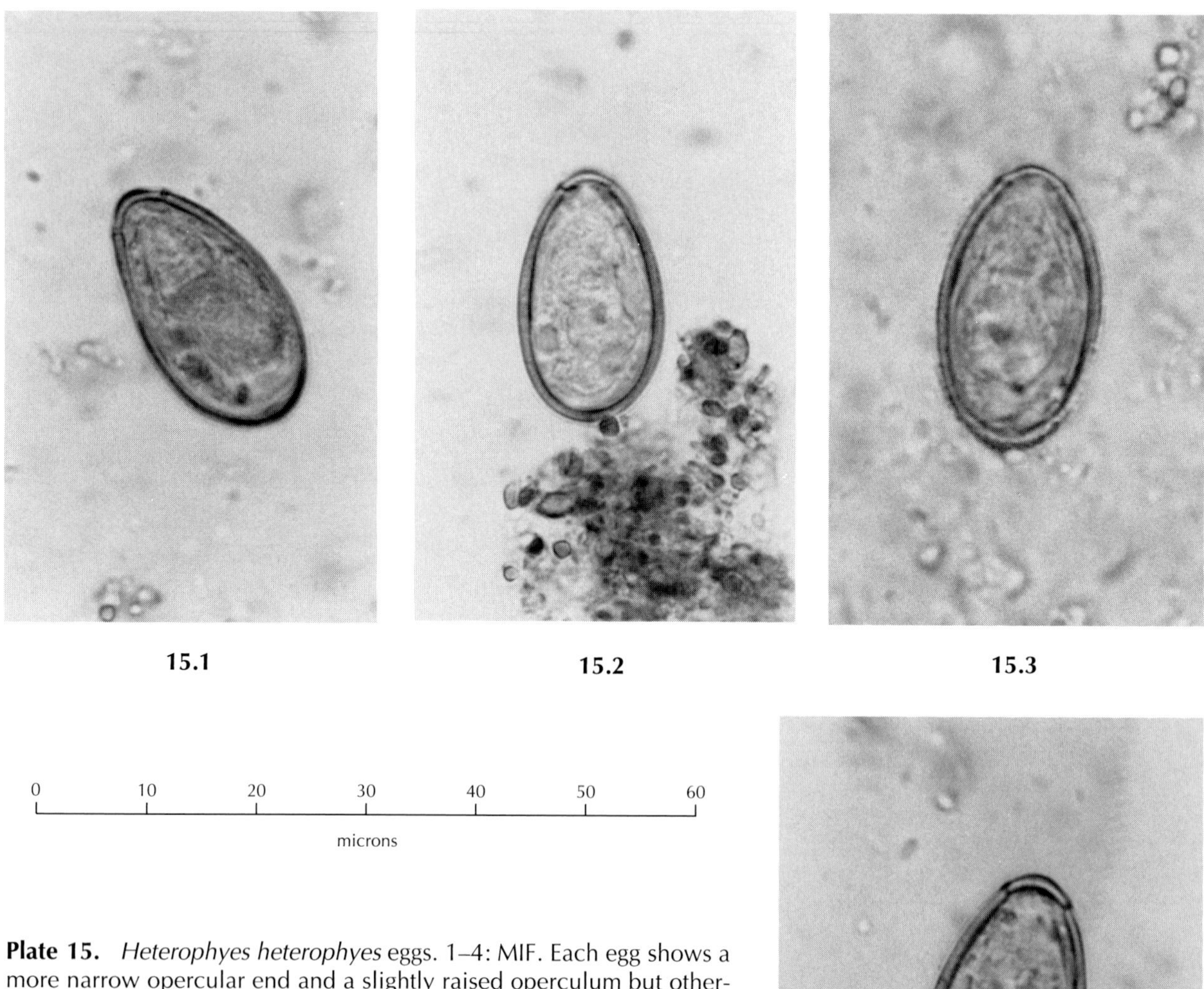

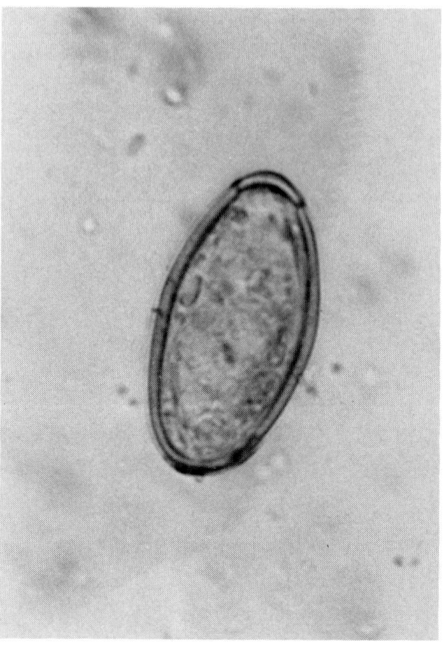

Plate 15. *Heterophyes heterophyes* eggs. 1–4: MIF. Each egg shows a more narrow opercular end and a slightly raised operculum but otherwise the eggs are similar to those of *Metagonimus yokogawai.* The symmetry of the miracidium can be seen in 1 (see also Plate 8). The absence of distinct shoulders separates *H. heterophyes* from the eggs of the liver flukes, *Clonorchis sinensis, Opisthorchis viverrini* and *O. felineus,* eggs 35 μm or less. The parasite is endemic in Central and South China, Japan, Taiwan, the Philippines, Palestine, and Egypt. The typical hosts are dogs, cats, foxes, man, and possibly other fish-eating mammals. The Nile delta of Egypt is one of the most important endemic areas. Infection occurs following the ingestion of infected raw or salted fish, especially mullet. In Egypt, the mullet inhabit fresh water but frequent brackish water during spawning season.

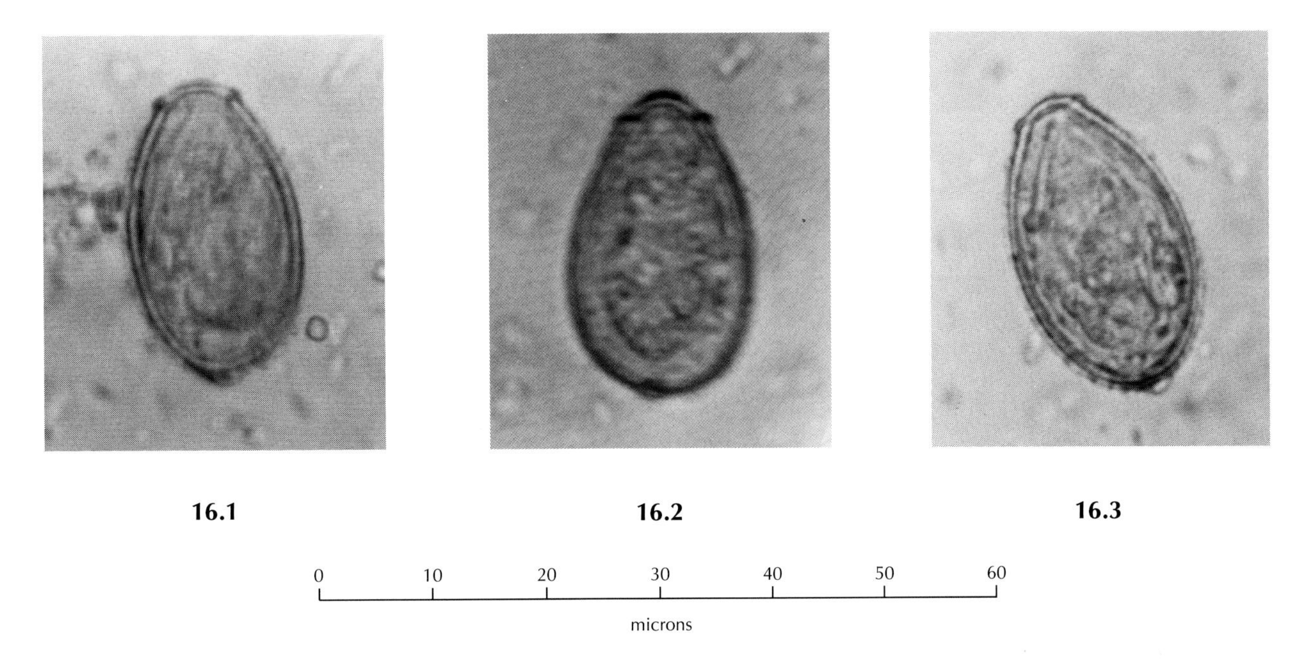

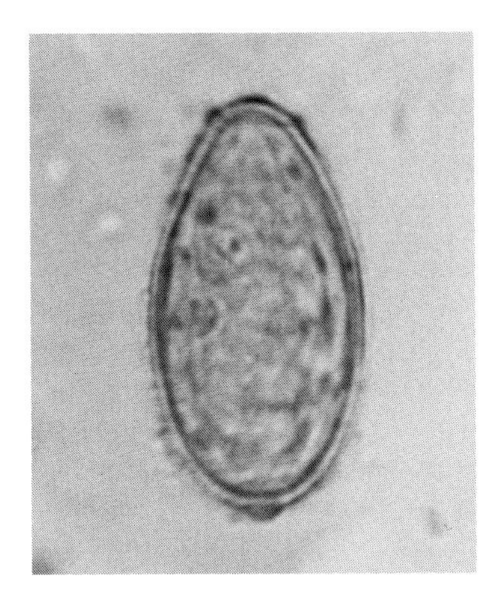

16.4

Plate 16. *Opisthorchis viverrini* eggs. 1–4: MIF. The distinct shoulders (1–4) separate the eggs of *O. viverrini* from those of the intestinal flukes, *Metagonimus yokogawai* and *Heterophyes heterophyes.* Shoulders may be less distinct, as in 3 and 4; separation of species may become more difficult and at times depend on recognition of the asymmetry of the cephalic glands of the miracidium. The surfaces of the eggs often appear rougher than other small eggs of trematodes (3, 4). There is usually a small knob (papilla) on the abopercular end of the egg as seen in all the photomicrographs and the operculum is often somewhat broader and flatter than that of *Clonorchis sinensis.* The primary endemic areas are Thailand and Laos in southeast Asia but its distribution may be far more extensive. *O. viverrini* is a typical parasite of the biliary passages of the civet cat and other fish-eating mammals. Man is considered an accidental host. The miracidium does not emerge from the egg until it is ingested by a suitable snail, intermediate host. Like the other members of this group of trematodes, the second intermediate hosts are fresh-water fish and infection occurs when infected, uncooked fish are ingested.

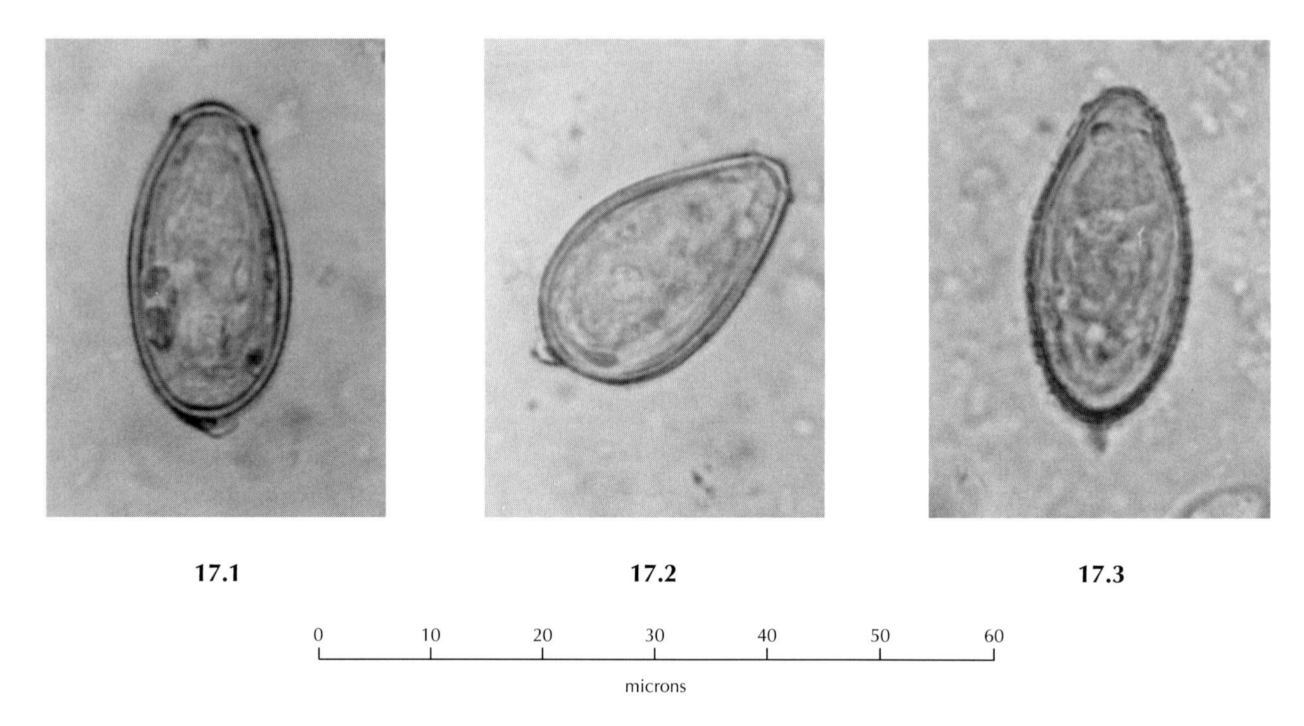

17.1 17.2 17.3

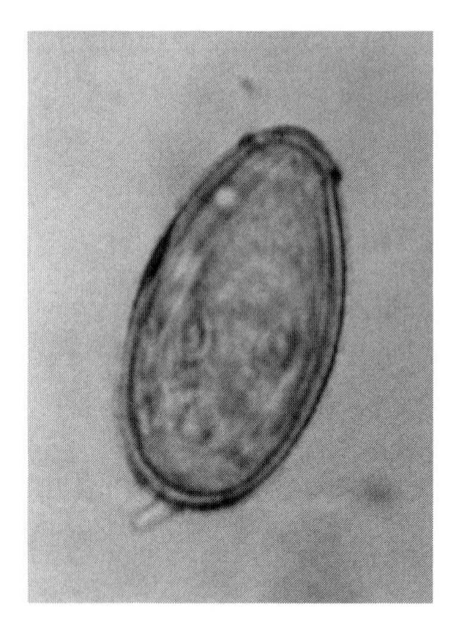

17.4

Plate 17. *Clonorchis sinensis* eggs. 1–4: MIF. The eggs of *C. sinensis* are distinct in that they have a buttonhook-like projection on the abopercular end (1, 2), which separates them from eggs of a size and shape similar to those of other species. Unfortunately, the buttonhook is not always the same shape (3, 4) and may not be seen clearly. The general shape of the eggs varies from ovoidal (2) to a broader, more oval-shape with a broader operculum (1, 4). The shoulders are distinct. The asymmetry of the miracidium can often be seen if eggs are lying correctly and are carefully examined (1, 4, and Plate 8). *C. sinensis* is endemic in many areas of the Far East. In some areas, such as Hong Kong, *C. sinensis* has become almost domesticated. Carp are raised in private water impoundments and human feces is used as fertilizer to enhance growth of the water plants. Fish harvested from the ponds are used to prepare raw fish dishes such as "yue-shan chuk" (a soup made with rice and raw carp). If the appropriate snails are present in the pond and fecal specimens from infected individuals are used as fertilizer, all elements necessary for completing the life cycle are present. Since the endemic areas of two or more of the species of trematodes that produce these small eggs may overlap, patients infected with more than one species are occasionally seen.

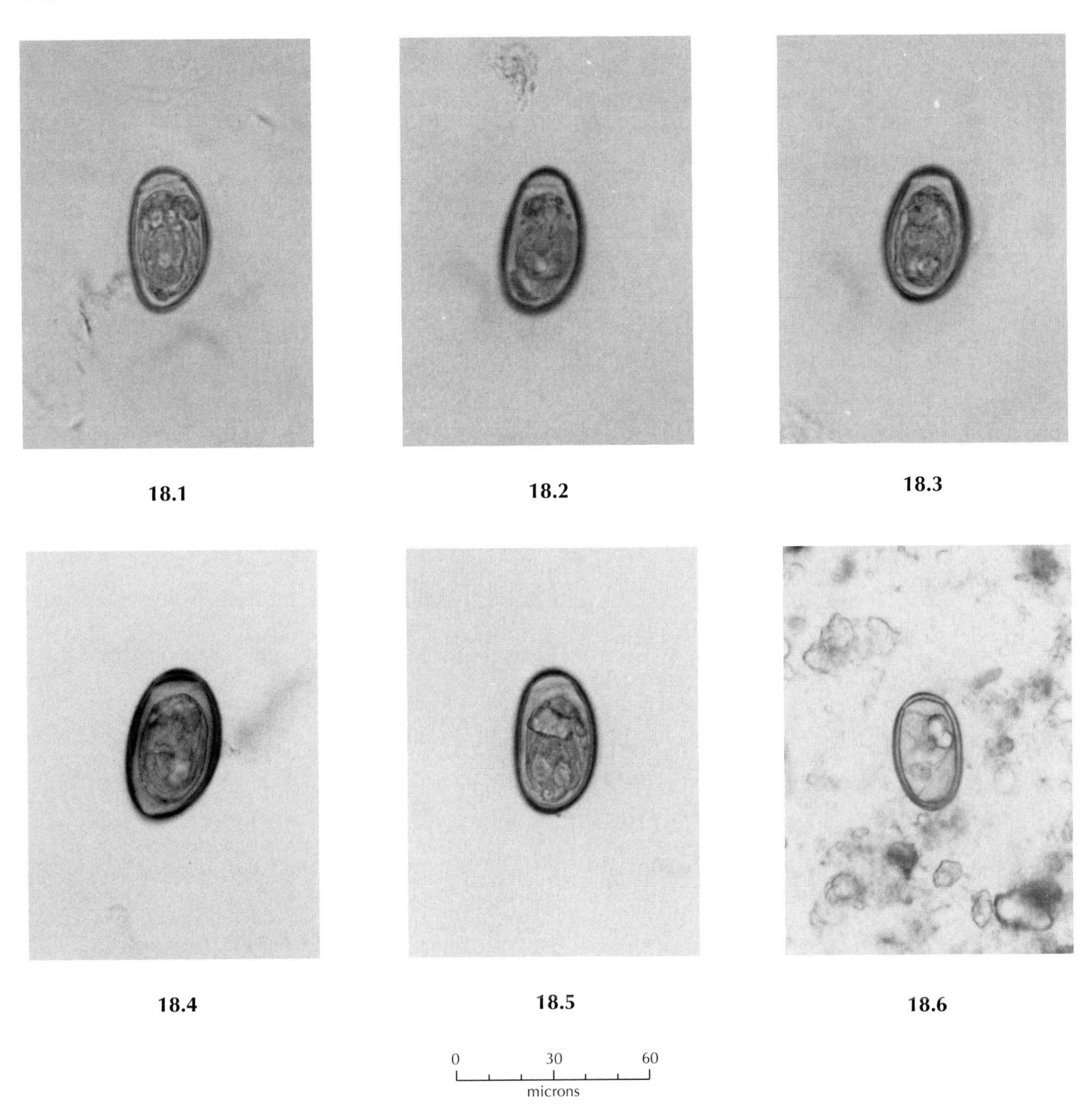

Plate 18. *Dicrocoelium dendriticum* eggs. 1–6: MIF. The eggs of *D. dendriticum* are thick-shelled, slightly flattened on one side, with a broad, bluntly rounded operculum which is tilted to one side (1, 4; less in 2). Eggs lying with the flat side up or down appear oval with a slightly flattened opercular end (3). The fully developed miracidium contained in the egg has symmetrically arranged cephalic glands (5). Eggs with dead disintegrating miracidia are occasionally seen (6). *D. dendriticum* is a parasite of the livers of sheep and many other herbivores and is endemic in most sheep-raising areas of the world. The first intermediate hosts of the parasite are land snails. The cercariae emerge from the snails in slime balls which may be eaten by certain species of ants. The metacercariae develop in the body cavity of the ant. Infection is acquired by ingesting an infected ant either accidentally or as food. The immature fluke emerges from the egg in the small intestine, penetrates the intestinal wall, migrates through the peritoneal cavity to the liver, enters the liver through its capsule, and eventually enters a bile duct. Because of the route of migration of the immature parasite through the peritoneal cavity of its definitive host, parasites are frequently found in an ectopic location.

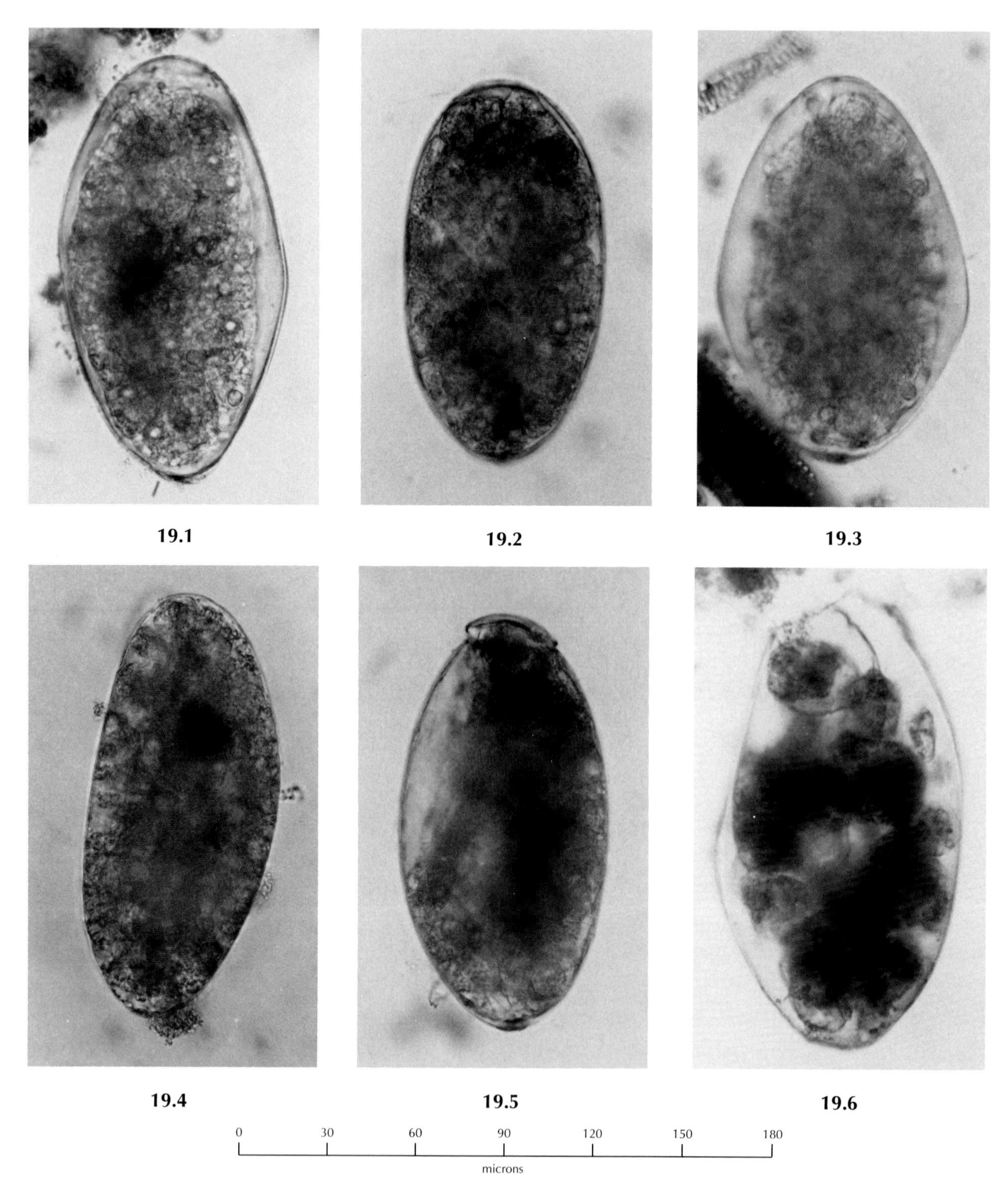

Plate 19. *Fasciola hepatica* eggs. 1–6: MIF. The eggs of *F. hepatica* are among the largest found in fecal specimens. They are thin-shelled, generally hen's-egg shaped (1, 2) or oval (5) and sometimes irregular (4). Occasionally eggs appear swollen (3). The operculum is indistinct, and there is often a slight thickening of the egg shell at the abopercular end (1, 3, 5). The operculum may be opened in older specimens (5) or even be broken off (6) when pressure is put on the egg in preparing a slide. The embryo is undeveloped when the egg is passed and develops into a miracidium in a moist or aquatic environment in from 9 to 15 days, or longer in a less suitable environment. Usually, the eggs cannot be differentiated from those of *Fasciolopsis buski*. The fluke is a common parasite of sheep and is present in most sheep-raising areas of the globe. The trematode may infect a number of other herbivorous mammals or man in endemic areas. Man becomes infected when eating watercress or other fresh aquatic grown plants with attached metacercariae.

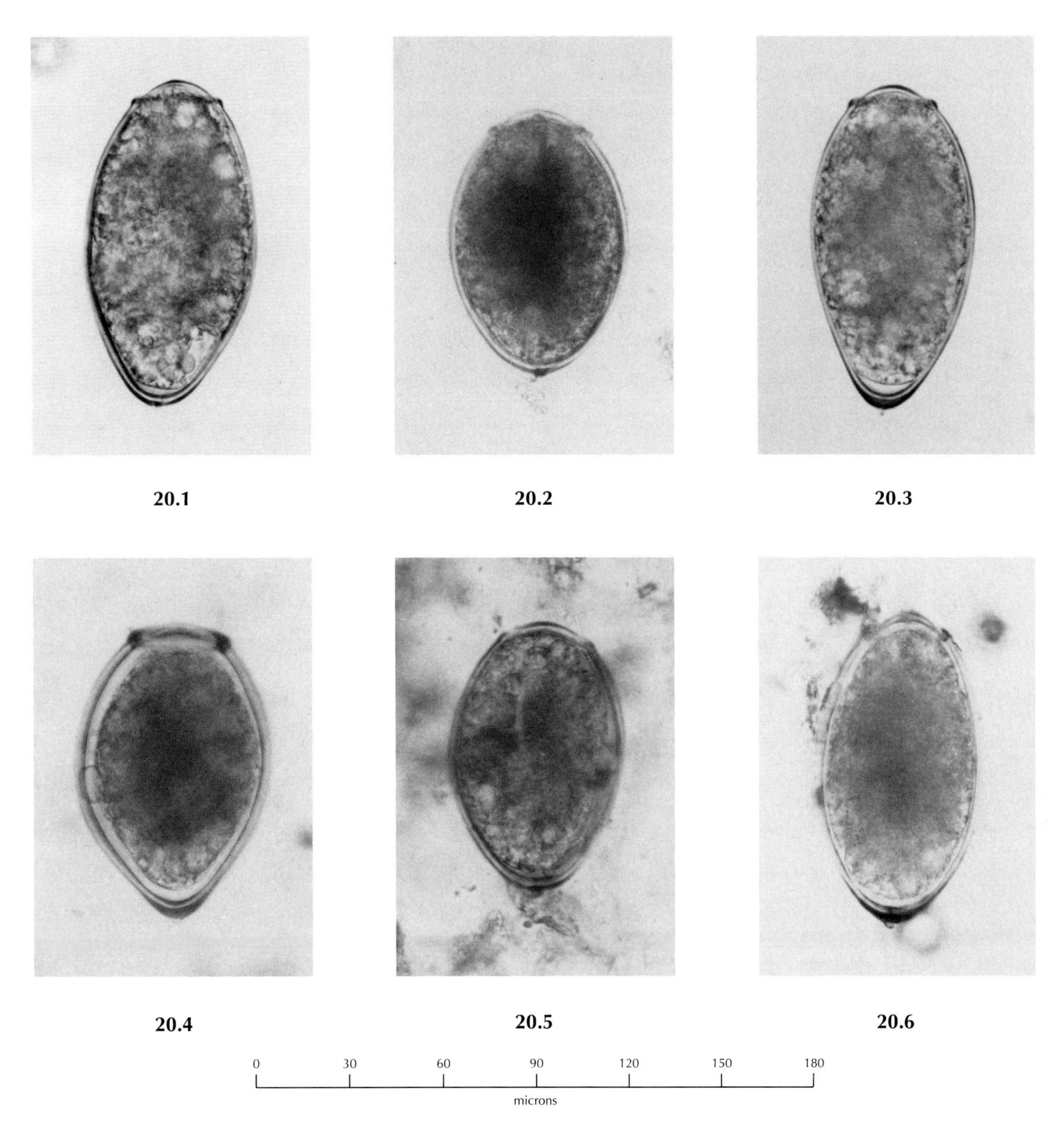

Plate 20. *Paragonimus westermani* eggs. 1–6: MIF. There are considerable variations between eggs of *P. westermani*, especially between eggs found in sputum and those found in feces. The photomicrographs reflect some of the variation seen in eggs found in feces. Generally eggs are ovoidal, tapering somewhat toward the abopercular end. In most eggs, the operculum is distinct and the shoulders pronounced (1–4; less so in 5, 6). The operculum is generally rounded (much less in 4). Eggs are sometimes broadest toward the abopercular end (1), at the center (2, 4–6), or toward the opercular end (3). The egg shell is usually thickened somewhat at the abopercular end, more in some eggs (3, 4, 6) than in others. The embryo is undeveloped when the eggs are passed in the feces and eggs must reach an aquatic environment to develop further. *P. westermani* has a wide geographic distribution but is usually found in small, limited foci within the endemic areas. Infection is acquired by eating freshwater crustaceans in which metacercariae are present, especially certain crayfish and crabs marinated in brine, vinegar, or wine and eaten as a delicacy. In addition to man, many wild mammals are naturally infected and many others have been infected experimentally.

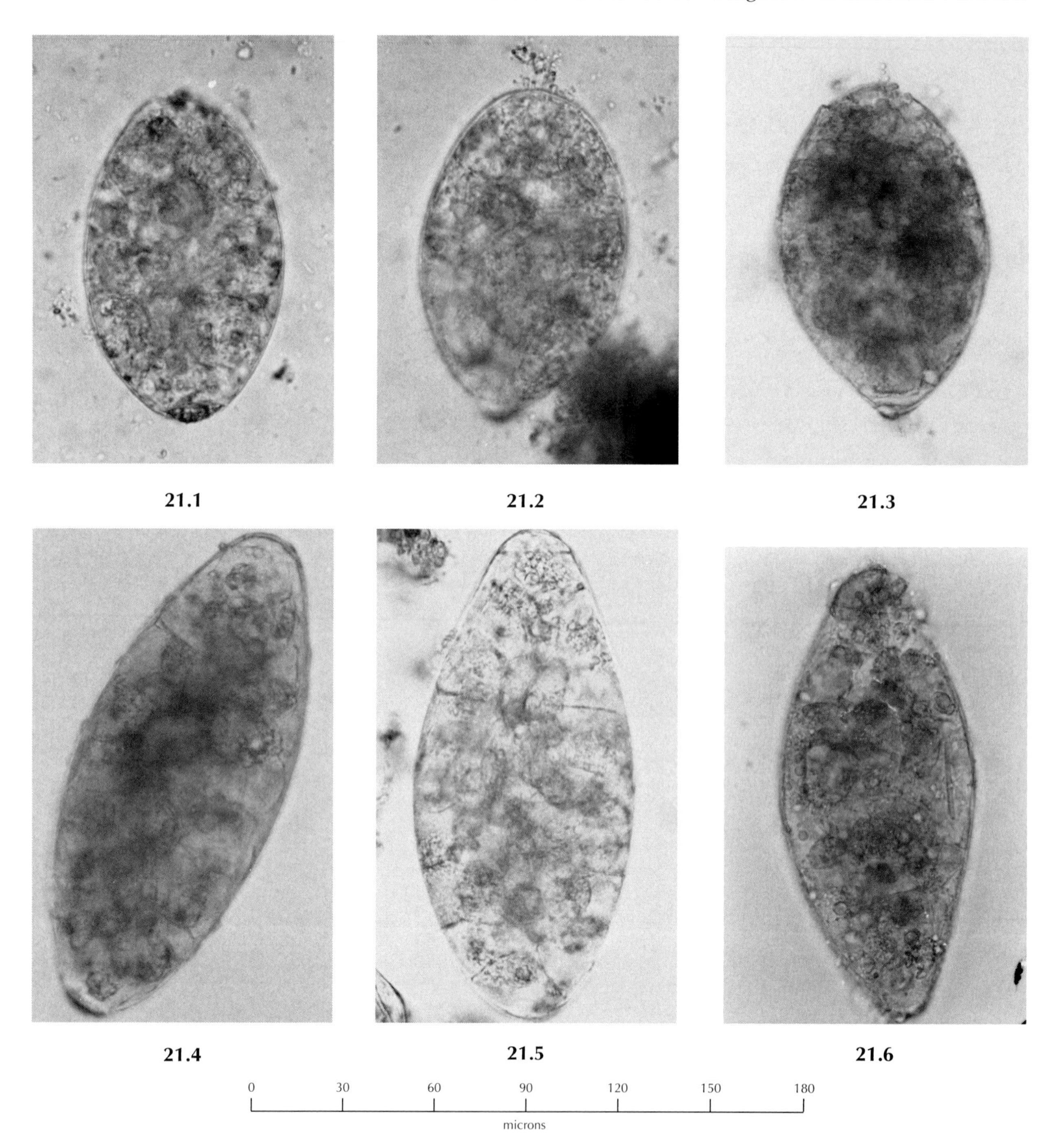

Plate 21. Eggs of the intestinal trematodes. *Echinostoma ilocanum.* 1–3: MIF. *Gastrodiscoides hominis.* 4–6: formalin. Eggs of *Echinostoma* spp. overlap in size, making it difficult to determine the species from the egg alone. Since *E. ilocanum* is most commonly found infecting man, its eggs are usually seen. The egg is thin-shelled, with an indistinct operculum, and often has a small concentration of dense material at the abopercular end (1–3). The embryo is undeveloped when the egg is passed. Eggs must reach fresh water for embryo development to continue. Both the first and second intermediate hosts are snails. In the Philippines and some other Asian areas, certain large snails that may act as second intermediate hosts are prized as food and are eaten raw. Infection occurs when a snail infected with metacercariae is ingested. The eggs of *G. hominis* are rhomboidal in shape, tapering from the center toward both ends (4–6). They have an indistinct, slanted operculum (4) and are thin-shelled. The operculum is sometimes forced open when preparing a slide (6). The embryo is undeveloped when passed and the egg must reach fresh water for its development to continue. Pigs, the typical hosts, act as a reservoir host for man. The parasite has a wide but spotty distribution being endemic in Assam and India and has been introduced into other areas by migrants. Infection occurs when metacercariae clinging to water plants are ingested.

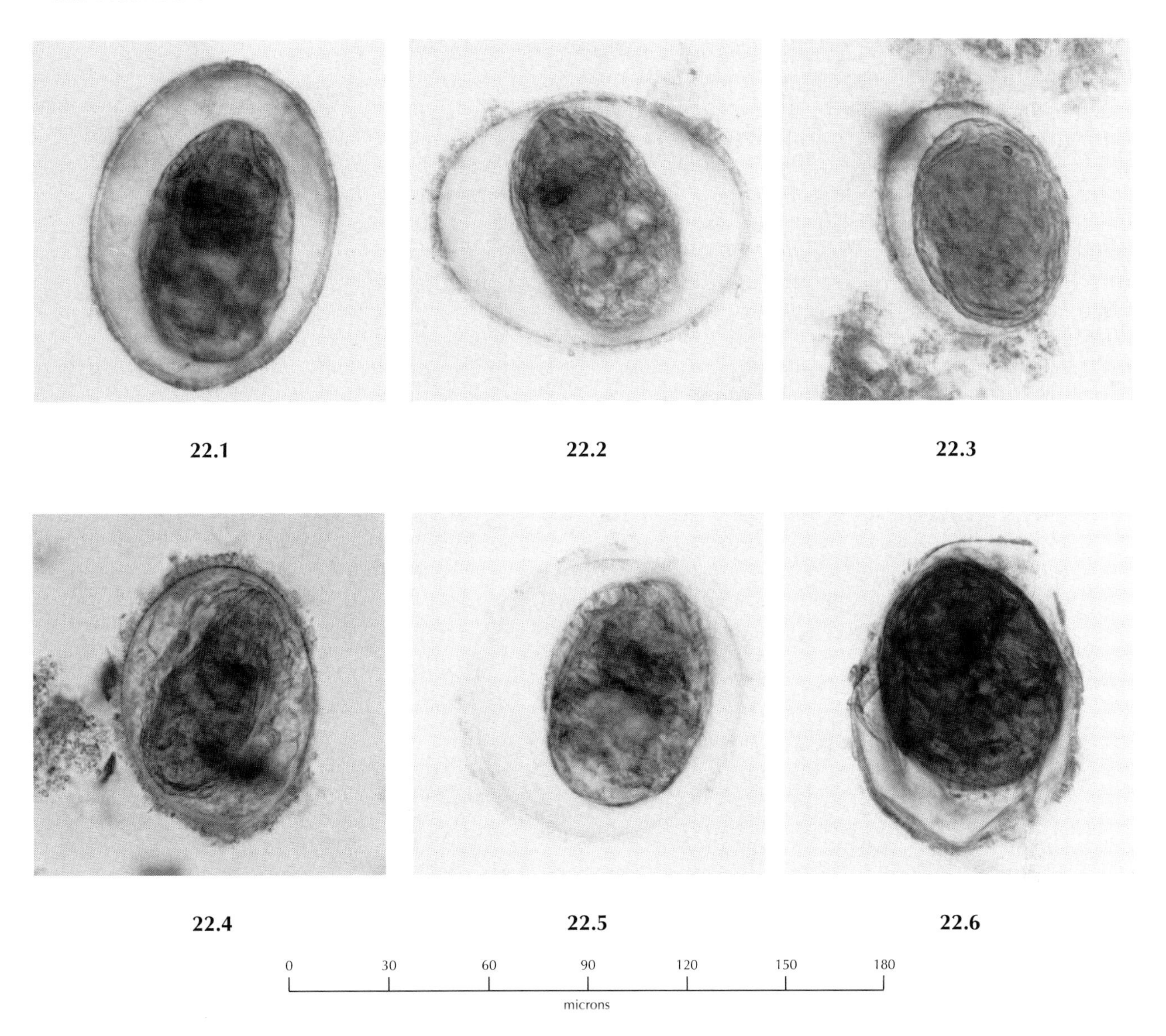

Plate 22. *Schistosoma japonicum* eggs. 1–6: formalin. The eggs of *S. japonicum* are broadly oval and contain a fully developed miracidium. A small, knob-like spine may be seen when the egg is lying in certain positions (not shown). Eggs vary considerably in size (1 large and 3 small). The shell may appear clean (1) or coated to various degrees with debris (2–4). Shells may become nearly transparent after sitting for a long time in formalin (5) and may collapse under pressure of slide preparation (6). The schistosomes inhabit the vascular system of the definitive host. *S. japonicum* inhabits veins associated with the intestine and eggs are usually passed with the feces. *S. japonicum* occurs mainly in Japan, eastern China, Taiwan, and some of the Pacific islands including the Philippines. Man appears to be the principal definitive host. In China, dogs, cats, cattle, sheep, and goats have been found to be naturally infected and, in the Philippine Islands, infection occurs in horses and hogs. The cercariae of schistosomes are free-living, and are the infective stage. They can directly penetrate the body surface of the definitive host to initiate infection. Cercariae most often emerge from snails in the afternoon on sunny days and contact with infested water is the mode of infection.

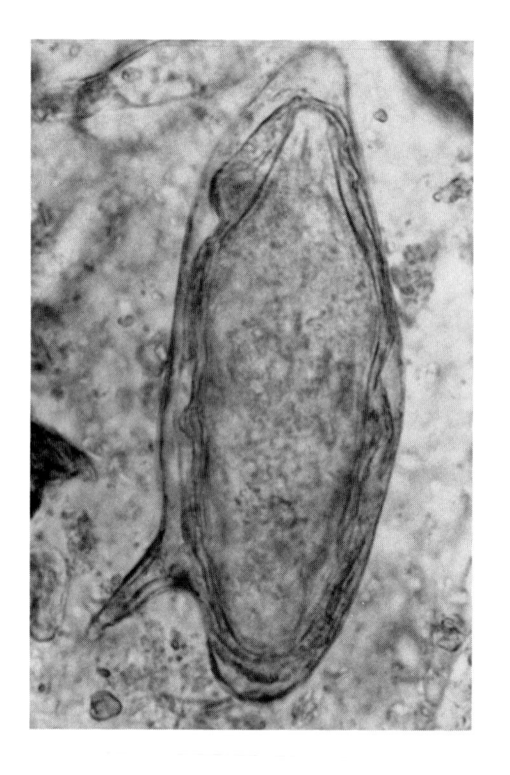

23.1

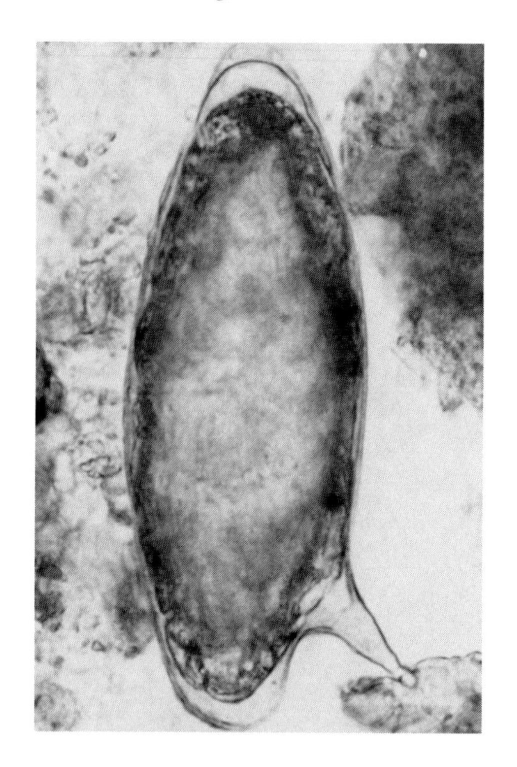

23.2

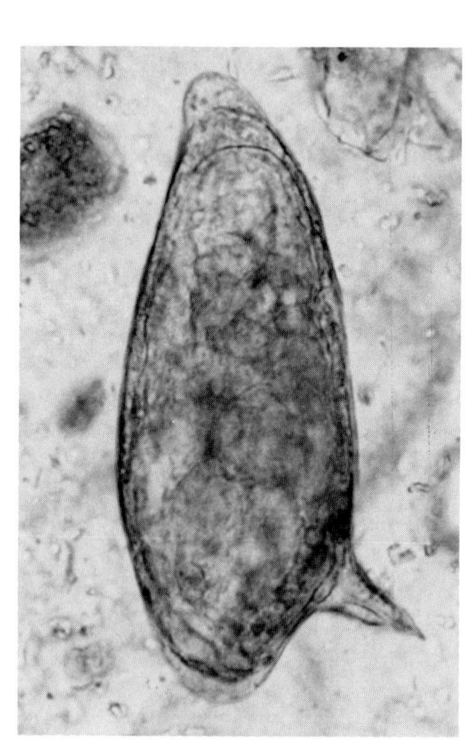

23.3

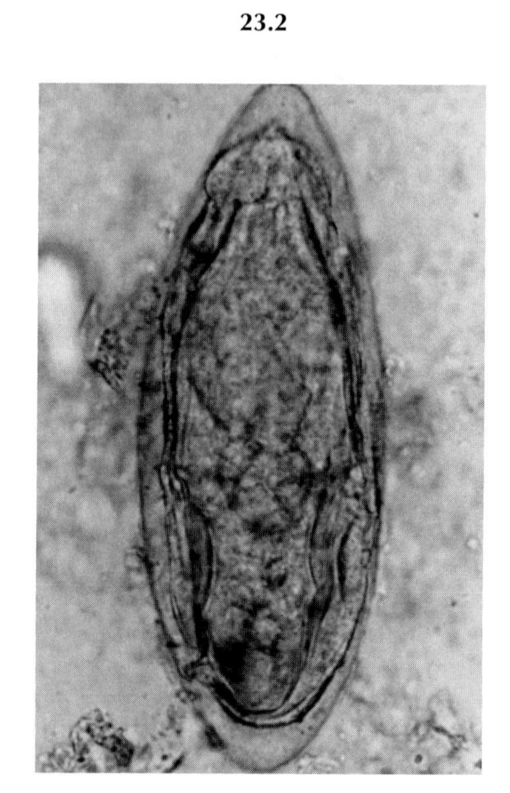

23.4

0 30 60 90 120 150 180

microns

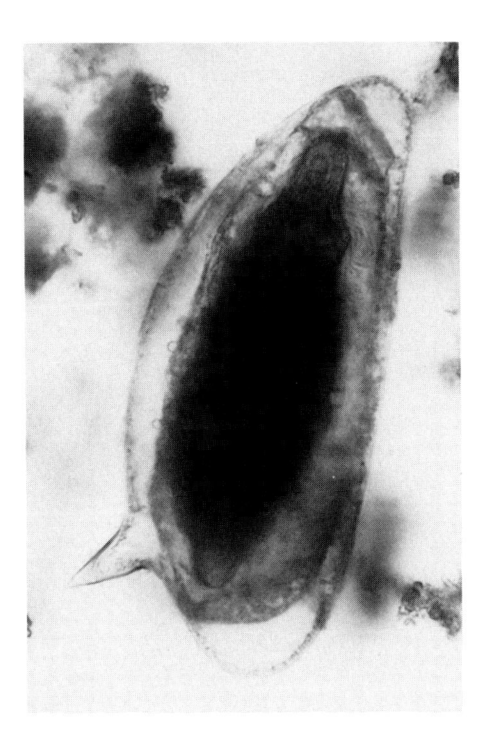

23.5

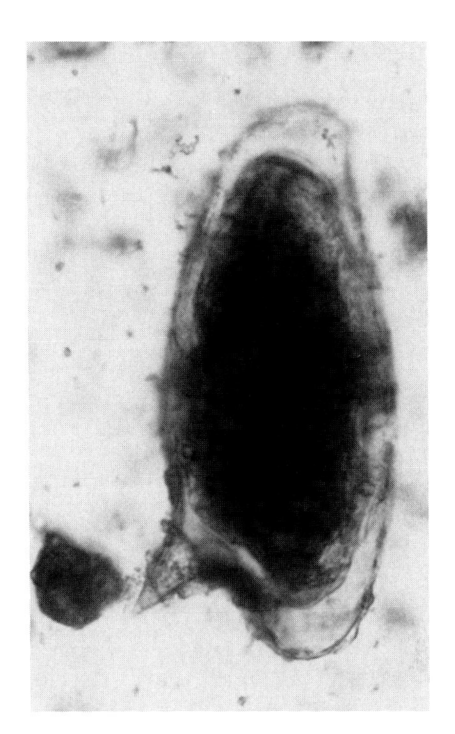

23.6

Plate 23. *Schistosoma mansoni* eggs. 1–5: MIF; 6. formalin. The eggs of *S. mansoni* are very large and distinctive because of their lateral spine (1–3). The miracidium is fully developed when the egg is passed and may be lying with its anterior end away from the spine end (1) or at the spine end (2). When the egg shell is in clear focus, the miracidium may be out of focus (3) because of the size of the egg and the depth of focus of the microscope. The spine may not be seen clearly when the egg is lying with the spine up or with the spine down (4). Eggs are often seen with dead miracidia in various stages of decomposition (5, 6). *S. mansoni*, like *S. japonicum*, inhabits venules associated with the intestine so eggs are usually passed in the feces. Infection occurs when exposure to fresh water brings the body surface of the definitive host into contact with free-swimming cercariae. It is endemic in much of Africa between the 30th degree parallels north and south and in pockets along the Nile river, especially in the delta of Egypt. In the western hemisphere, it is common in the eastern costal regions of South America and is endemic in most of the islands of the West Indies from Cuba eastward and south to Trinidad. Infections are found in individuals in many localities where the parasite is not endemic due to migration from endemic areas.

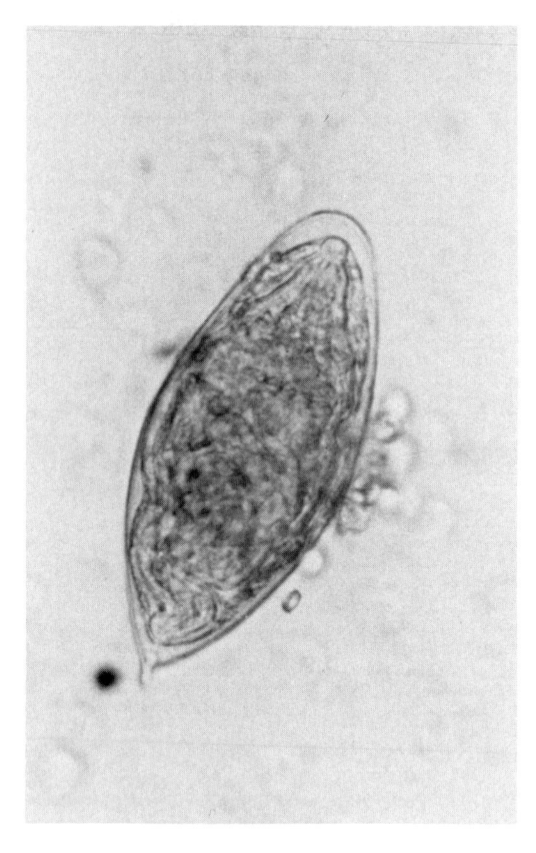

24.1

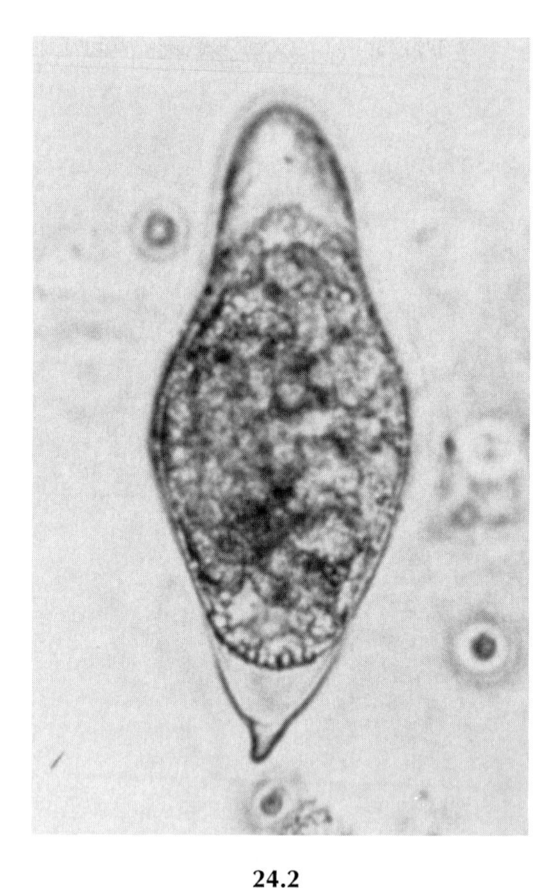

24.2

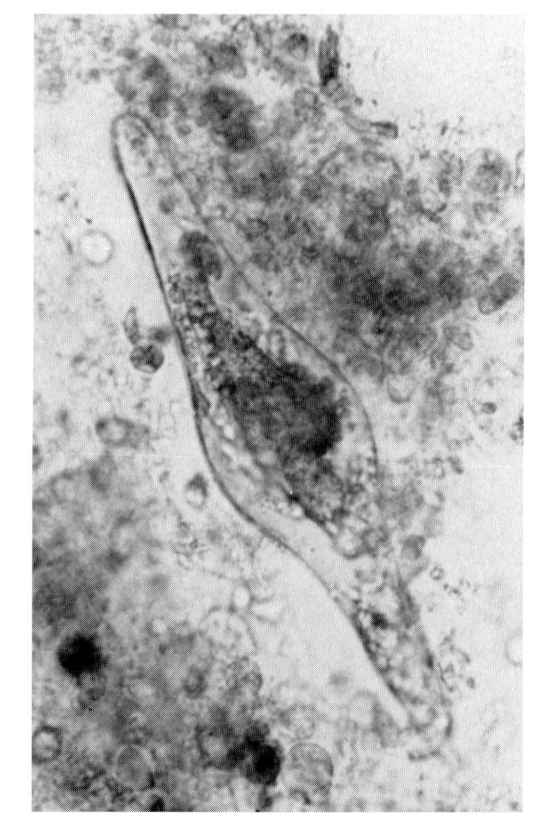

24.3

0 30 60 90 120 150 180
microns

Plate 24. Eggs of *Schistosoma haematobium, S. intercalatum,* and *S. bovis.* 1. *S. haematobium,* from urine, formalin; 2. *S. intercalatum,* from feces, formalin;* 3. *S. bovis,* from feces, MIF. The eggs of these three species vary considerably in length and shape. Eggs of *S. haematobium* are the smallest, appear in feces in small numbers and infrequently, and show the least size variation. The egg shown here is typical of those seen in urine specimens. Eggs of *S. intercalatum* vary from about 140 to 220 µm in length. They are broadest in the middle and taper toward both ends. Since it has been reported in seven counties in Africa south of the Sahara, it may be seen more frequently in travelers. *S. bovis* is the longest of schistosome eggs appearing in human feces measuring from 158 to 257 µm in length and may represent either a true or spurious infection. The egg shown has a dead, decomposing miracidium and was recovered from human feces.

* Photomicrograph courtesy Dr. H. Mehlhorn, Department of Parasitology, Ruhr-University, Germany.

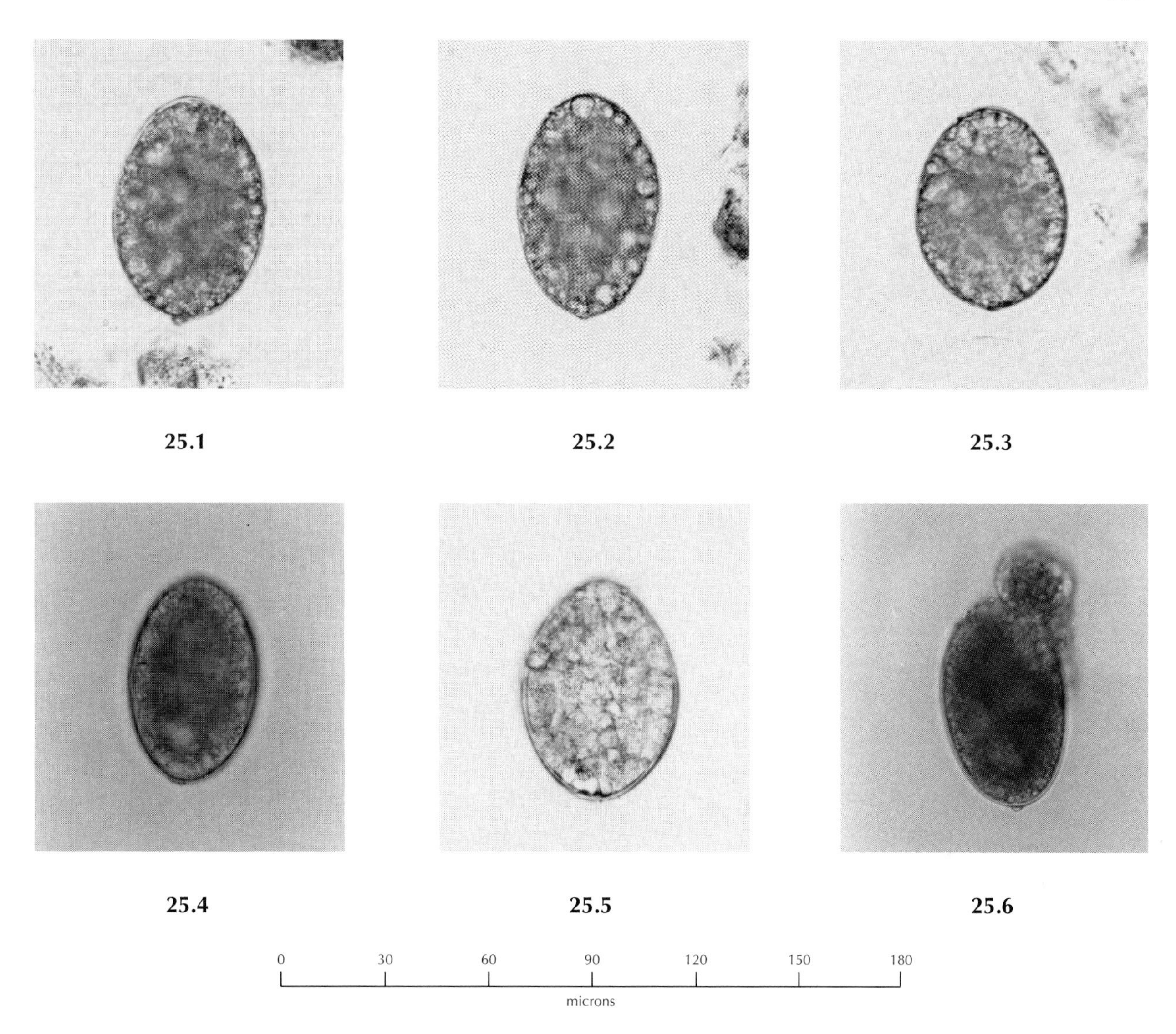

Plate 25. *Diphyllobothrium latum* eggs. 1–6: formalin. The eggs are thick-shelled, broadly barrel-shaped and more round (1, 3, 5) or more oval (2, 4, 6). The operculum is broad and often difficult to see. The embryo is undeveloped when the egg is passed. The egg usually has a flattened, button-like thickening (2, 4, 5) or a small papillae (1, 6) at the abopercular end just to one side of center. In fecal preparations, the operculum is often broken open and may or may not remain attached to the egg (6). *D. latum* is a pseudophyllidian tapeworm with a scolex having two bothria as organs of attachment and is often referred to as the broad fish tapeworm. It is prevalent in cooler temperate zones where fresh-water fish are an important part of the diet. It is endemic in the lake areas of Europe, Middle East, Far East, and parts of Africa. It was apparently introduced into the Great Lakes of North America and continues to be endemic is some localities. The strobila may range from 3 to 10 m in length and be comprised of as many as 4000 proglottids. Unlike the cyclophyllidian tapeworms, eggs are passed singly from the genital pore of each gravid proglottid. Man and fish-eating mammals are the definitive hosts. Infection is acquired when certain fish, or parts of fish such as roe, are eaten uncooked.

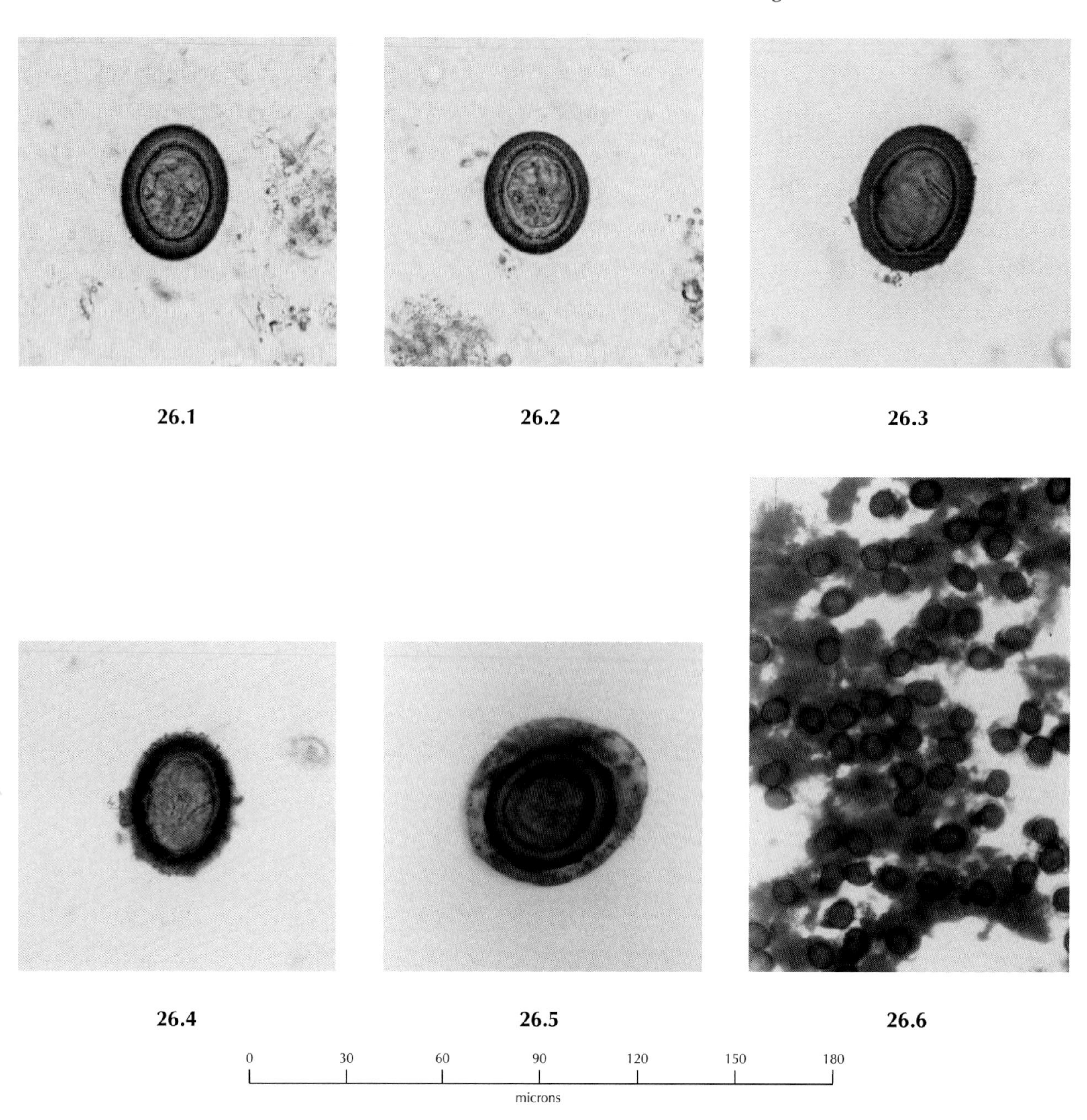

Plate 26. Eggs of *Taenia* spp. 1–6: MIF. The eggs of *Taenia* spp. are subspherical and vary from 30 to 40 μm at their longest axis. The embryo has 6, small, refractive hooklets, some of which may be clearly seen in some eggs (3, 4) but not in others (2). The outer shell is made up of truncated prisms cemented together, giving the appearance of striations (2). The surface may be clean (1–3) or may be coated with debris (4). Rarely seen are eggs surrounded by the primary membrane (5). Sometimes, a portion of a disintegrating proglottid with clusters of eggs (6) is seen in a fecal preparation (about 1/4 the magnification of other photomicrographs). Man is the only typical, definitive host of *T. solium* and *T. saginata*. Intestinal infection is initiated by ingestion of a cysticercus larva embedded in inadequately cooked pork or beef, respectively. Eggs found in human feces cannot be differentiated. Specific identification is made by examining the proglottid or scolex of the adult parasite. Humans may become infected as aberrant intermediate hosts by ingestion of eggs of *T. solium*. When *Taenia* eggs are found, extreme care should be taken to prevent contamination until the species can be determined. When a cysticercus is ingested, the scolex evaginates, attaches to the wall of the small intestine, and develops into the adult parasite. Some terminal, fully gravid proglottids may break off and be passed in the feces or disintegrate releasing eggs in the feces.

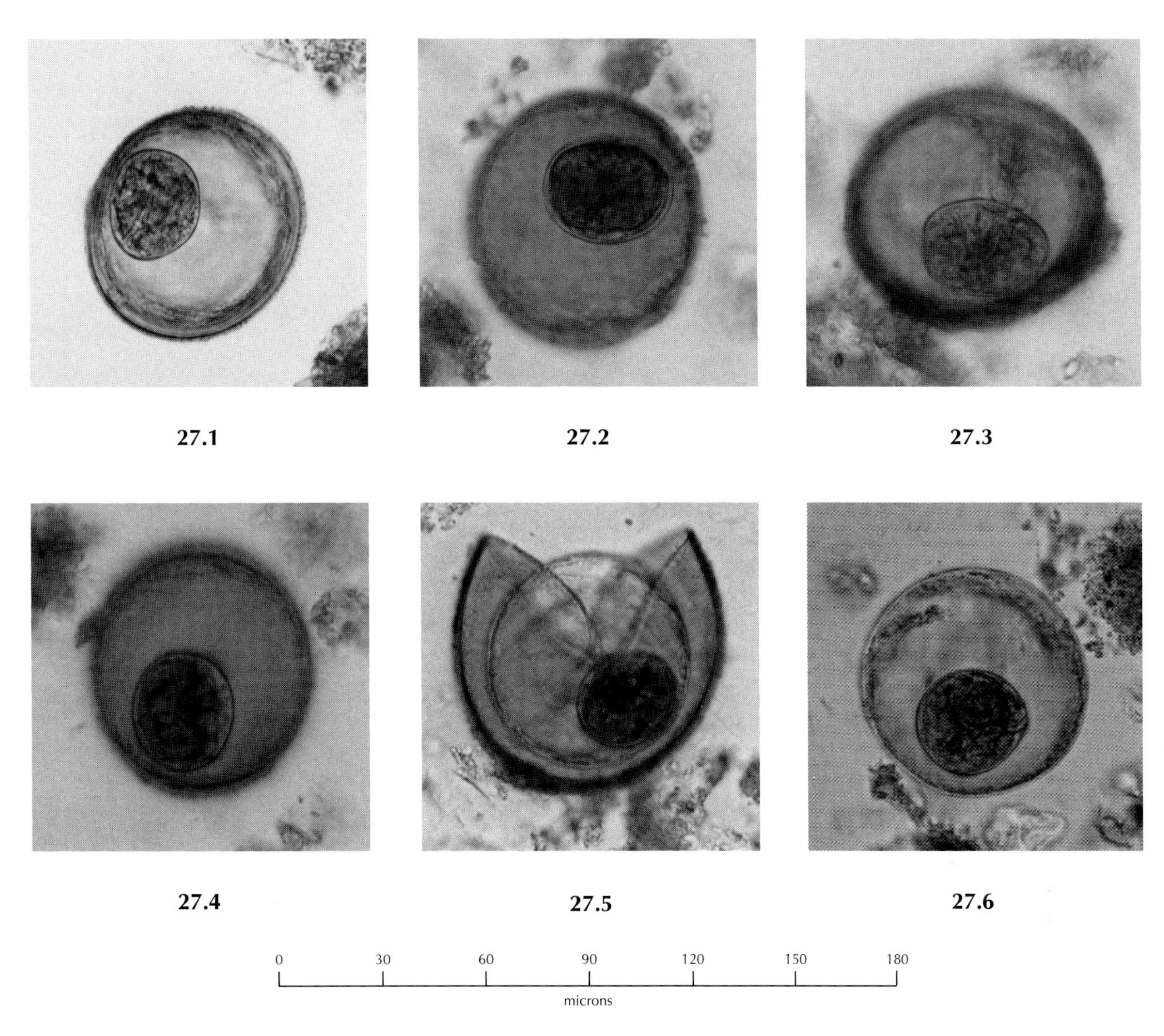

Plate 27. Eggs of *Hymenolepis diminuta*. 1–6: MIF. The ovoidal eggs of *H. diminuta* vary in diameter from 60 to 85 µm. The onchosphere is about one third the diameter of the egg, usually lies close to one side of the inner shell, and has six hooks some of which may be seen (1, 3, 4). When the hooks are visible, they often lie in the shape of an open fan (4). The thickenings of the embryophore at the poles are visible only when the onchosphere is lying in a lateral position (not shown). The shell has two layers that are not discernible in intact eggs. The outer shell may be covered by a loose coating (2, 3). In older specimens or in the preparation of slides, the thick outer shell may be broken, revealing the thin inner shell (5), or missing entirely (6). The adult parasites are cosmopolitan in rats, mice, and other murine species. Man and dogs are accidental definitive hosts. A variety of larval and adult insects can act as intermediate hosts, including beetles, fleas, roaches, and myriapods. Man becomes infected by accidentally ingesting a small beetle (or other insect), usually from dry cereals or grains.

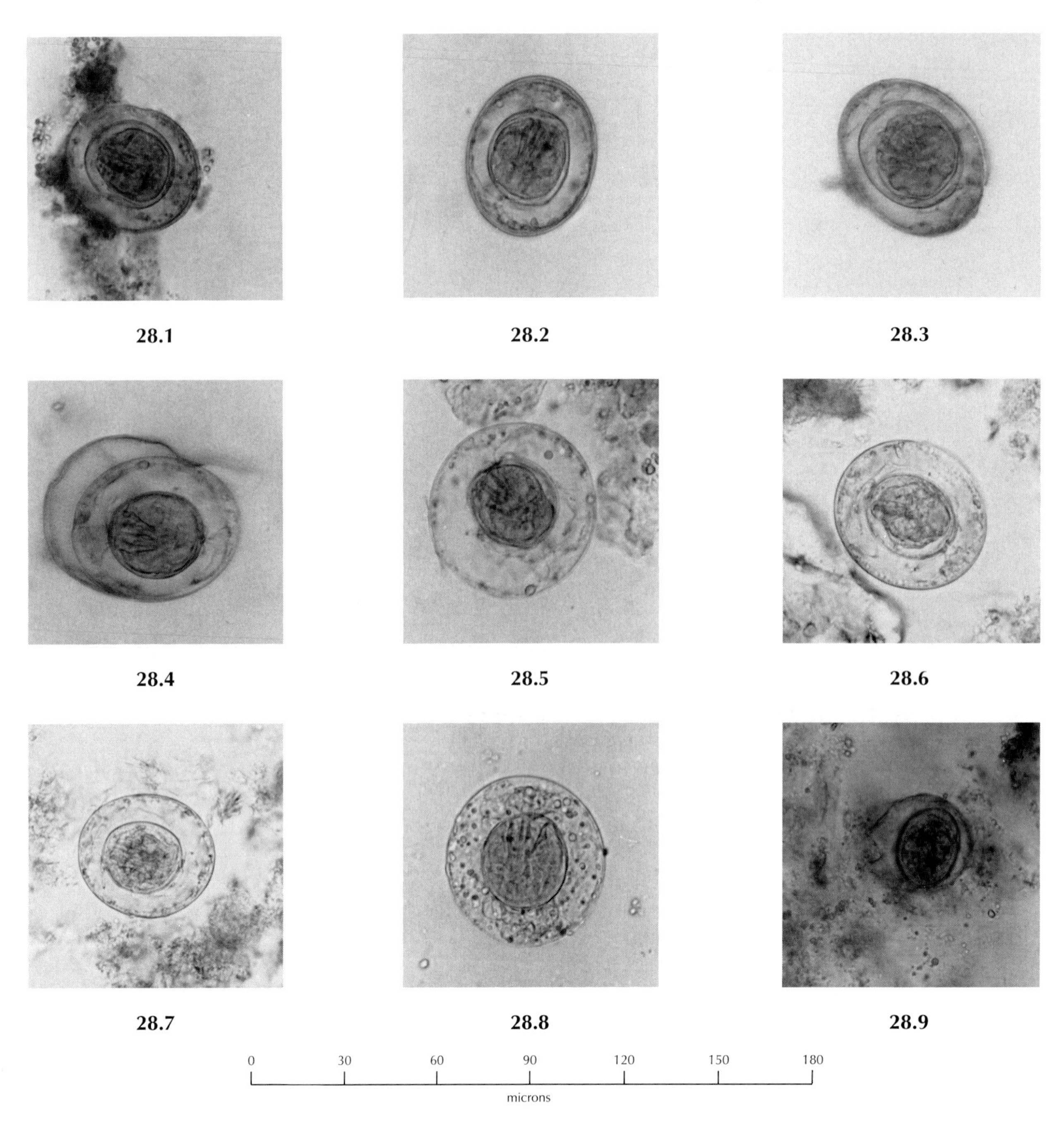

Plate 28. Eggs of *Hymenolepis nana*. 1–9: MIF. Eggs of *H. nana* are similar to those of *H. diminuta* but are smaller, varying from 30 to 60 µm in diameter. The onchosphere has six hooks some of which are usually visible (1–4, 8). It is surrounded by the thin embryophore, which is clearly visible, and is about one half the diameter of the egg, although the ratio may vary (compare 1 and 5). In *H. nana* eggs, there are fine filaments originating at the poles of the embryophore and extending into the space between the membrane and the inner shell (2–8) which is a distinguishing feature. As in *H. diminuta*, there are outer and inner shells. The outer shell often ruptures (3), pulls away from the inner shell (4), leaving the egg devoid of an outer shell (4–7). In slide preparations or in old, preserved specimens, egg shells of some eggs may collapse around the onchosphere (9). *H. nana* is cosmopolitan and is the most common tapeworm of man in the U.S. Because of its size, it is referred to as the dwarf tapeworm. It requires no intermediate host and is acquired by ingestion of the eggs from a contaminated source. Some of the gravid, terminal proglottids may break off and disintegrate in their passage down the intestinal tract releasing the eggs to be passed with the feces. If passage through the intestine is slowed, freed eggs may hatch and the onchosphere may attach to the mucosa establishing an autoinfection.

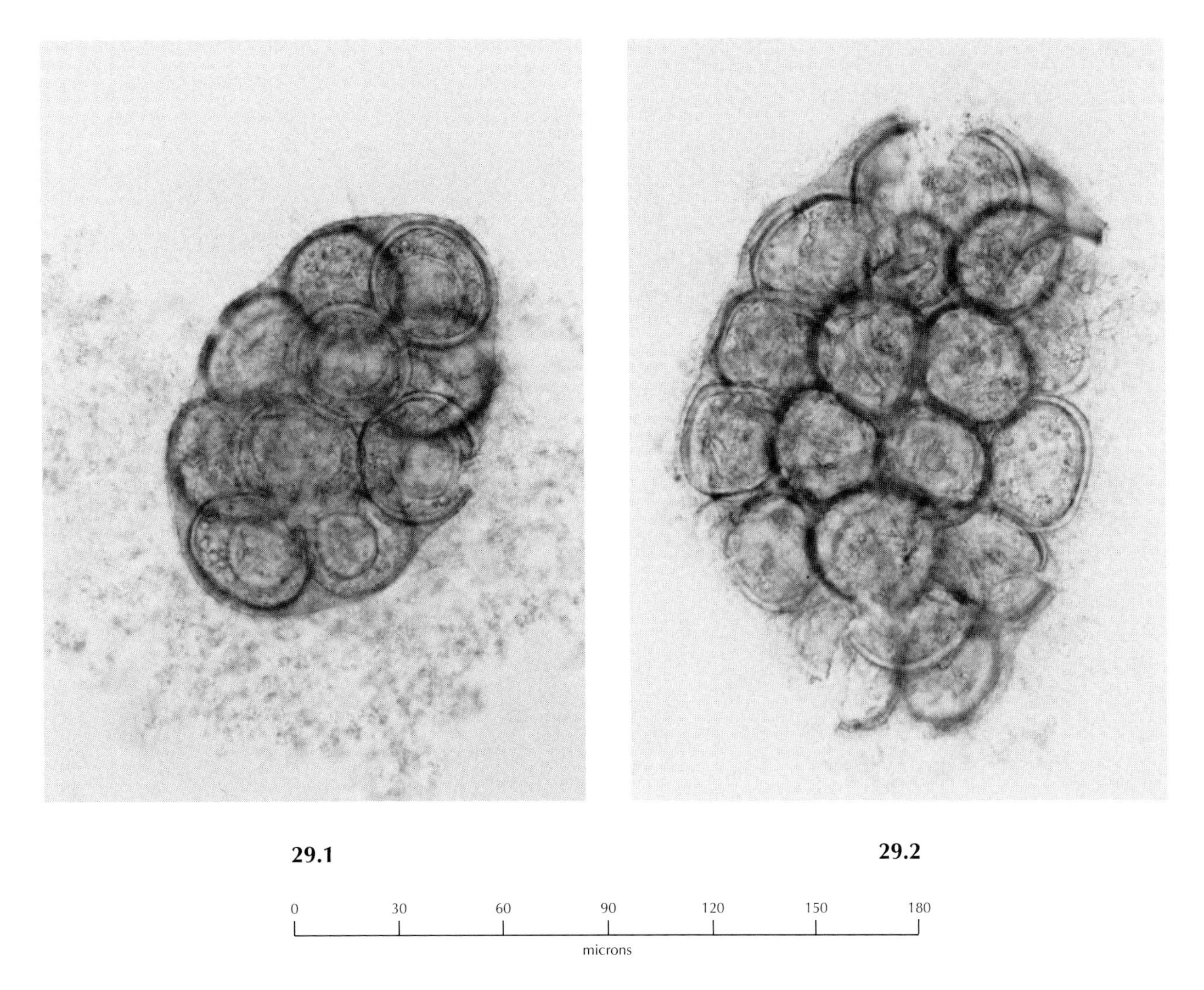

Plate 29. Eggs of *Dipylidium caninum.* 1 and 2: MIF. The eggs of *D. caninum* are within packets of 10 to 25, surrounded by an embryonic membrane in the mature and gravid proglottids. Individual eggs are spherical and vary from about 30 to 60 µm in diameter. They are similar to eggs of *H. nana* with an onchosphere with six hooks but with no outer shell and no filaments in the space between the embryophore and shell. Usually, the terminal, gravid proglottids break from the strobila and either migrate out of the anus or are passed in the feces. Proglottids commonly become attached to hairs around the anus of animals where they become dried and break open to release the eggs. Some egg packets may be freed from the proglottid within the intestine and are passed in the feces. The smaller of these usually remain pretty much intact (1) but larger packets tend to break up (2) in fecal preparations. *D. caninum* is a cosmopolitan parasite of canines and felines. Man is an accidental definitive host. The biting dog louse and dog, cat, and human fleas are the usual intermediate hosts. Infection occurs when fleas containing one or more cysticercoid larvae are ingested.

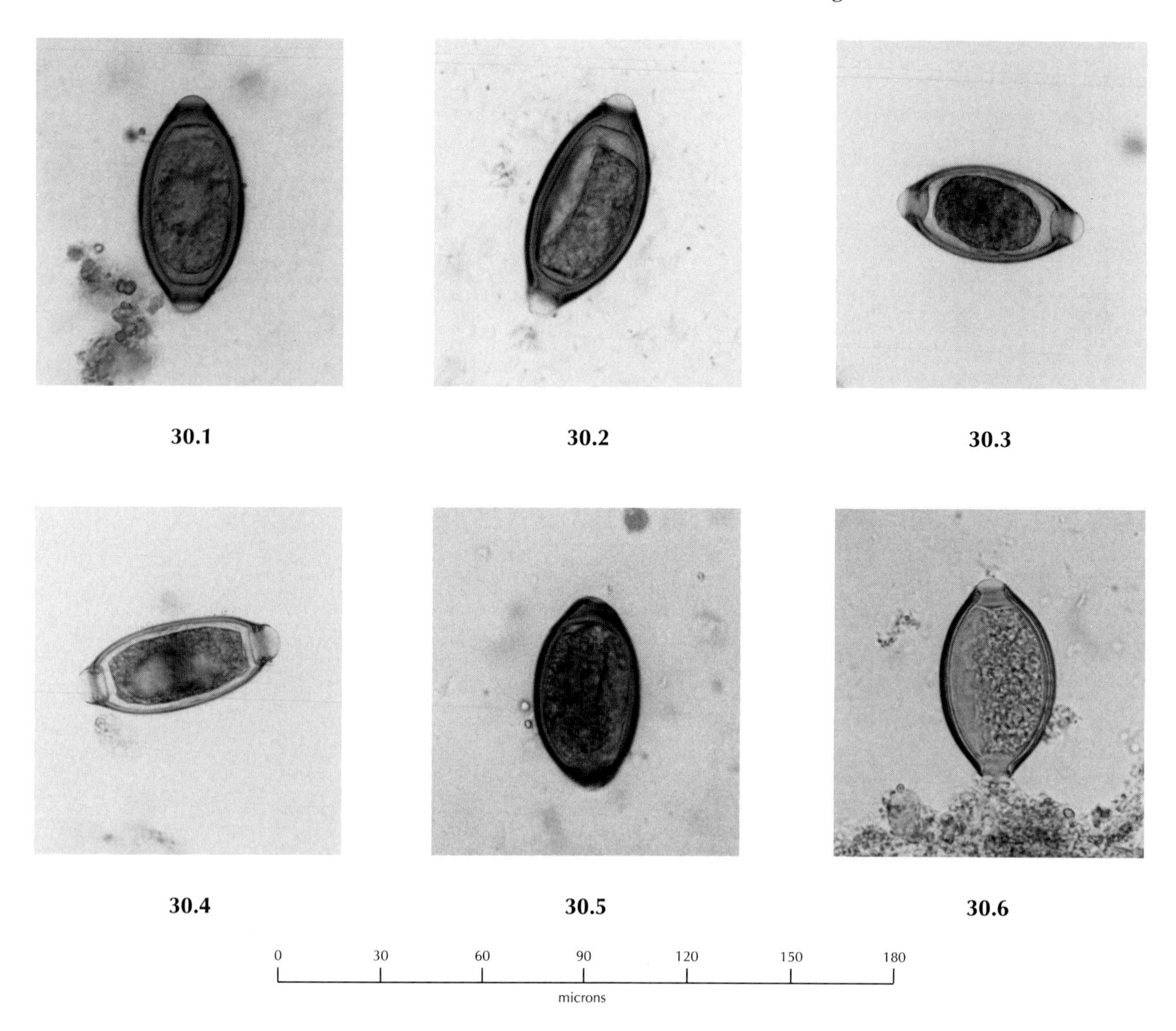

Plate 30. *Trichuris trichiura* eggs. 1 and 2: MIF; 3: PIF; 4–6: MIF. The eggs of *T. trichiura* are ellipsoidal or football shaped, from 50 to 65 μm long × 22 to 30 μm wide, and have mucoid plugs at each end (1–6). They vary somewhat from their typical ellipsoidal shape (1, 2), to those that are thinner (4), or fatter (6). They are rarely short with unclear plugs (5). The embryo may be undeveloped (1, 3, 4), partially developed (2), or dead and decomposing (6). Man is the typical, definitive host of *T. trichiura*. The parasite is cosmopolitan and is an important parasite in tropical agricultural areas, especially among poorer classes. In some areas, it competes with *Ascaris* as the most common parasite. Like *Ascaris*, the eggs must incubate in the soil to become infective. Infection occurs by ingestion of eggs containing infective juvenile worms. Adult worms are referred to as whipworms because they have a whip-like appearance with a long, thin, anterior end and a much thicker and shorter posterior end. The adults inhabit the colon, particularly in the area of the caecum, where their anterior end is threaded into the villi and the posterior end extends into the lumen.

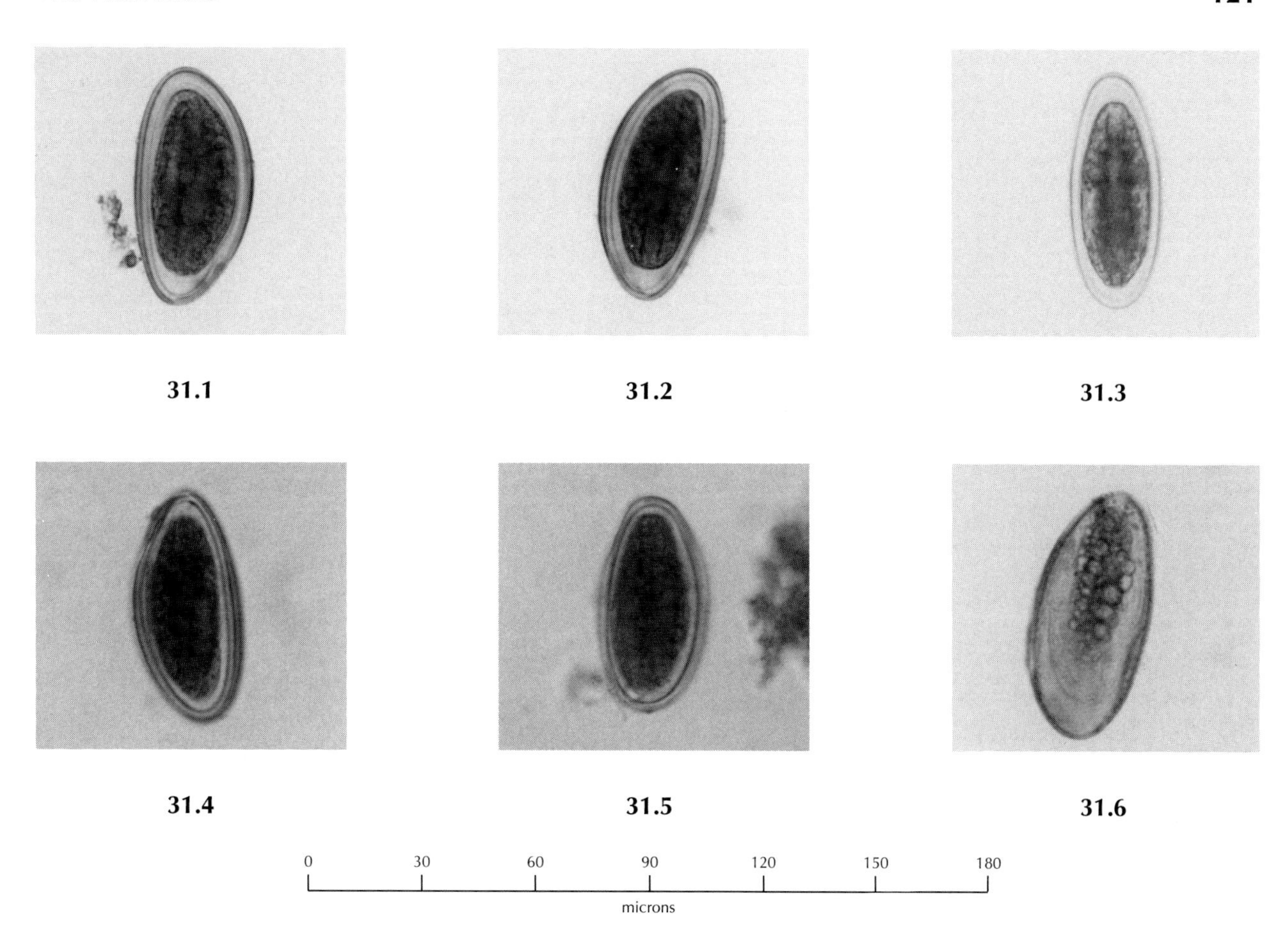

Plate 31. *Enterobius vermicularis* eggs. 1–6: MIF. The eggs of *E. vermicularis* found in fecal specimens do not vary greatly and usually do not become fully embryonated as do the eggs laid outside of the intestine in the folds of the anus. A small segment of the tail of the juvenile in an early stage of development often can be seen (1, 2, 4). The eggs measure 50 to 65 × 20 to 30 µm and are flattened on one side and rounded on the other. When the eggs are lying with the flat side up or down, the egg appears symmetrical (3, 5). Since eggs passed in the intestine do not develop, the embryo in eggs laid in the upper part of the colon may die before the eggs are passed in the feces (6). The parasite is cosmopolitan, especially in temperate and warm climates. Infection is initiated by ingestion of the egg containing an infective juvenile, either hand-to-mouth or from a contaminated source such as unwashed underclothes, bed sheets, or even dust under humid conditions. Sandboxes are an ideal source of infection for small children. Eggs are not killed by the usual household disinfectants but are killed at high temperatures. To eliminate the parasite after it has been introduced into a household often requires heroic measures.

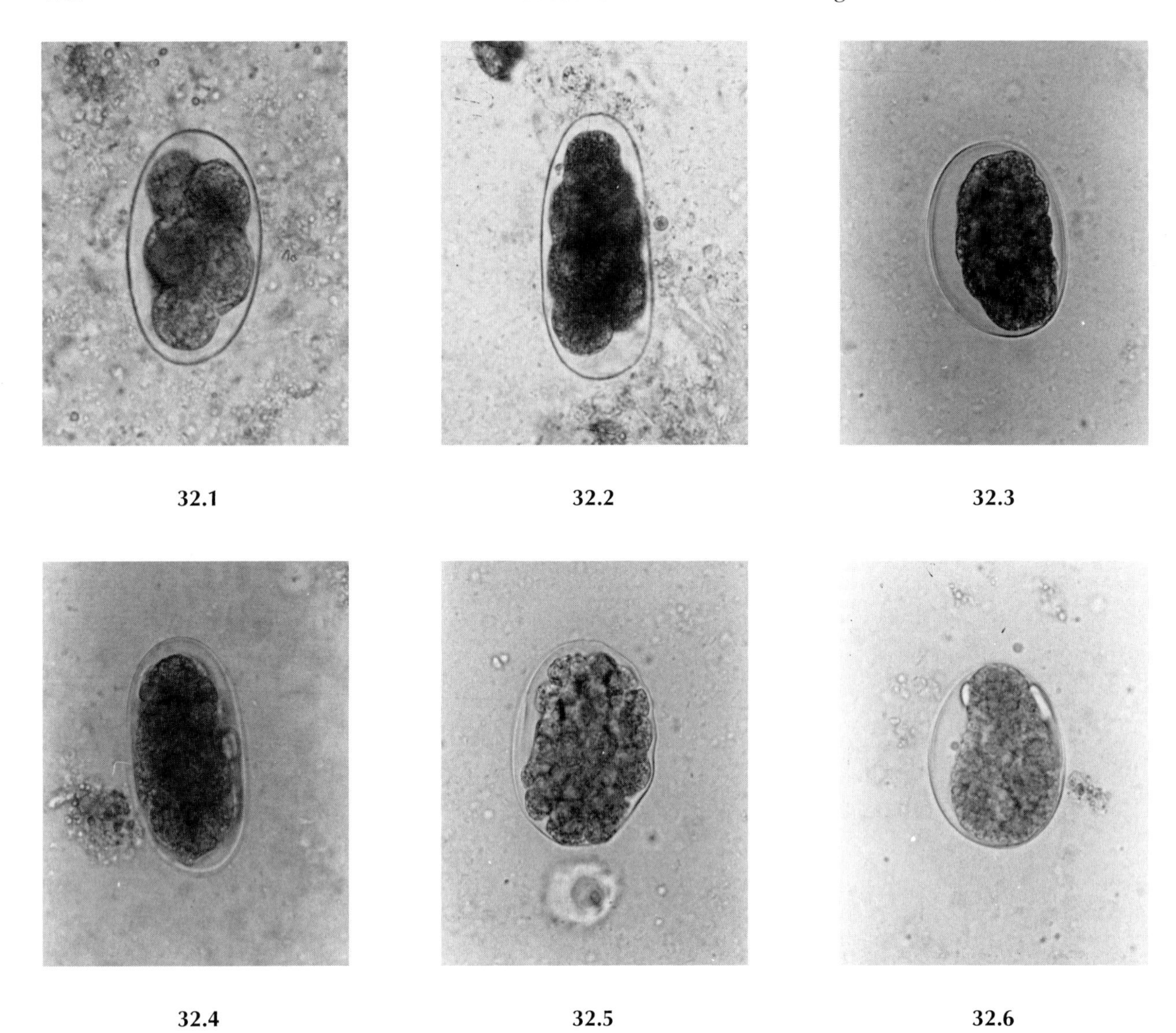

32.1

32.2

32.3

32.4

32.5

32.6

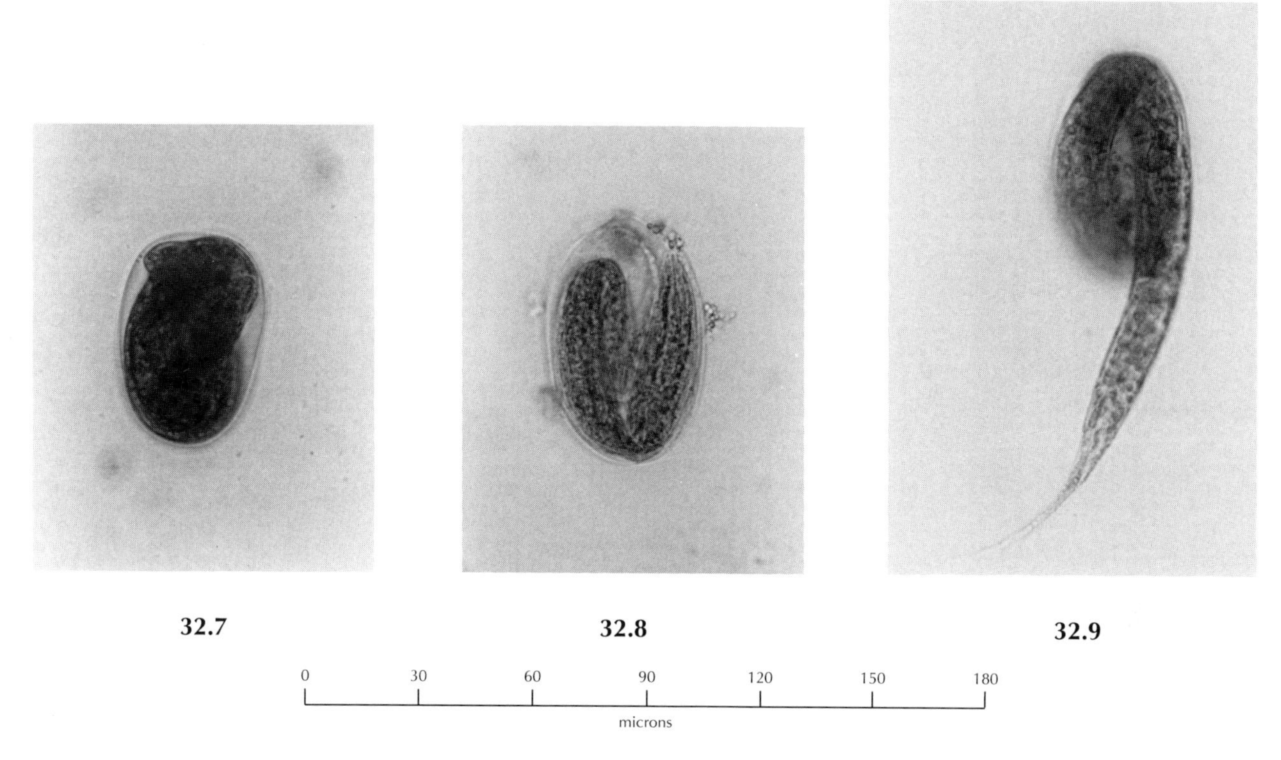

Plate 32. Hookworm eggs, *Necator americanus* or *Ancylostoma duodenale.* 1 and 2: MIF; 3–9: PIF. Hookworm eggs are thin-shelled, may be oval (1), or elliptical (2) and if passed shortly after being laid by the adult, the embryo may have advanced to the 4 to 8 cell stage (1, 2). If evacuation from the intestine is delayed, the embryo may be well advanced (3–6) or even may develop to the embryo stage (7, 8). If evacuation is delayed further, the juvenile worm may break out of the egg (9). Some juvenile worms may be found in fecal preparations (Plate 12, Plate 35). Hookworm infection is worldwide and is most prevalent at low altitude levels in tropical or subtropical areas, especially where soils are sandy or loosely packed. Hookworm disease is considered to be separate from hookworm infection and depends on the number of adult worms in the intestine. The severity of hookworm disease is related directly to the blood loss in the patient; blood leaks from unhealed lesions where worms have attached. Light infections where blood loss is compensated may not produce overt symptoms. In infected patients where blood loss cannot be compensated, the disease must be treated or the patient's health will deteriorate. Infections with *A. duodenale* are considered more serious because of the greater blood loss per worm.

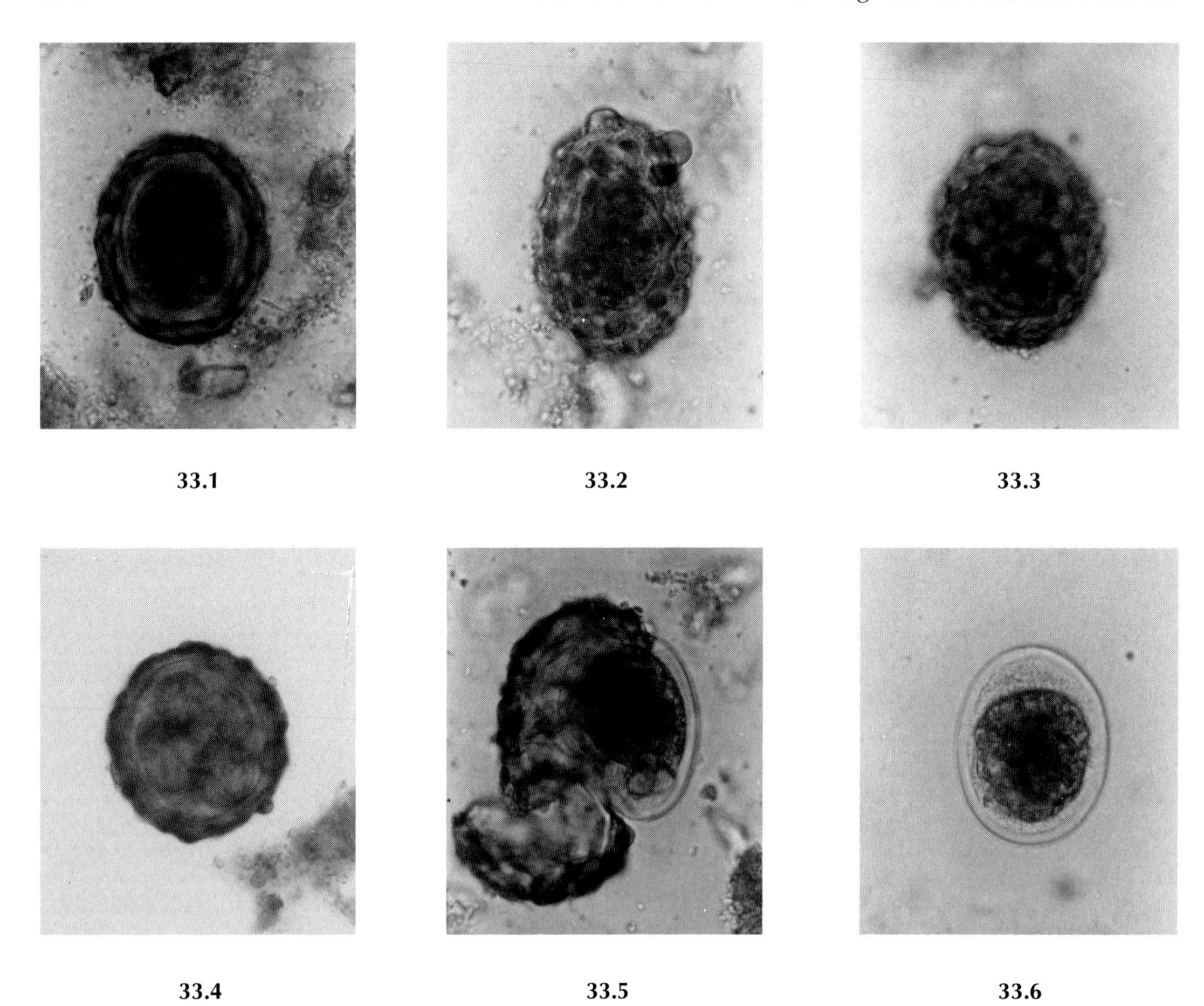
33.1
33.2
33.3
33.4
33.5
33.6

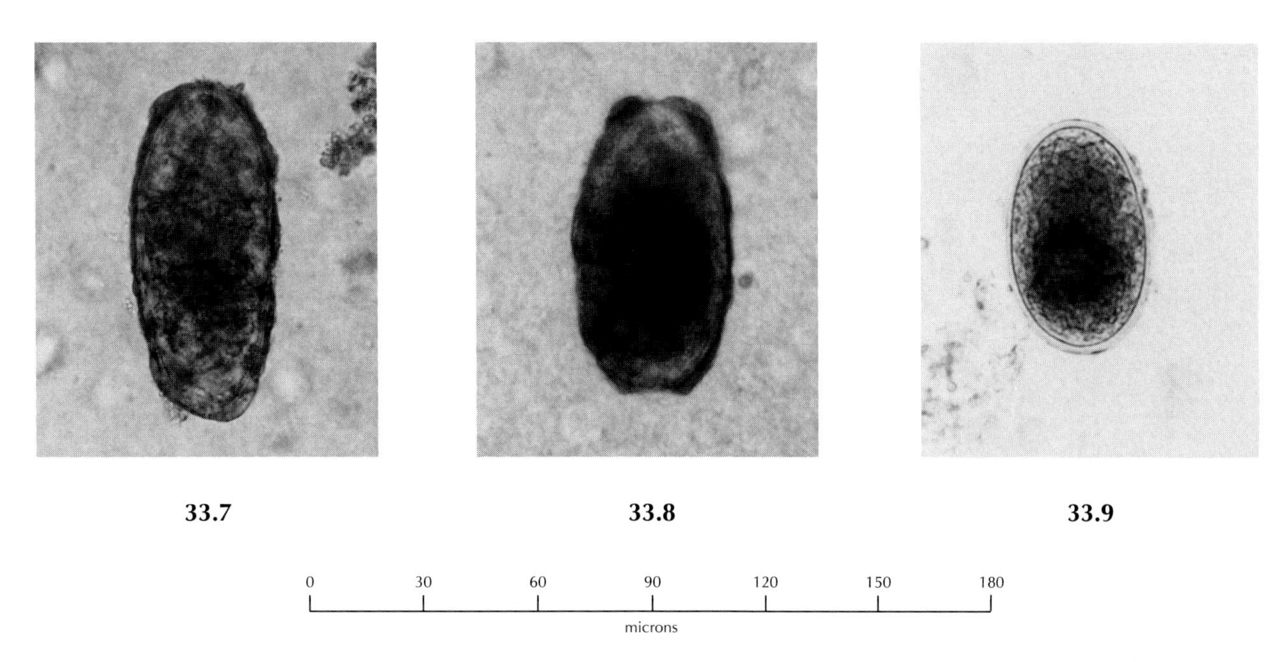

Plate 33. *Ascaris lumbricoides* eggs. 1–7: PIF; 8 and 9: MIF. The typical eggs of *A. lumbricoides* have a thick shell with a rough mammillated coating, the cortex. Eggs are usually broadly oval, 45 to 75 × 35 to 50 μm, and have an embryo of 1 to 4 cells (1–4). Under certain circumstances, the eggs may lose their cortex (5) revealing the thick shell (6). Unfertilized eggs may also be present (7–9). These are long and rhomboidal in shape and vary from 88 to 95 μm in length. They are usually covered with a cortex (7, 8). When the cortex is missing, the absence of an embryo becomes more obvious and the thick shell differentiates it from hookworm (9). *A. lumbricoides* is the largest and most common of the intestinal nematodes. It is most prevalent in warmer climates where sanitation is poor. It is more common where soils are dense, e.g., clay soils, where eggs remain on the surface. Infection with *Ascaris* can be mild or serious, usually related to the intensity of the infection. Adult worms tend to migrate when disturbed and can cause serious complications. The infection is easily treated and, whenever possible, infected individuals should be treated before medication for other conditions is given.

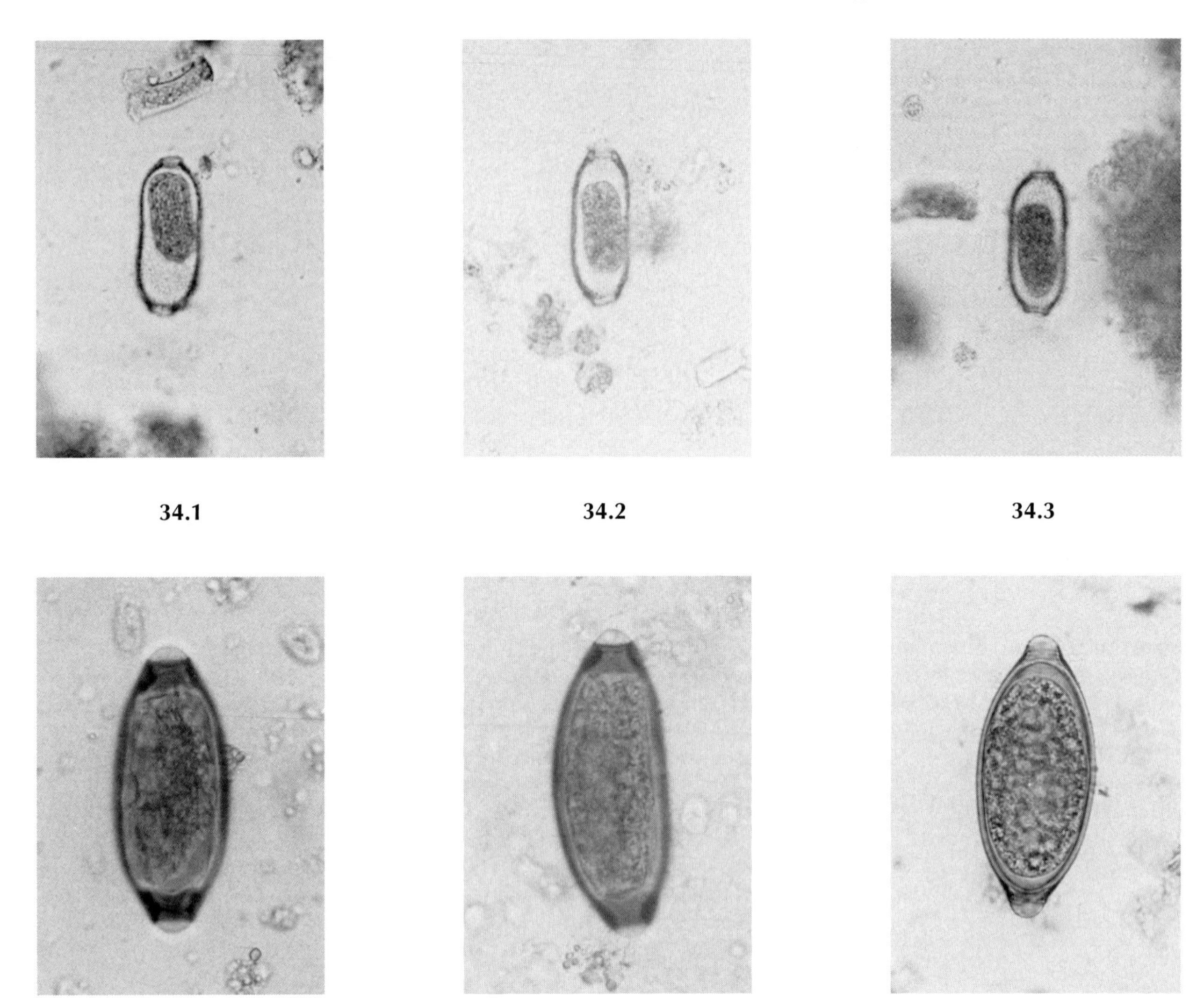

34.1 34.2 34.3

34.4 34.5 34.6

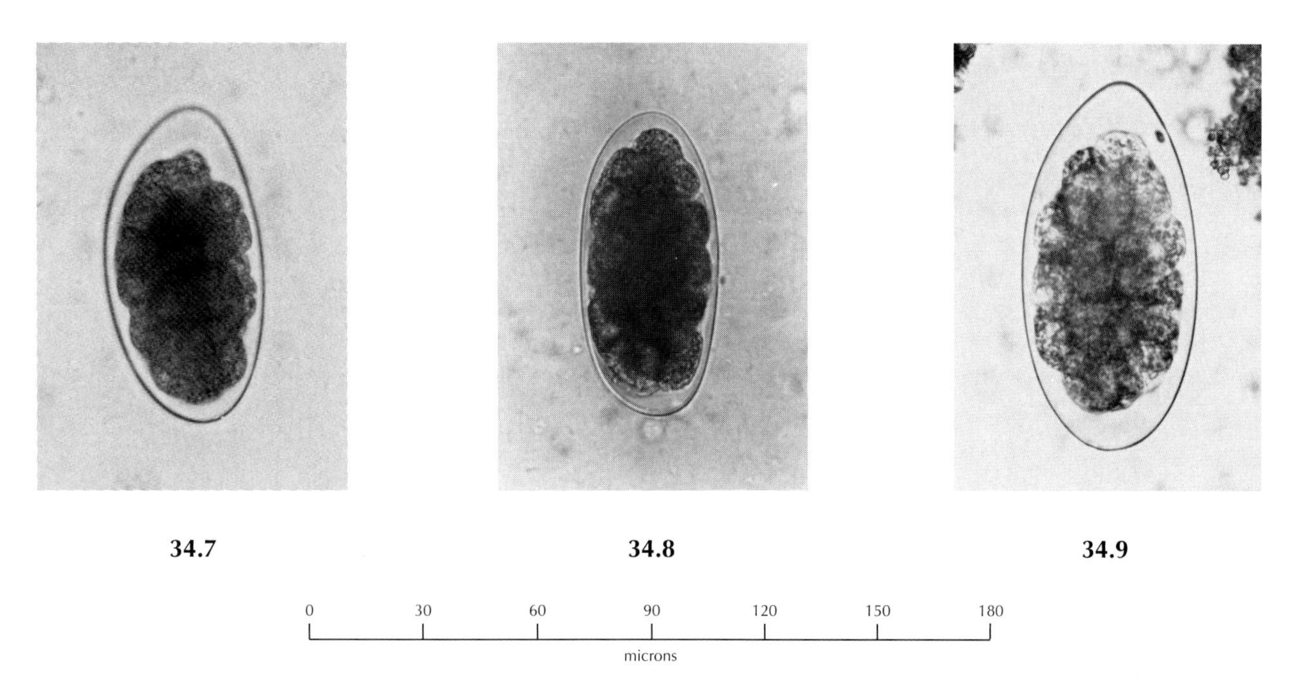

Plate 34. Three nematode eggs less commonly seen, 1–3 *Capillaria philippinensis,* 4–6 *Trichuris vulpis,* and 7–9 *Trichostrongylus orientalis.* 1: MIF; 2 and 3: formalin; 4–7: MIF; 8: PIF; 9: MIF. The eggs of *C. philippinensis* are relatively small, 36 to 45 × 19 to 22 μm, with flat sides and small flattened polar plugs (1, 3). The parasite inhabits the human intestine and causes serious disease in infected individuals. The complete life cycle is unknown. The eggs of *T. vulpis* are similar to those of *T. trichiura* but larger, 70 to 80 × 25 to 30 μm. Eggs may have slightly flattened sides (4, 6) or may be more football shaped (5). The embryo is undeveloped. *T. vulpis* is a parasite of the large intestines of canines. Eggs require 2 weeks or more incubation outside of the host to develop to the infective stage. Man becomes an accidental definitive host following ingestion of eggs containing infective juveniles. The eggs of *T. orientalis* are ellipsoidal, usually tapering toward one end, and average 85 × 35 μm. The shell is thin and the embryo is usually near the morula stage of development (7–9). The parasite occurs in man in agricultural areas of the Orient and *T. orientalis* and/or related species extend westward to Indonesia, India, Arabia, and Egypt.

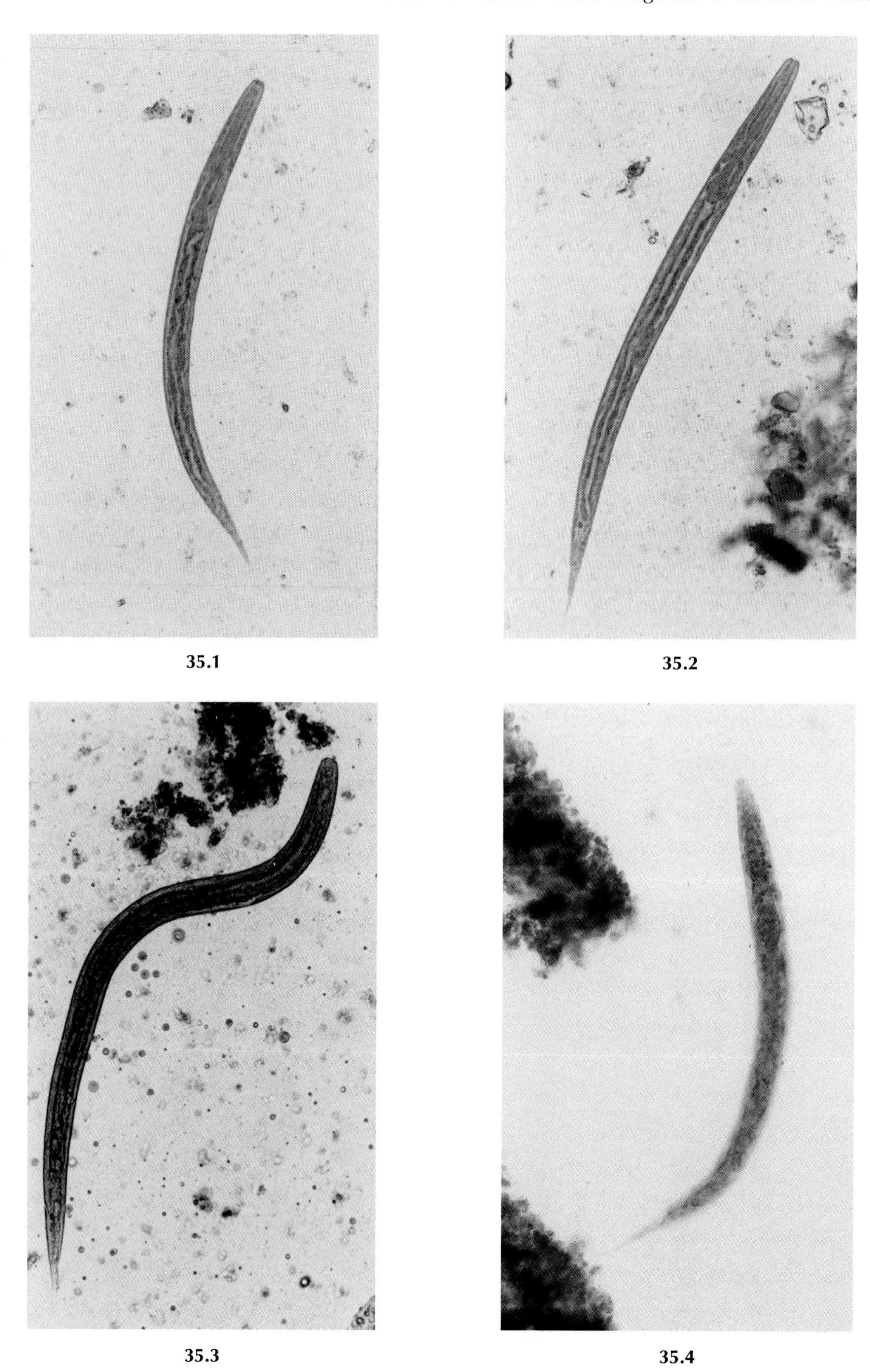

35.1

35.2

35.3

35.4

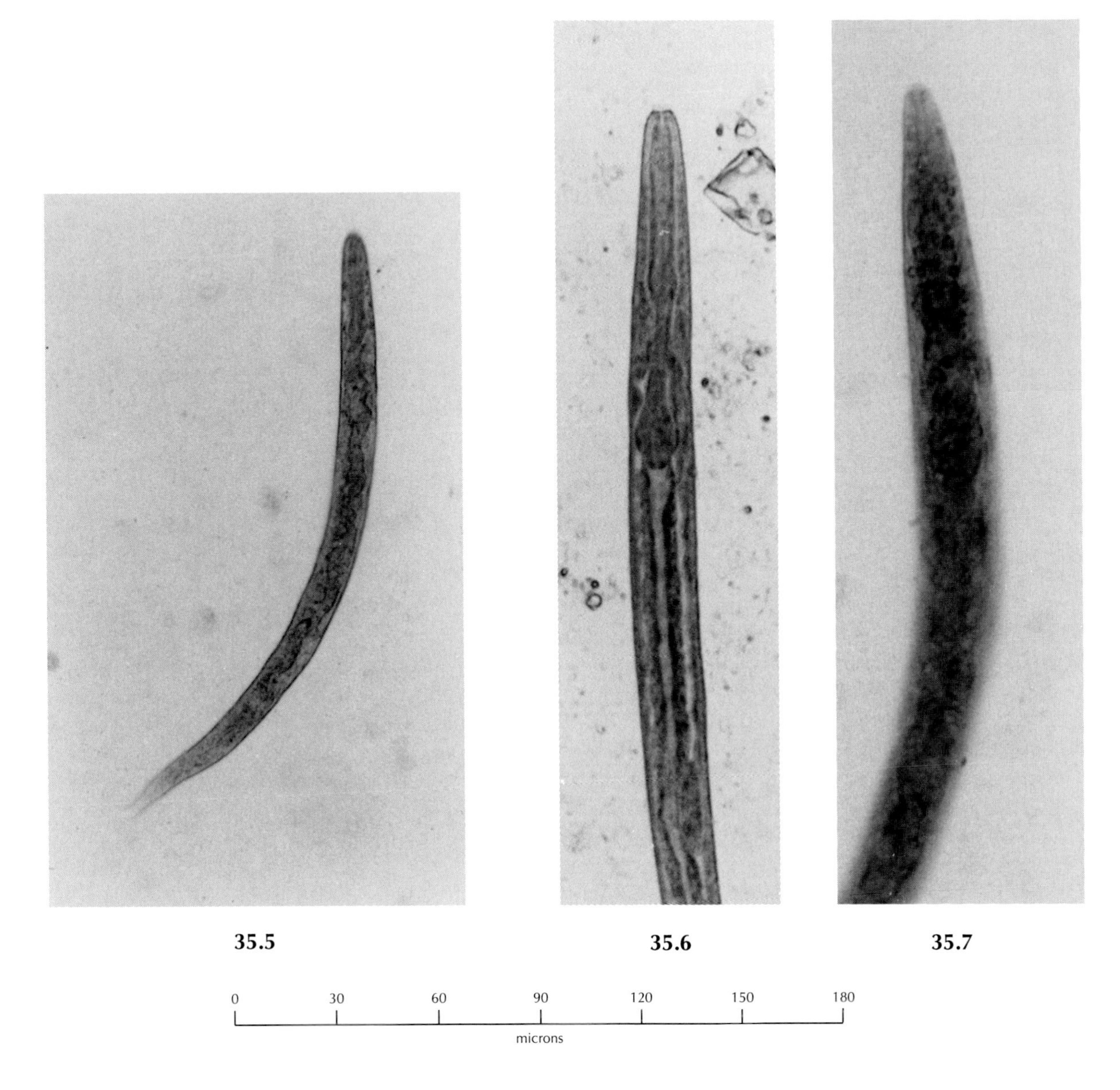

Plate 35. Juvenile nematodes, *Strongyloides stercoralis* 1, 2, 3, and 6; and Hookworm, 4, 5, and 7. 1–7: MIF (photomicrographs 1 through 5 are to scale). The short buccal cavity, 3 to 4 μm long, and relatively large genital primordium are the identifying characteristic of the first-stage (rhabditoid) juvenile of *S. stercoralis* (1, 2, 6). Other features are distinguishable in rhabditoid juveniles including a usually clearly defined body cavity, esophagus, esophageal bulb, nerve ring, intestine, and a distinct anal opening (see Plate 12). A juvenile that has developed to the second stage, which may occur in 12 to 24 hours (3). In the second-stage juvenile (3), except for the short buccal cavity, diagnostic features are not usually definable. Cultivation in fecal preparations allows development to the third stage in which other identifying features become discernible. The buccal cavity of hookworm juveniles is much longer, about 8 μm (4, 5, 7), and the genital primordium is small but internal structures are not clearly seen because of the highly refractile, granular-appearing structures that tend to hide the body cavity.

Chapter 9

The Protozoa

As stated in the introduction to the parasites, four major groups of protozoa are found in the gastrointestinal tract of man: amoebae, flagellates, ciliates, and coccidia. Since 1967, *Blastocystis hominis* has also been classified as a protozoan (Zierdt et al., 1967) but it has not been associated with any of the other taxonomic groups of protozoa. Most of the species are cosmopolitan and are found wherever fecal specimens are examined for parasites. Only a few members of these groups cause physical illness. Among the amoebae, *Entamoeba histolytica* is the only species considered to be pathogenic, but anecdotal experience suggests that *Dientamoeba fragilis* may also cause illness under certain circumstances. *Giardia lamblia* is the only flagellate of the intestinal tract that causes disease and *Balantidium coli* is the only ciliate. There is still controversy concerning the coccidia of the intestinal tract of man, but most workers recognize three genera: *Isospora, Sarcocystis,* and *Cryptosporidium* and members of each genus can cause disease.

STRUCTURE OF THE INTESTINAL PROTOZOA

The protozoan cell does not differ in any essential way from cells of multicellular animals except that they are usually independent and may possess structures not seen commonly in cells of multicellular animals. They have a nucleus, a cell membrane encompassing the cytoplasm, some organelles within the body, and others extending from the body, the latter usually associated with food gathering or motility. Some multicellular animals have cells with organelles related to motility, e.g., pseudopodia (white blood cells) and cilia (cells lining the bronchi of the respiratory tract).

Amoebae that may inhabit the intestinal tract of man are presented in Chapter 7, Table 3. These species are not considered to be natural parasites of other animals, so man is the primary reservoir for the intestinal amoebae. They are the simplest of the protozoa inhabiting the intestinal tract. There are differences in their structure that allow specific identification but in general they are made up of the same components.

The three major groups of protozoa, other than the amoebae, that inhabit the intestinal tract of man are presented in Chapter 7, Table 3. The Mastigophora (flagellates) are *Giardia lamblia*, trophozoites and cysts; *Chilomastix mesnili*, trophozoites and cysts; and *Trichomonas hominis* trophozoites only. Several other species of flagellates are rarely found in fecal specimens including *Enteromonas hominis* and *Retortamonas intestinalis* both of which have trophozoite and cyst stages.

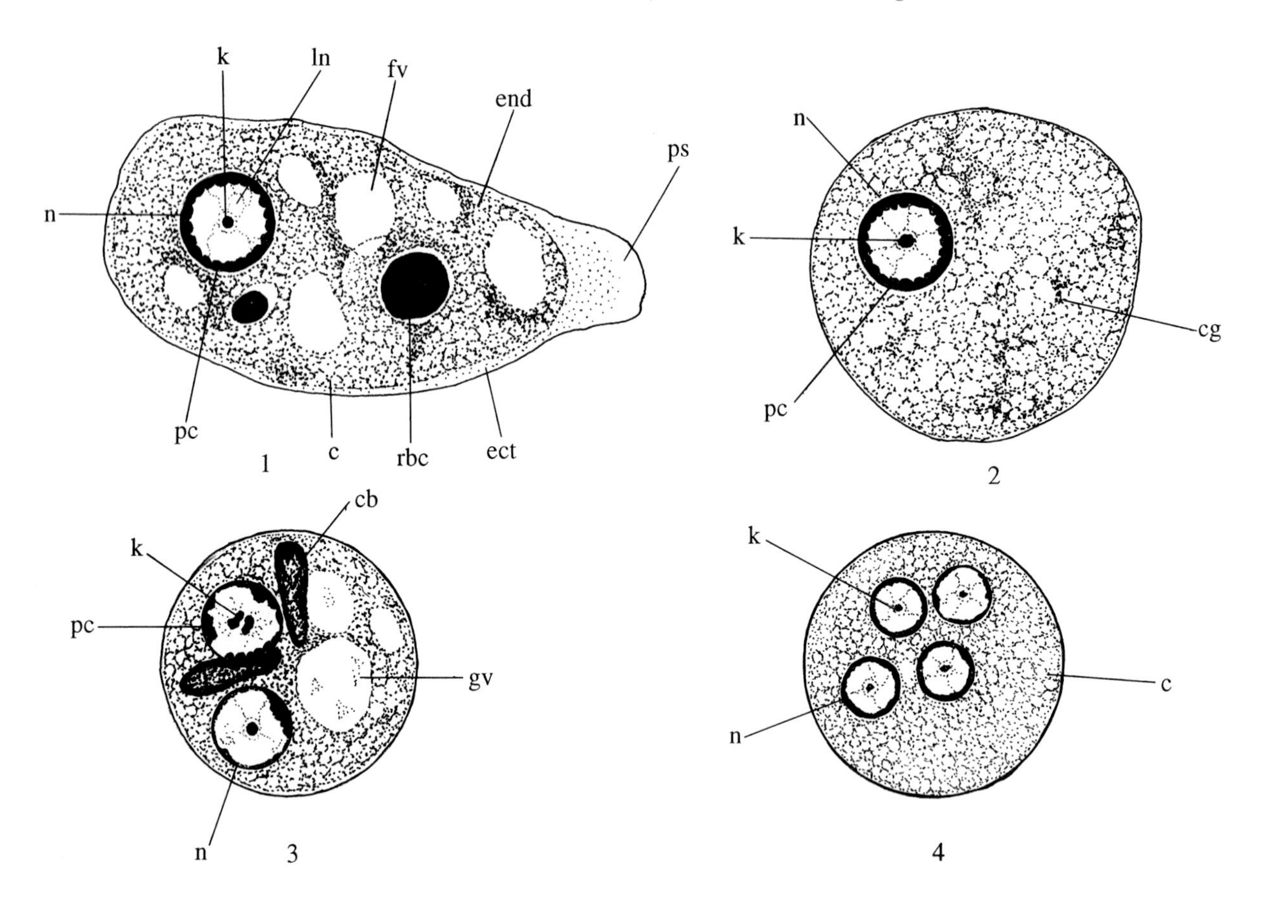

Plate 36. Schematic drawings of amoebae showing structures used in identification, represented by *Entamoeba histolytica.* 1: Trophozoite; 2: precyst; 3: immature cyst; 4: mature cyst. cb: chromatoid bar (body); c: cytoplasm; cg: cytoplasmic granules; ect: ectoplasm; end: endoplasm; fv: food vacuole; gv: glycogen vacuole; k: karyosome; ln: linin network; n: nucleus; pc: peripheral chromatin; ps: pseudopodium; rbc: red blood cell.

Balantidium coli is the only ciliate and the largest of the intestinal protozoa known to infect man. This ciliate can cause severe intestinal ulceration and attendant symptoms.

Blastocystis hominis is another organism seen frequently in fecal specimens. The organism was generally accepted as a nonpathogenic yeast by most investigators; however, in 1967 Zierdt et al. reclassified it as a protozoan. The organism is cosmopolitan and has been reported as the cause of diarrhea in immunosuppressed patients.

Certain of the morphological features and organelles of protozoa are used in determining their taxonomic position. Some of these structural characteristics are especially useful in making an identification of protozoa found in fecal preparations. Diagrams of members of each of the major groups have been selected to show and name these various structures. The protozoan must first be placed in a group, e.g., the amoebae or the flagellates, and then be identified.

The conditions in which protozoa live are dynamic and they are continually changing. Protozoans grow, divide, and encyst so at any particular time some of the structures shown in the diagrams may or may not be present.

STRUCTURE OF THE INTESTINAL AMOEBAE

Entamoeba histolytica was selected to represent the intestinal amoebae (Plate 36). Four stages are shown: the active vegetative form, the trophozoite; the precyst, a stage in which inclusions and vacuoles are absent; the immature cyst, in which chromatoid bars and glycogen vacuoles may be present; and the mature cyst, in which there are four nuclei and the inclusions are no longer present. These different stages are most commonly seen in *Entamoeba histolytica* and *Entamoeba coli* (see Plate 47).

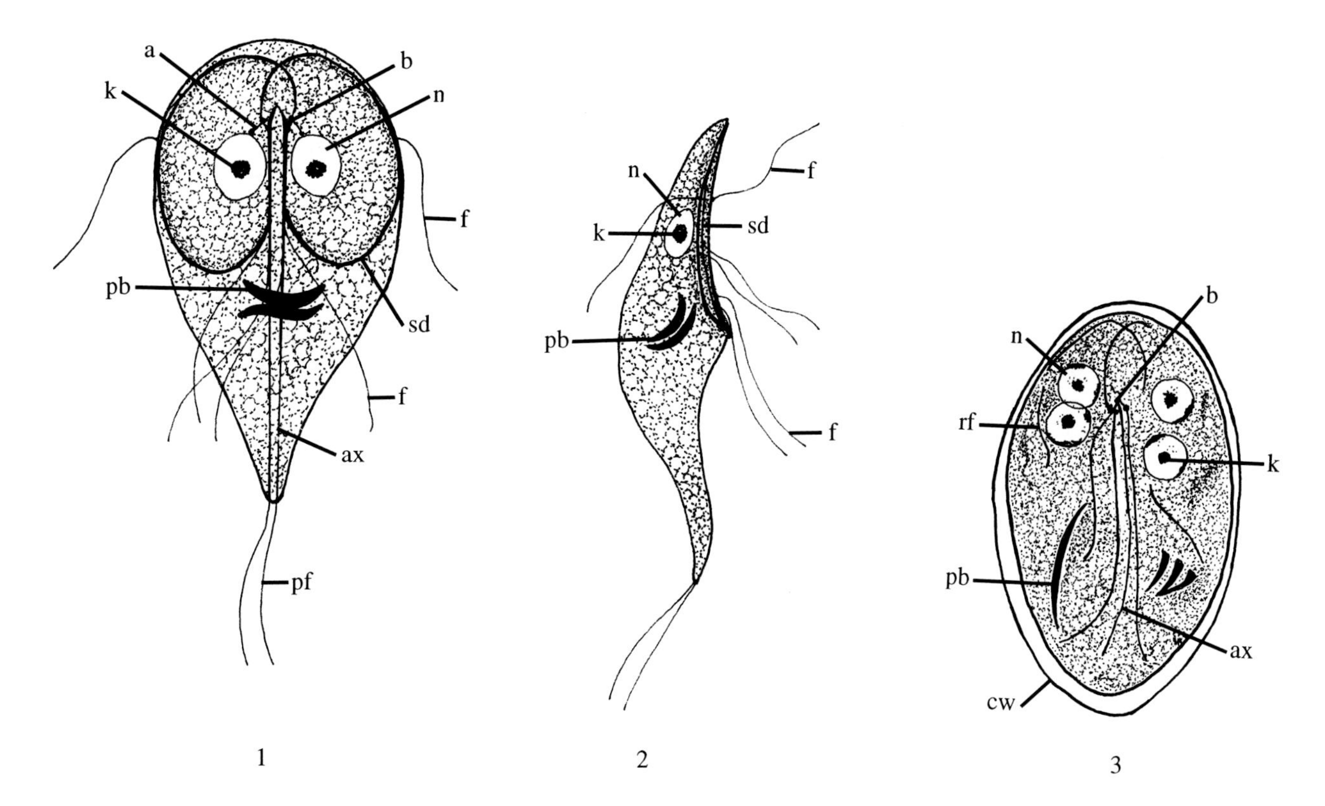

Plate 37. Schematic drawing of the flagellate *Giardia lamblia* showing structures used in identification. 1: Trophozoite, dorsal view; 2: trophozoite, lateral view; 3: cyst. a: axoneme; ax: axostyle; b: blepharoplast; cw: cyst wall; f: flagellum; k: karyosome; n: nucleus; pb: parabasal body; pf: posterior flagellum; rf: retracted flagellum; sd: sucker disk (ventral).

The cytoplasm has two components, the outer tougher and more hyaline ectoplasm that forms the outer layer of amoebae (and other protozoa) and the inner, more fluid component, the endoplasm, that is granular and contains many of the organelles and any inclusions. In Plate 36:1, a clear pseudopodium of ectoplasm stretches out and, when actively motile, the amoeba flows into it. For food gathering, the pseudopodium simply surrounds its prey and draws it inside where it is surrounded by the cytoplasm and transferred to a vacuole such as the one surrounding the red blood cell in the drawing. The ingested material is digested while in the vacuole (see the smaller included body in a vacuole, Plate 36:1) and the wastes are extruded in much the same manner as the prey was included only in reverse.

The nucleus of each species is somewhat different. In *Entamoeba histolytica*, a relatively even, granular layer of chromatin lines the inner surface of the nuclear membrane. A similar chromatin layer is also present in *E. hartmanni*, *E. polecki*, and *E. coli* but is not present in other species. Each species also has a karyosome but the size and structure differs in some (see Plates 42–50 for details on the various species). Delicate strands of material (linin network) can often be seen between the peripheral chromatin and karyosome in *E. histolytica.* These structures may be an artifact related to fixing and staining the amoebae but, since they appear only in certain species, they are sometimes helpful in identification.

When an amoeba is ready to encyst, it rounds up and loses its inclusions (Plate 36:2). A cyst wall is produced and the internal structure begins to change. The nucleus divides, the chromatin appears irregular, the karyosome may appear broken up, and chromatoid bars and glycogen vacuoles may appear in the the cytoplasm (Plate 36:3). Nuclear division in the cyst of *Entamoeba histolytica* produces four nuclei. When the cyst becomes mature, the inclusions and vacuoles disappear and the cytoplasmic granules usually form a pattern of small circles giving the cytoplasm the appearance of a lace-like network (Plate 36:4).

STRUCTURE OF THE INTESTINAL FLAGELLATES

Because each major genus of flagellate that infects man has structures differing from those seen in another genus, all three species of the commonly seen

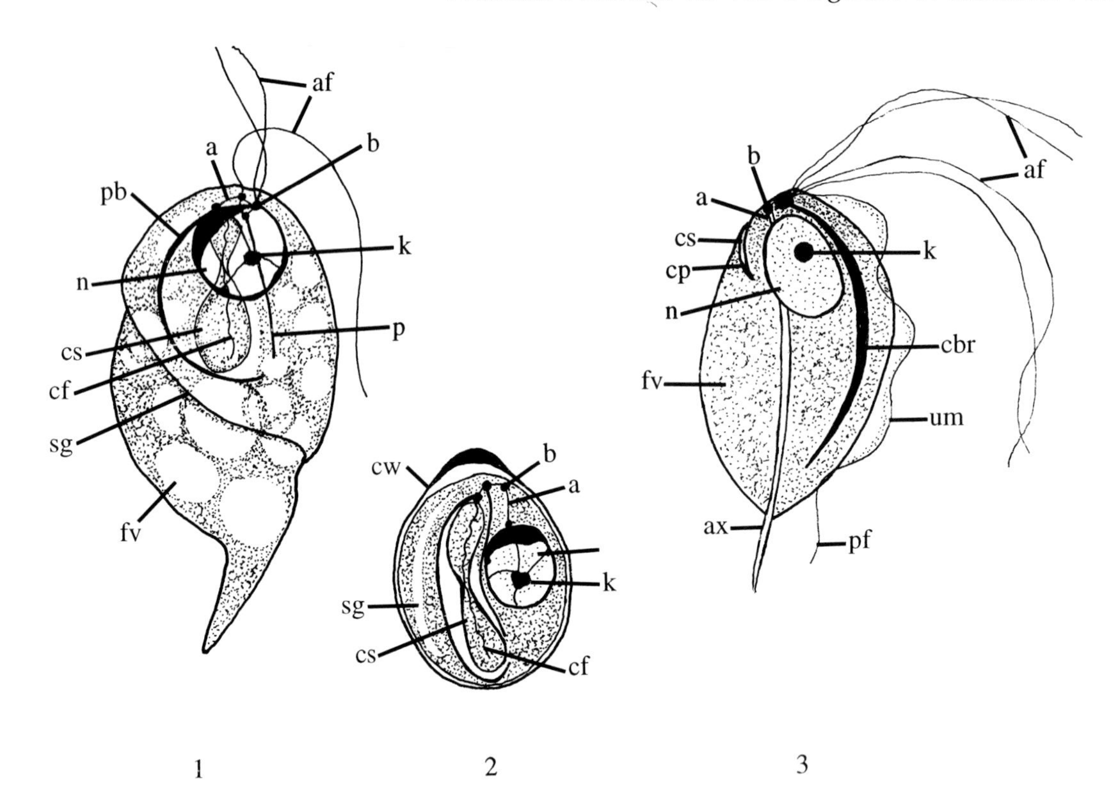

Plate 38. Schematic drawings of the flagellates *Chilomastix mesnili* and *Trichomonas hominis* showing structures used in identification. 1: *C. mesnili*, trophozoite; 2: *C. mesnili*, cyst; 3: *T. hominis*, trophozoite, no cyst known. a: axoneme; af: anterior flagella; ax: axostyle; b: blepharoplast; cbr: chromatoid basal rod (costa); cf: cytostomal flagellum; cp: cytopharynx; cs: cytostome (mouth cavity); cw: cyst wall; fv: food vacuole; k: karyosome; n: nucleus; p: parastyle; pb: parabasal body; pf: posterior flagellum; sg: spiral groove; um: undulating membrane. (Redrawn after Wenyon, CM, p. 121 (after Kofoid and Swezy, 1920), *Protozoology, A Manual for Medical Men, Veterinarians and Zoologists*. New York: Hafner Publishing Co., 1965.)

flagellates are selected, *Giardia lamblia* (Plate 37), and *Chilomastix mesnili* and *Trichomonas hominis* (Plate 38), in order to illustrate the various structures.

The morphology of intestinal flagellates is very complicated. Like the amoebae, they often have a nucleus with a karyosome and chromatin (in some species), and food is digested in food vacuoles. In most flagellates, however, the food is brought into the body through a peristome that functions as a mouth opening. From the peristome, food enters into the cytostome (mouth cavity) and moves into the cytopharynx (esophagus). At the base of the cytopharynx vacuoles form and food is transferred to the vacuoles which carry it through the body of the organism.

Their means of locomotion are flagella, and the shape of the body affects the movement pattern through the medium in which the flagellate resides. *Chilomastix* swims in a spiral motion, *Trichomonas* moves in circles, and *Giardia* moves with a rocking motion in a more or less straight line. When looking at a fresh specimen in which there are active trophozoites of a species of flagellate, the type of movement helps in making an identification.

The presence of a flagellum usually places a protozoan in this group; however, some protozoa that appear without a flagellum may also be classified as a flagellate based on other morphological features (e.g., *Dientamoeba fragilis*). Each flagellum arises from a minute granule called a blepharoplast. The flagellum begins as an axoneme (rhizoplast, a fibril) which passes through the cytoplasm and extends beyond the surface of the body carrying with it a thin sheath of cytoplasm.

Axonemes (rhizoplasts, fibrils) may have other functions. The primary blepharoplast may be connected to the nuclear membrane by a rhizoplast. In *Chilomastix*, two fibrils passing along the margins of the cytostome hold it rigid (Plate 38). One margin is formed by a more coarse fibril, the parabasal body,

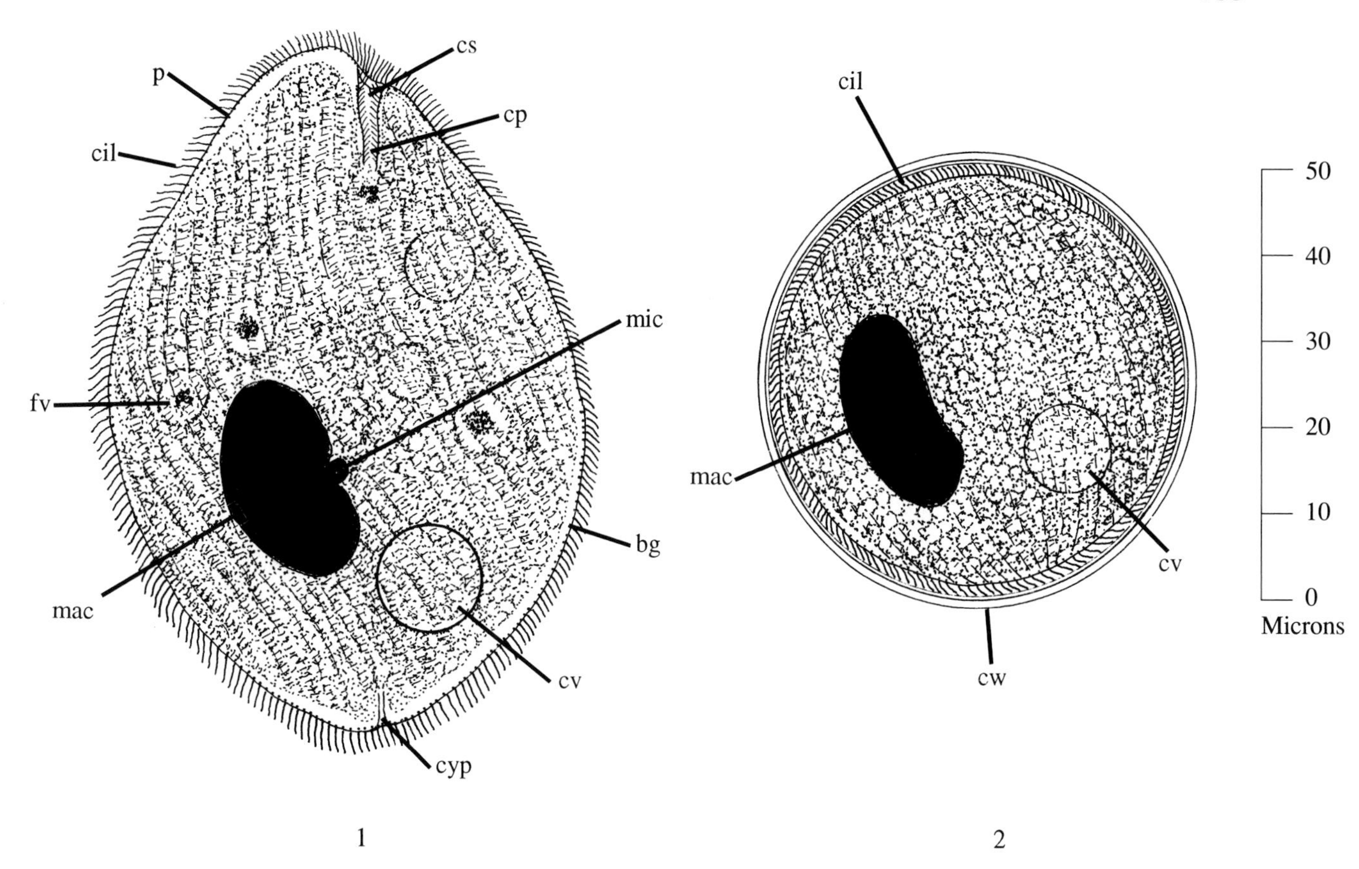

Plate 39. Schematic drawings of *Balantidium coli* showing structures used in identification. 1: Trophozoite; 2: cyst. bg: basal granule; cil: cilia; cp: cytopharynx; cv: contractile vacuole; cw: cyst wall; cs: cytostome; cyp: cytopyge; fv: food vacuole; mac: macronucleus; mic: micronucleus; p: pellicle.

and the other by a more delicate fibril, the parastyle. In *Trichomonas*, an axoneme passes over the body surface and raises the periplast to form an undulating membrane and a rigid fibril runs along its base. The undulating membrane also functions in locomotion. The more rigid fibril that supports the cytostome in *Chilomastix*, the basal fibril of the undulating membrane of *Trichomonas*, and two structures of unknown function that occur in the posterior region of *Giardia* are called parabasals (parabasal bodies). *Trichomonas* has a rod-like structure, the axostyle, that tends to hold the body more rigid. The organelle apparently originates in the area of the anterior blepharoplasts, then passes through the body to the posterior end where it projects as a pointed rod (Plate 38).

Chilomastix and *Giardia* form cysts that are their means of protection during the period of transmission, i.e., after being evacuated from one host and while waiting to be ingested by another host. Some structures persist in the cyst stage and can be recognized (Plates 37, 38). The cyst forms are characteristic of the species. *Trichomonas hominis* has no cyst form and their means of transmission from one host to another is unclear.

Unlike the other intestinal flagellates of man, the trophozoite of *Giardia* has two nuclei and is bilaterally symmetrical with a thin axostyle dividing the two sides (Plate 37, Figure 1). A large sucker disk is located on the ventral side of the trophozoites by means of which they attach to the microvilli of the small intestine. The sucker disk is more clearly seen in a lateral view (Plate 37:2). There are four pairs of flagella arising at opposite points of the body.

The flagella are withdrawn during encystation. After encystation, the nuclei divide once producing a cyst with four nuclei. The blepharoplasts and parabasal bodies also divide (Plate 37:3).

STRUCTURE OF THE INTESTINAL CILIATES

Balantidium coli, the only ciliate, is the largest of the intestinal protozoa of man. It inhabits the large in-

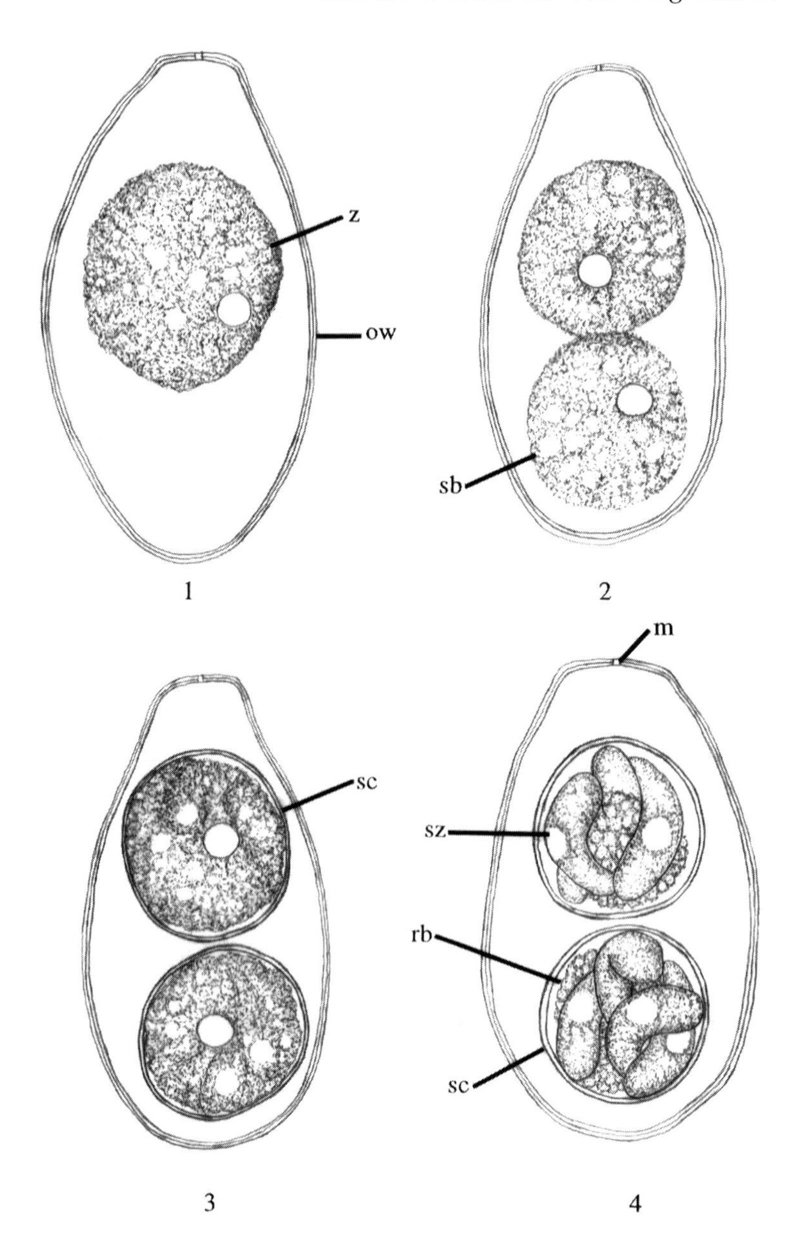

Plate 40. Schematic drawings of *Isospora belli* showing developmental stages of the oocyst and identifying structures. 1: Oocyst with unsporulated zygote (sporont); 2: oocyst with two sporoblasts; 3: oocyst with two sporocysts; 4: oocyst with two sporocysts each with four sporozoites. m: micropyle; ow: oocyst wall; rb: residual body; sb: sporoblast; sc: sporocyst; sz: sporozoite; z: zygote (sporont).

testine and may be found in either the trophozoite or less frequently in the cyst stage in feces. The trophozoite (Plate 39:1) usually measures from 50 to 80 μm in length and is covered with cilia arranged in longitudinal rows in grooves between ectoplasmic ridges. Each cilium arises from a basal granule (similar to a blepharoplast).

At the anterior end is the cytostome (mouth cavity) lined with cilia that spiral into the cytostome which opens into a cytopharynx (esophagus). As in some other flagellates, the food vacuoles are formed at the base of the cytopharynx. Often, food vacuoles are numerous and contain highly refractile globules. At times, some may contain red blood cells.

A kidney bean-shaped or sausage-shaped macronucleus usually lies near the middle of the body and a micronucleus lies close to it, often in a slight groove or pouch in the macronucleus.

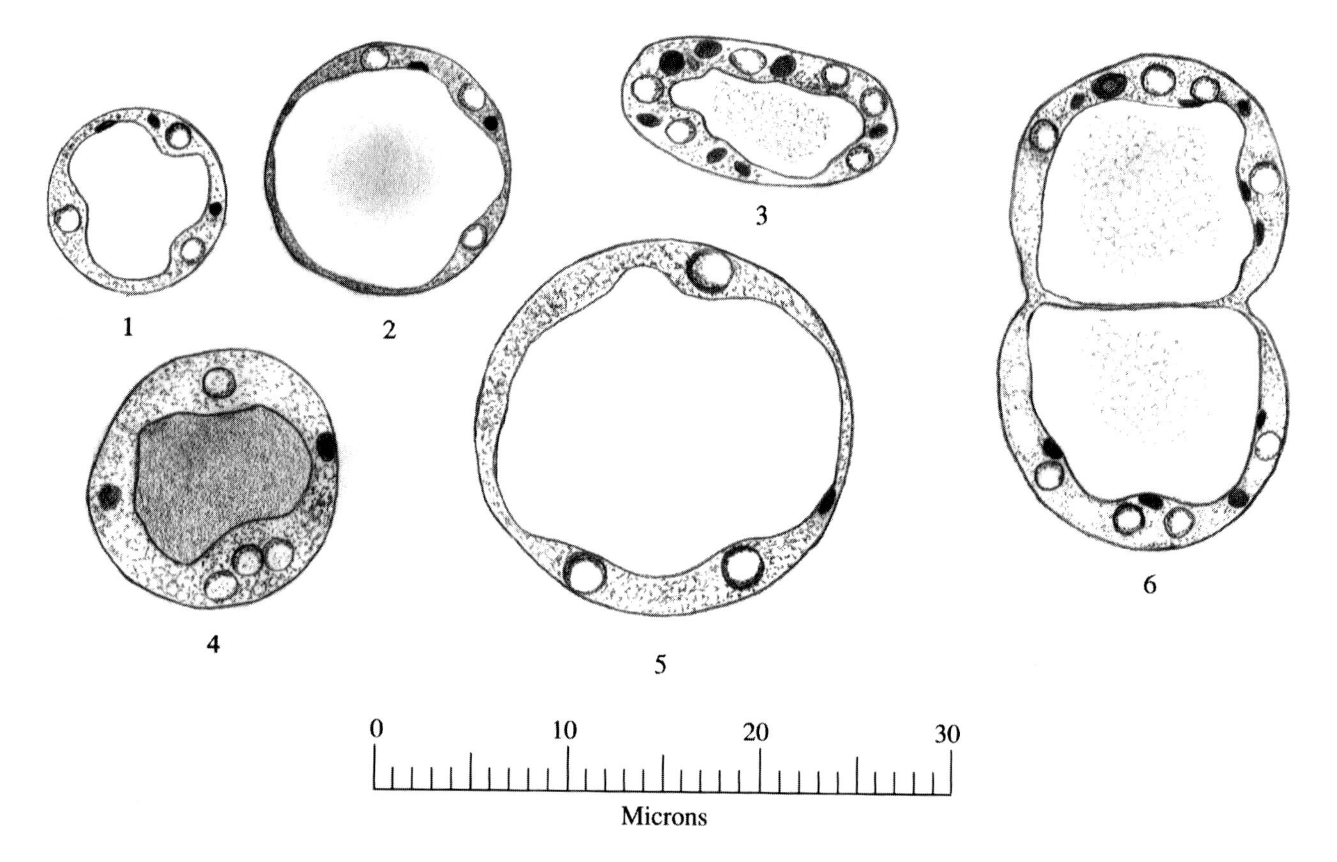

Plate 41. Diagrams of *Blastocystis hominis*. 1, 2, 5: Common cyst-like forms of different sizes; 3: early developing cyst-like stage; 4: cyst-like stage with dense center; 6: dividing stage. The cyst-like forms found in fecal specimens vary greatly in size from about 6 μm to as much as 40 μm. They divide by binary fission and may produce the amoeboid form (not shown).

The excretory system consists of two contractile vacuoles that carry liquid wastes to the body surface for discharge. The contractile vacuoles fill, move to the surface, and empty through the ectoplasm. Food vacuoles carry solid wastes to the posterior end of the body where wastes are passed out through the cytopyge that functions as an anus.

Reproduction takes place by transverse fission of the trophozoite, which occurs after the division of the two nuclei and other specific structures.

Some individual organisms become nearly spherical and encyst (Plate 39, Figure 2). They secrete a cyst wall which consists of two distinct layers. The cyst is purely protective since no multiplication takes place within it. The cysts usually measure from 45 to 65 μm. The cilia can be seen in the early cyst and the organism may move around inside the cyst wall. In older cysts, cilia may disappear. The macronucleus and sometimes the micronucleus, along with a contractile vacuole may be visible (see also Plate 69).

STRUCTURE OF THE INTESTINAL COCCIDIA

The three genera of coccidia that infect man, *Isospora, Sarcocystis*, and *Cryptosporidium*, are quite different in their life cycle and in their structure (refer to Plate 6). All three species form oocysts. In the definitive host, the zygote produced by the union of the male and female gametes rounds up and secretes a resistant cyst wall consisting of outer and inner portions, the oocyst. Within the cyst, the zygote may develop into a sporoblast and then directly into four sporozoites (*Cryptosporidium*), or it may divide to form two sporoblasts (*Isospora* and *Sarcocystis*) or four sporoblasts (*Eimeria*). A wall is formed around each sporoblast to produce a sporocyst. Within each sporocyst, the sporozoites are formed. In both *Isospora* and *Sarcocystis* of man, there are two sporocysts each with four sporozoites.

In Plate 40, the general structure of coccidia is represented by *Isospora belli*. A transparent-walled oocyst is usually passed in the feces. The wall is composed of a thicker outer layer and a very thin inner membrane. The oocyst of *I. belli* measures from about 25 to 35 × 12 to 16 µm and is usually more narrow at one end. There is a small plug at the narrower end called a micropyle which is the point where the oocyst opens to release sporozoites.

Within the oocyst, the zygote appears as a spherical, cytoplasmic body which contains a globule of refractile material (Plates 6, 40, 70). As development takes place, the nucleus of the zygote divides to form a sporont which then divides to form two sporoblasts. Each sporoblast secretes a cyst wall to become a sporocyst that measure 10 to 14 × 8 to 12 µm. Within the sporocysts, four elongated sporozoites with a residual mass develop by an asexual process called sporogony (schizogony).

Oocysts of *I. belli* may be missed in routine fecal examination because the oocyst wall is so nearly transparent and not as refractive as in most other species. The oocyst wall and internal structures are acid-fast, so they may be more readily seen in fecal smears stained with acid-fast stains. If a coccidial infection is suspected but not found by direct examination or after concentration, an acid-fast stain should be done on the concentrate.

Two sporocysts, each containing four sporozoites, are formed in the oocysts of *Sarcocystis* but these are usually released from the oocyst in the intestinal lumen and are passed in the feces as sporocysts, often in pairs (see Plates 6, 71). Single sporocysts average about 15 × 9 µm.

The oocysts of *Cryptosporidium parvum* (Plates 6, 54, 72) do not produce sporocysts but four sporozoites are produced free in the oocyst. Oocysts are extremely small, measuring from 4 to 6 µm. They do not take iodine stain, as most yeast cells do, and they are acid-fast, whereas most yeast cells are not. It is often possible to distinguish the oocysts of cryptosporidia in wet mounts with the microscope set at high contrast using the 100× oil objective (Plate 72) especially in specimens fixed in MIF or PIF, but usually special stains or monoclonal antibody procedures are needed to confirm the diagnosis.

Organism of Unknown Taxonomic Position

Blastocystis hominis, now classified as a protozoan, does not fit into any of the established taxonomic groups of protozoa. (Zierdt et al., 1967; Zierdt, 1991). The classical form is readily recognized in human fecal preparations (see Plates 6, 41, 73, 84). It is highly refractile, usually has a large central area that appears similar to a vacuole which is surrounded by a thin layer of cytoplasm containing several light-staining nuclei and dark-staining volutin granules. The ameboid form also may be present in wet preparations but requires high contrast to differentiate it from other materials and it is rarely recognized in routine examinations of specimens.

THE INTESTINAL AMOEBAE

Seven species of amoebae are commonly found in fecal specimens from man. These include *Entamoeba histolytica* [large and small forms (races)], *E. coli*, *E. polecki*, *E. hartmanni*, *Endolimax nana*, *Iodamoeba butschlii*, and *Dientamoeba fragilis*. Some free-living species also have been found in feces but these are not considered to be intestinal parasites. Other genera are opportunistic and can cause meningoencephalitis, *Naegleria* and *Acanthamoeba*; members of the latter genus also invade the cornea of the eye causing serious eye disease.

The intestinal amoebae vary in size, as shown in Table 1. They are differentiated on the basis of structure and size. Trophozoites divide by binary fission. Most amoebae that form cysts go through nuclear division within the cyst and then divide again after excystation in a

Table 1. Size Ranges of the Intestinal Amoebae Found in Feces

	Usual size range in microns	
Species	**Trophozoite form**	**Cyst form**
Entamoeba histolytica (large race or form)	12–50	10–16
E. histolytica (small race or form)[a]	7–12	5–10
E. coli	15–50	15–30
E. polecki	10–25	8–18
E. hartmanni	5–12	5–10
Endolimax nana	6–12	5–10
Iodamoeba butschlii	6–20	5–15
Dientamoeba fragilis[b]	4–20	Unknown

[a] Most workers do not acknowledge the existence of a small form or race of *E. histolytica* but separate *E. histolytica* from *E. hartmanni* on the basis of size. Burrows (1959) described the small form as separate from *E. hartmanni* (see Plates 42–44 and 55–58).

[b] Considered to be a flagellate (Camp et al., 1974) but appearing in feces in the amoeboid form.

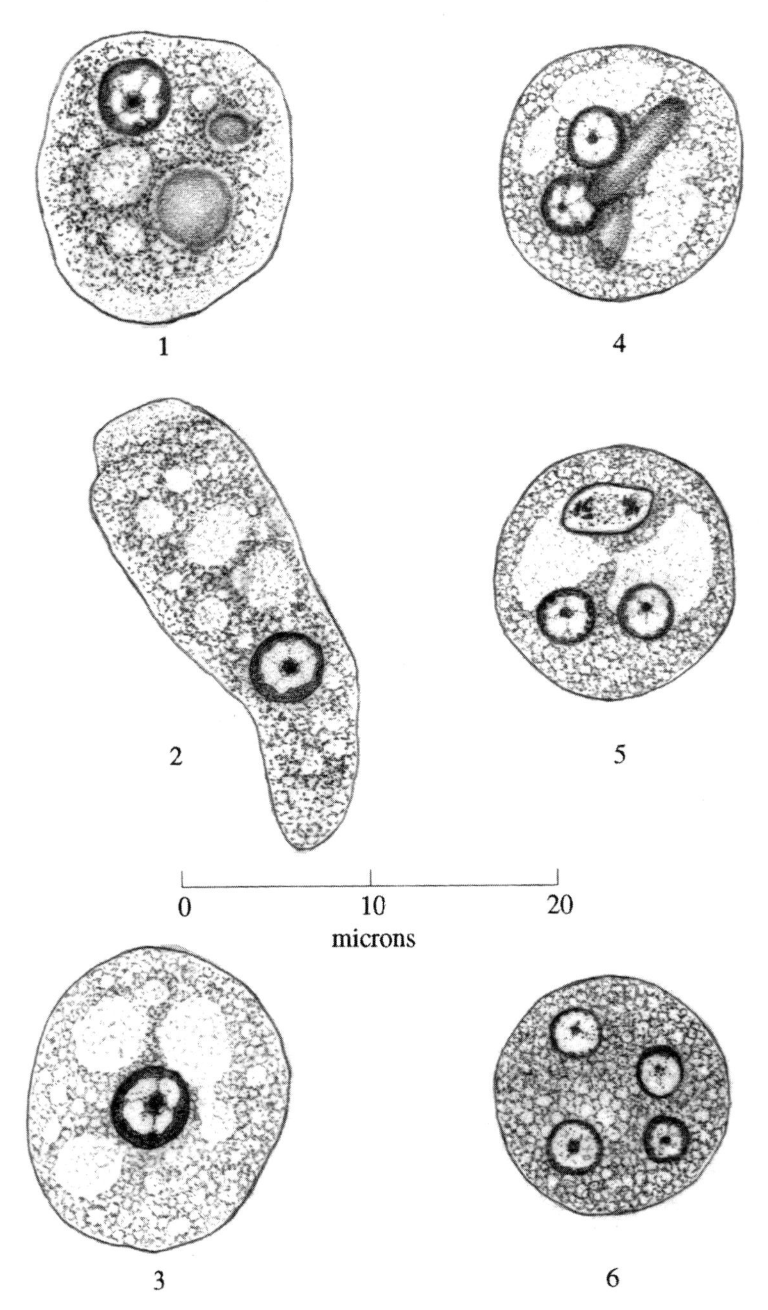

Plate 42. Diagrams of *Entamoeba histolytica* large form (race). 1, 2: Trophozoites; 3: precyst; 4, 5: immature cysts; 6: mature cyst. The active trophozoite (1, 2) has a delicate, finely granular cytoplasm usually with relatively few vacuoles, occasionally with ingested red blood cells (1). The nucleus has a light-density nucleoplasm, evenly beaded peripheral chromatin, and a distinct karyosome. There are often fine granules in the nucleoplasm extending from the karyosome to the peripheral chromatin. When moving, the extended pseudopodium is relatively free of granules (2, upper end). In preparation for encystment, the vacuoles are emptied and the trophozoite rounds up (3), then it forms a cyst wall. Once encysted, the cytoplasm goes through a number of changes, glycogen vacuoles may form, and the nucleus begins to divide (5). Chromatoid bars may be formed in the cytoplasm (4) but are reabsorbed by the time the mature cyst (6) is formed. The cytoplasmic granules in the mature cyst are usually arranged to form small circles and there is usually nothing else present (6). The size of the nucleus is usually larger than one fifth the cyst diameter. If the mature cyst is 15 μm in diameter, the nucleus is 3 μm or greater in diameter. The nuclei are relatively much smaller in mature cysts of *E. coli*. *E. histolytica* can usually be identified in the trophozoite or mature cyst stages, or when chromatoid bars with rounded ends are present in immature cysts.

new host. Cysts of *Iodamoeba butschlii* and *Entamoeba polecki* usually have only one nucleus.

Dientamoeba fragilis has been classified as a flagellate, although the flagellate form has not been found (Camp et al., 1974). Its ameboid trophozoites are destroyed in water, so it is doubtful that they can withstand passage through the stomach. There must be a stage, yet unknown, that is the infective form. Since it is found only in the ameboid form in the intestine of man, it is included here with the amoebae.

Each species of amoeba has some characteristics in common with some other species in its group, making identification somewhat difficult. It is a combination of certain characteristics that are found in each species that allows one to make an identification. It follows that the entire organism should be evaluated. One should observe the kind of cytoplasm. Are the granules coarse or fine; dispersed evenly, irregularly, or in a pattern? Are the vacuoles large, small, numerous, with or without inclusions, and what is included? What type of nucleus is present? Is the nuclear membrane distinct? What is the size of the nucleus in relation to the size of the organism? Is peripheral chromatin present; is it beaded; is it dispersed evenly or irregularly around the inner periphery of the nucleus? Is the karyosome

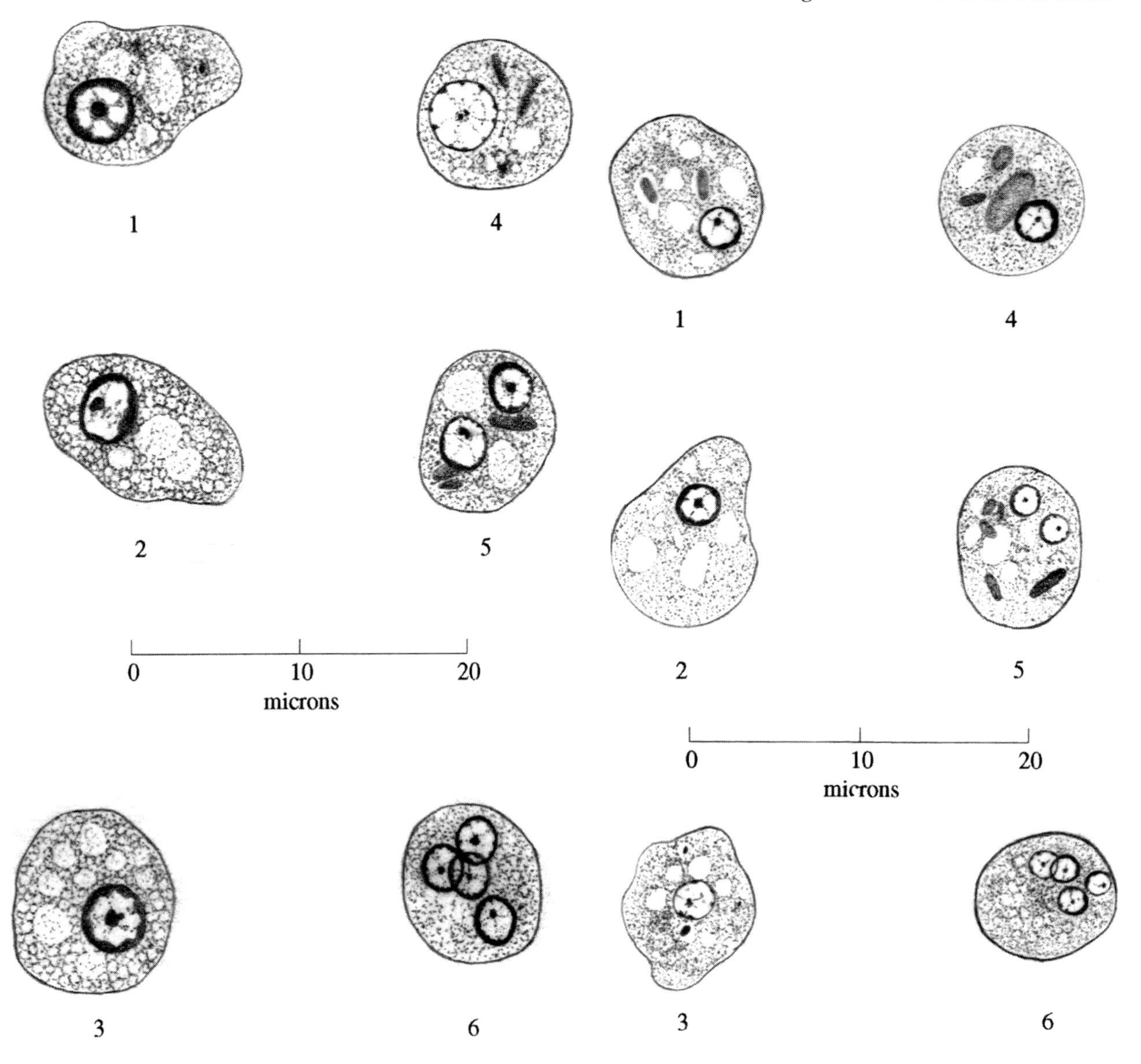

Plate 43. Diagrams of *Entamoeba histolytica*, small form (race). 1, 2: Trophozoites; 3: precyst; 4, 5: immature cysts; 6: mature cyst. Trophozoites of the small form of *E. histolytica* are usually less than 10 μm in diameter. Their nuclei are large in relation to the size of the amoeba, usually greater than one fourth, often as large as one third, the diameter of the amoeba (1, 2). There are relatively few vacuoles and only rarely do they have inclusions so that, except for the vacuoles, a rounded-up amoeba is similar to a precyst (3). In an immature cyst with a single nucleus, the diameter of the cyst is hardly greater than twice that of the nucleus (4). There may be small chromatoid bars or granules in the cytoplasm and small glycogen vacuoles (4, 5). The mature cyst of the small form is much like that of the large form except that the nuclei appear to fill a greater part of the amoeba (6).

Plate 44. Diagrams of *Entamoeba hartmanni*. 1–3: Trophozoites; 4, 5: immature cysts; 6: mature cyst. Trophozoites of *E. hartmanni* vary in size from 5 to 12 μm in diameter (usually under 10 μm). The cytoplasm is finely granular, often with many small vacuoles, occasionally with inclusions, usually bacteria (1). The nucleus is small in relation to the size of the trophozoite, usually one fifth or less than the diameter (2). It has evenly dispersed, granular, peripheral chromatin and a central karyosome. Once they encyst, chromatoid bars are usually formed which may be large (4) or small (5). The immature cyst may have a number of glycogen vacuoles (4, 5). When mature, the vacuoles disappear and the four small nuclei fill less than one third the area of the cyst (6).

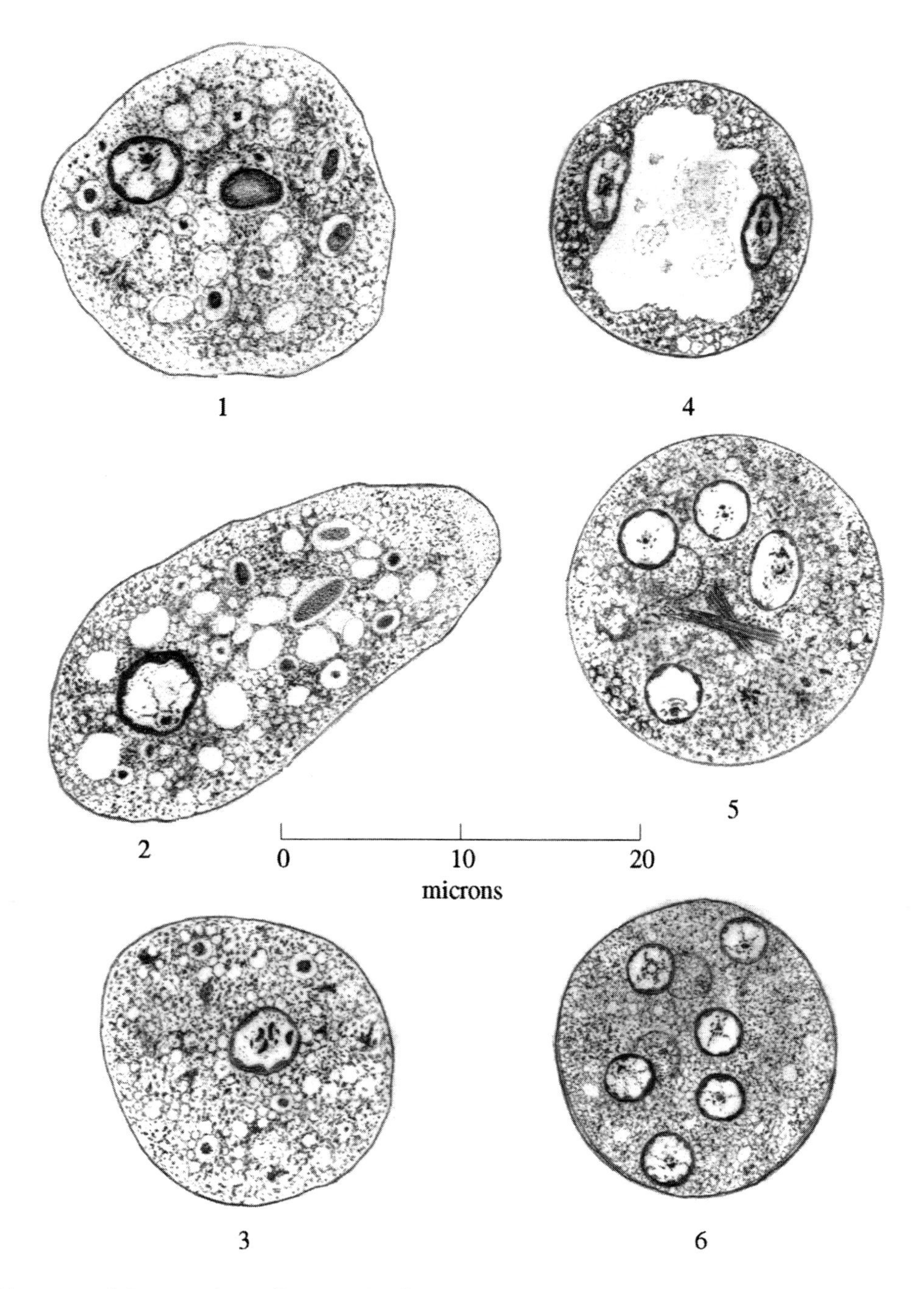

Plate 45. Diagrams of *Entamoeba coli.* 1, 2: Trophozoites; 3: precyst; 4, 5: immature cysts; 6: mature cyst. The trophozoites of *E. coli* range from 15 to 50 μm in diameter but are usually between 18 and 28 μm. The nuclei are relatively small, usually about one sixth the diameter of the trophozoite with irregular peripheral chromatin and an eccentric karyosome (1, 2). The karyosome is often seen in division in the trophozoites and immature cysts and appears to be made up of several particles (3). The cytoplasm of the trophozoite is dense, heavily granular, and appears cluttered with many small vacuoles, often with included bacteria; vacuoles are usually only large enough to accommodate the included material (1, 2). In the precyst, inclusions and vacuoles disappear, encystment begins, and the nucleus begins division (3). The binucleate cyst usually has a large glycogen vacuole that may appear to press the nuclei against the cyst wall (4). As the cyst matures, splinter-like chromatoid bodies occasionally form in the coarsely granular, irregular cytoplasm (5). Nuclear division continues until there are eight nuclei, producing a mature cyst having a cytoplasm with evenly dispersed coarse granules and without vacuoles (6). On occasion, further nuclear division will take place in individual cysts producing large cysts with up to 32 nuclei.

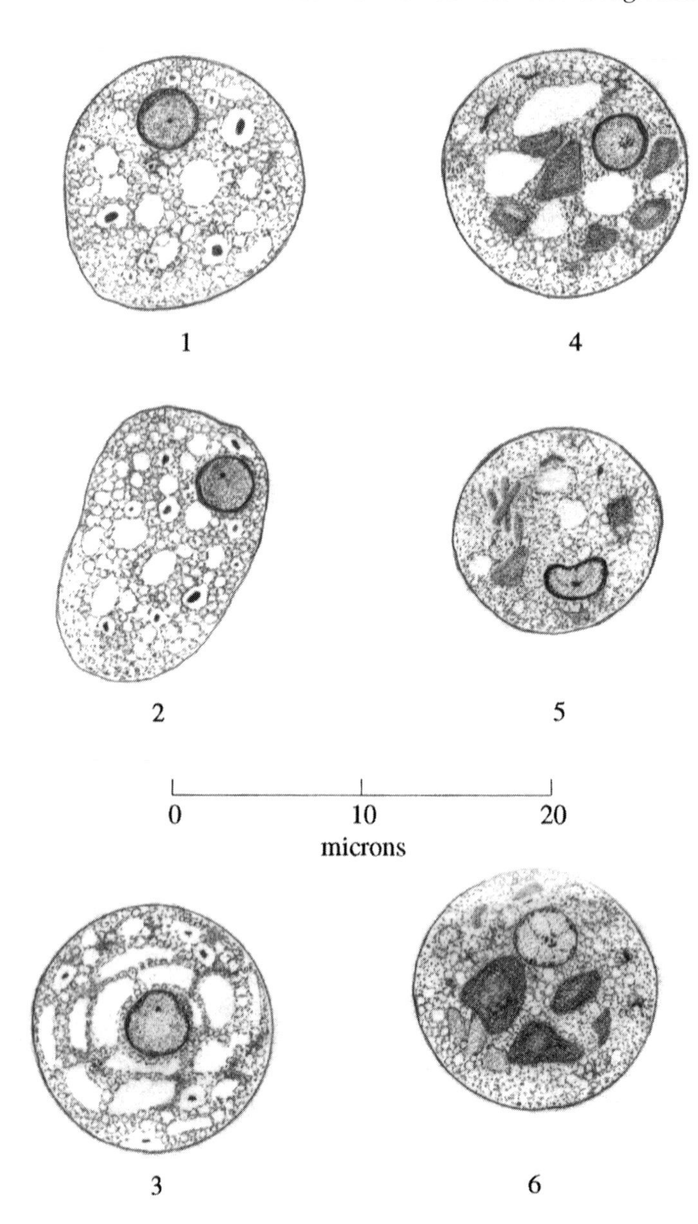

Plate 46. Diagrams of *Entamoeba polecki.* 1–3: Trophozoites; 4–6: cysts. The trophozoite of *E. polecki* usually has many small vacuoles in a dense cytoplasm much like that of *E. coli.* Often, a few of the vacuoles have inclusions, usually small bacteria (1, 2). Sometimes trophozoites have several large clear areas (vacuoles) and many small vacuoles giving them an unusual appearance (3). The nucleus has relatively regular, usually thin, peripheral chromatin and the karyosome is small and often indistinct (1–6). The nucleus is much like that of *E. histolytica* except that the nucleoplasm is very dense and the karyosome is very small. The nuclear membrane is often misshaped. The cyst of *E. polecki* may or may not have a less-dense cytoplasm, several large or small vacuoles, and several to many large dense masses or chromatoid-like bodies (4–6). The cytoplasm in diagram 5 is irregular with two dense masses and several small cigar-shaped ones that appear similar to chromatoid bars. The nuclei of the cysts (4–6) are typical, with relatively even peripheral chromatin, a pinpoint-like karyosome, and a dense nucleoplasm. Nuclei are often irregular in shape (5).

large, small, divided into parts, and is there a linin network of fine granules that radiate from the karyosome to the peripheral chromatin? Many characteristics should be observed before a specific identification is made. An identification based on only one or two features may not be accurate.

The sizes of the various amoebae (see Table 1), the presence or absence of certain structures, and the size of such structures when present are used to differentiate species. An example of the use of these criteria is demonstrated in Table 2, which compares the characteristics of *Entamoeba histolytica* and *Entamoeba coli*. It should be obvious that there are few hard and fast rules for identifying species. The approach that appears to result in the fewest errors is to identify an amoeba by the process of elimination. A "key" for the identification of intestinal amoebae in wet preparations, especially PIF- and MIF-fixed specimens, based on this approach, has been developed and is presented later in this Chapter. The Key can be used for amoebae fixed by other methods but works best for those seen in wet mounts.

The intestinal amoebae can be divided into two groups based on whether the nucleus has or does not have peripheral chromatin. The four species having a nucleus with peripheral chromatin — *E. histolytica, E. hartmanni, E. coli,* and *E. polecki* were presented in Plate 3, Chapter 7, and each is described separately below.

Most drawings of protozoa in textbooks were drawn from fecal films fixed in Schaudinn's and stained with hematoxylin. The drawings in Plates 42–50 are of amoebae seen in fecal films prepared from specimens fixed in MIF or PIF and stained with Polychrome IV, and may appear differently.

Entamoeba histolytica

Entamoeba histolytica, *Large Race*

There are two forms or races of *E. histolytica*. The trophozoites of the large race of *E. histolytica* (Plate 36, 42, 55) are from 12 to 30 µm in diameter, have finely granular cytoplasm, and usually few vacuoles, rarely with included erythrocytes, in various stages of digestion (Plate 42:1, 2). Pseudopodia, when present, are nearly free of granules (Plate 42:2). The size of the nucleus is about one fifth to one sixth the diameter of the rounded up trophozoite. The amoeba rounds up, evacuates all inclusions, and absorbs all the vacuoles and other structures when it is ready to encyst (Plate 42:3).

Cysts of *E. histolytica* (large race) may be seen in various stages of development (Plate 42:4–6). Mononucleate and binucleate cysts usually have at least one large glycogen vacuole and may have chromatoid bars (Plate 42:4, 5). After four nuclei are formed in the cyst, the chromatoid bars are reabsorbed, glycogen vacuoles disappear, and the cyst is mature (Plate 42:6). The cytoplasm of the mature cyst is finely granular with granules forming a pattern of fine circles. Only mature cysts, or immature cysts with chromatoid bars, are diagnostic (see also Plates 56, 74).

Entamoeba histolytica, *Small Race*

The trophozoites of the small race of *E. histoytica* (Plates 43, 57) vary from 7 to 12 µm in diameter and have very few vacuoles (Plate 43:1, 2). The cytoplasmic inclusions and vacuoles are absorbed in the precyst stage (Plate 43:3). The nucleus of the trophozoites of both the large and small race are similar in size so that in the small race it is one third to one half the diameter of the trophozoite. It usually can be differentiated from *E. hartmanni* on the basis of the size of the nucleus in relation to the size of the amoeba and the type of cytoplasm which has few large vacuoles, usually without inclusions. In 1959, Burrows published a paper showing the differences between the two species which should be consulted (see also Plates 57, 58).

The cysts are usually from 6 to 8 µm in diameter. The nucleus in the early cyst appears large in relation to the size of the parasite (Plate 43:4, 5). Some cysts have vacuoles and chromatoid bars as do cysts of *E. hartmanni* and the large race (Plate 43:4, 5) but these disappear in the mature cyst (Plate 43:6).

Entamoeba hartmanni

The trophozoites of *Entamoeba hartmanni* (Plates 44, 58, 76) range in size from 5 to 12 µm, usually 6 to 8 µm in diameter. The cytoplasm has irregular cytoplasmic granules; many small vacuoles, some with inclusions; and may appear cluttered (Plate 44:1, 2). The nucleus has relatively regular, peripheral chromatin and a near central karyosome. The nucleus is relatively small, about one fourth to one sixth the diameter of the trophozoite (Plate 44:1). The precyst is similar to that of the small race of *E. histolytica* but with a smaller nucleus.

Cysts of *E. hartmanni* range in size from 5 to 10 µm and often are not round. Immature cysts, like those of *E. histolytica*, may have glycogen vacuoles and one to several chromatoid bars (Plate 44:4–5). The mature cyst has four small nuclei usually occupying about one fourth to one third of the cyst (Plate 44:6) which helps in distinguishing it from the small race of *E. histolytica.*

Entamoeba coli

The cytoplasm of the trophozoites of *Entamoeba coli* (Plates 45, 59, 75) is coarsely granular with many vacuoles that vary in size. The vacuoles often have included bacteria or occasionally yeast cells, giving the trophozoite a cluttered appearance (Plate 45:1, 2). The nucleus has irregular peripheral chromatin and a karyosome that may be either small or appear to be made up of several parts as in division. The size of the trophozoite can vary from 15 µm to as much as 50 µm. The vacuoles are emptied and the cytoplasm loses its cluttered appearance in the precyst stage (Plate 45:3). In this stage, it is often indistinguishable from *E. histolytica.*

The cysts may be seen in various stages of development. It is common to see cysts with two nuclei and a large glycogen vacuole that appears to press the cytoplasm and nuclei against the cyst wall (Plate 45:4). As the cyst develops, splinter-like chromatoid bundles may form and the cytoplasm may take on a more lace-like appearance (Plate 45:5). The nuclei of mature cysts have a more regular peripheral chromatin and a small eccentric karyosome. The cytoplasmic granules are evenly dispersed but vary in size (Plate 45:6). Usually the mature cyst has eight nuclei, but nuclei may continue to divide. Cysts with 12, 16, or even 32 nuclei are somtimes seen.

Entamoeba polecki

The trophozoites of *Entamoeba polecki* (Plates 46, 61, 77) have some characteristics of *E. histolytica* and some of *E. coli.* There are other characteristics seen only in *E. polecki.*

The cytoplasm appears less cluttered than *E. coli* but more dense and often more vacuolated than *E. histolytica* (Plate 46:1, 2). The peripheral chromatin of the nucleus is usually thin, the karyosome pin-point-like, and nucleoplasm dense, sometimes obscuring the karyosome (Plate 46:1, 2). The nuclear membrane of the trophozoite as seen in wet preparations is often irregular because of pressure when the slide is prepared.

The cysts of *E. polecki* usually have a single nucleus with thin peripheral chromatin, dense-staining nucleoplasm, and a small indistinct karyosome (Plate 46:4–6). Early cysts may have a number of vacuoles in a dense, irregular cytoplasm. Usually, plate-like, irregular, dense masses form in the cytoplasm which stain differently than the chromatoid bars seen in *E. histolytica* (Plate 46:4–6).

The differences between *E. polecki* and *E. histolytica* were outlined by Burrows in 1959. He demonstrated that *E. polecki* can be distinguished from *Entamoeba histolytica* by

- The more variable nuclear shape in the trophozoites
- The usually more dense nucleoplasm and small size of the karyosome
- The predominant mononuclear cysts
- The cytoplasmic condensations and dense masses in cysts
- The diffuse type of vacuolation seen in most cysts

Although there are similarities between the species, when an entire organism with all its components is examined, discernible differences become more obvious. Giving consideration to only one component, such as the nucleus, does not always allow a clear definition of a species and can lead to an incorrect identification. *E. histolytica, E. coli,* and *E. polecki* are often confused when only the nucleus is considered.

Stages in the Development of *Entamoeba histolytica* and *E. coli.*

Some of the developmental stages of *Entamoeba histolytica* and *E. coli* are shown in Plate 47. The cytoplasm of the trophozoites differs greatly. That of *Entamoeba histolytica* appears clean, with few fairly large vacuoles most of which have no inclusions. Trophozoites in specimens that have remained unfixed for several hours may become highly vacuolated which suggests the importance of fixing fecal specimens as soon as possible after evacuation. On the other hand, the cytoplasm of trophozoites of *E. coli* is usually filled with vacuoles, some fairly large and others very small, many with inclusions.

The structure of the nucleus of *E. histolytica* is very consistent, unless it is in division. In the trophozoite, peripheral chromatin is usually beaded and evenly

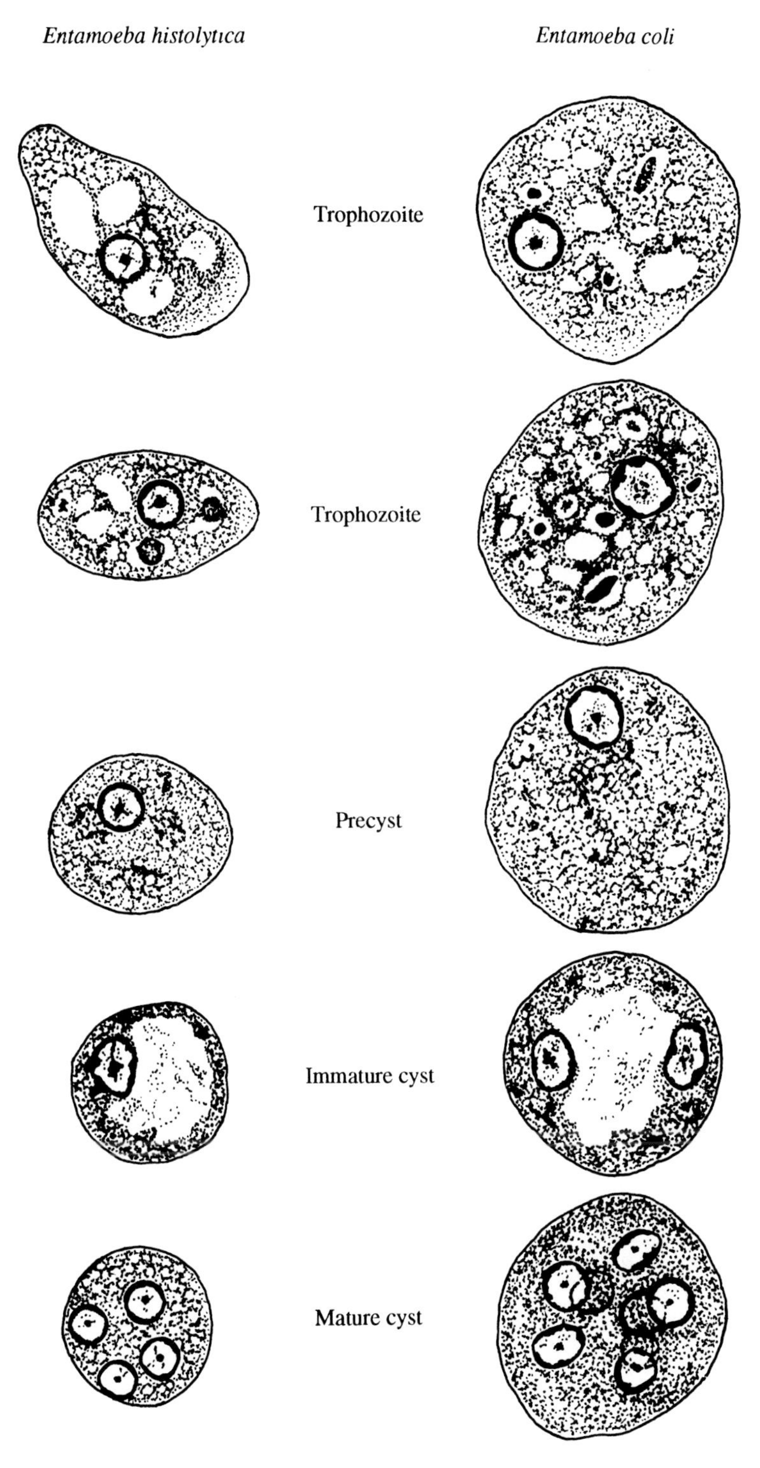

Plate 47. Comparison of the stages in the development of *Entamoeba histolytica* and *E. coli* in the intestinal tract of man. The diagrams (not to scale) show comparable stages of the two species. The structure is described for each species in Plates 42 and 45. The usual difference in size is shown here but one should note that a large *E. histolytica* may equal the size of *E. coli* (refer to Table 1). *E. histolytica* has a more delicate, less vacuolated, cleaner appearing cytoplasm, usually without inclusions. When there are inclusions, they are usually few in number and consist mostly of ingested, partially digested red blood cells whereas the materials ingested by *E. coli* are often numerous, include bacteria, and frequently give the trophozoite a cluttered appearance. The nucleus of *E. histolytica* is more sharply defined and more consistent in form in either wet mounts or permanently stained smears. In order to make an identification, the entire organism should be considered. A list of the various characteristics and how frequently they may occur in each species is given in Table 2.

Table 2. Comparision of Characteristics Seen in *Entamoeba histolytica* and *E. coli*[a]

Active trophozoite nucleus	*E. histolytica*	*E. coli*	Mature cyst nucleus	*E. histolytica*	*E. coli*
Peripheral chromatin present	Always	Always	Peripheral chromatin present	Always	Always
Peripheral chromatin beaded and regular	Usually	Occasionally	Peripheral chromatin beaded and regular	Usually	Often
Peripheral chromatin not beaded and irregular	Rarely	Usually	Peripheral chromatin not beaded and irregular	Occasionally	Usually
Karyosome large, about $1/4$ to $1/5$ diameter of nucleus[b]	Usually	Rarely	Karyosome large, about $1/4$ to $1/5$ diameter of nucleus	Usually	Rarely
Karyosome central	Usually; may be slightly out of center	Rarely; may appear central in some positions	Karyosome central	Usually	Rarely; may appear central
Karyosome made up of several particles	Rarely	Frequently	Karyosome made up of several particles	Rarely	Frequently
Nucleus with delicate strands of chromatin between periphery and karyosome	Frequently	Rarely	Nucleus with delicate strands of chromatin between periphery and karyosome	Frequently	Occasionally
			Usual number of nuclei	4	8[c]
Cytoplasm			**Cytoplasm**		
Heavily granular	Rarely	Always	Heavily granular, irregular in size	Occasionally	Usually
Many small vacuoles	Rare or when moribund	Always	Granules evenly dispersed as granules of powder dropped on water	Never	Usually
Vacuoles containing red blood cells	Frequently	Never	Granules forming a pattern as a lacework of small circles	Usually	Rarely
Vacuoles containing bacteria	Rarely	Usually	Chromatoid bars with rounded ends	Frequently[d]	Never
Vacuoles containing larger cells and/or starch	Occasionally	Frequently	Chromatoid bars splinter-like	Never	Occasionally
Pseudopods broad and nongranular	Always	Rarely			
Pseudopods small projections of cell membrane	Rarely	Usually			

Table 2. Comparision of Characteristics Seen in *Entamoeba histolytica* and *E. coli* (Continued)

Size range	*E. histolytica*	*E. coli*	Size range	*E. histolytica*	*E. coli*
20–30 µm	Usually	Usually	12–15 µm	Usually	Usually
>30 µm	Rarely	Frequently	>15 µm	Rarely	Frequently
<20 µm	Frequently	Rarely	<12 µm	Frequently[e]	Rarely

[a] Table 2 is modified from "Table for the Comparison of *Entamoeba histolytica* and *Entamoeba coli*", *Culturette Brand MIF Procedure Kit*, Price, D. L., Marion Merrell Dow, Kansas City, MO, 1978. With permission.

[b] In MIF-type preparations the area surrounding the karyosome takes stain and may appear larger than in hematoxylin- or trichrome-stained fecal films fixed in PVA.

[c] Cysts of *E. coli* with more than 8 nuclei (12–32 nuclei) are sometimes seen. These are referred to as hypernucleate or supercysts.

[d] When chromatoid bars are present, the cysts are not in fact mature even though they may occasionally have 4 nuclei (*E. histolytica*) or 8 nuclei *E. coli*).

[e] There are small forms of *E. histolytica* and cysts may be less than 8 µm. The nuclei of these small forms are usually equal in size to those in the larger forms. The size of the nucleus in relation to the size of the cyst separates the small form of *E. histolytica* from *E. hartmanni*.

dispersed around the nuclear membrane and the karyosome is distinct.

In *E. coli* trophozoites, as often seen also in *E. polecki*, the nuclear membrane may be irregular in shape, the karyosome may be indistinct, and often the peripheral chromatin is irregularly dispersed.

The precysts are quite similar and are not distinguishable one from the other. The immature cysts may also be very similar but differentiation becomes more clear when the cysts mature. As the cysts form, the structural appearance of the two species may be quite similar and, unless there are individual organisms with some distinguishing features such as chromatoid bars, the species may not be readily determined. When the cysts are mature, the two species are usually distinguishable. Table 2 lists various morphologic structures seen in *E. histolytica* and *E. coli*, and how frequently they may occur in the two species. Many of these features are used in identification. The table may be used with Plate 47 to gain a better appreciation of the overall structure and the differences between the two species.

The four species of intestinal amoebae having a nucleus without peripheral chromatin are shown in Plate 4, Chapter 7. As in the previous example, when the four species are viewed together, one gets a clearer impression of their differences.

Endolimax nana

The trophozoites of *Endolimax nana* (Plate 48, 62, 78) are usually round to oval (Plate 48:1, 2) but may be stretched out with no distinct pseudopodium (Plate 48:3). They range from 6 to 12 µm but are usually 6 to 8 µm in diameter. The cytoplasm is finely granular, usually with many small vacuoles, and often with included bacteria (Plate 48:1, 3). The nucleus has no peripheral chromatin but cytoplasmic granules adhere to its outer surface, usually making the nucleus stand out. The karyosome is large and slightly flattened, taking about one half the diameter of the nucleus. Often the karyosome is made up of several parts (Plate 48:2).

The cysts are round to oval. Early cysts may have a glycogen vacuole (Plate 48:4). As cysts mature the cytoplasm becomes very finely granular sometimes with dense condensations of granular-like material (Plate 48:5). The mature cyst has four nuclei that appear as punched-out holes in the cytoplasm containing relatively large granules, the karyosomes. The cytoplasm is clean and usually free of the granular-like material (Plate 48:6).

Iodamoeba butschlii

The trophozoites of *Iodamoeba butschlii* (Plate 49, 63, 79) vary greatly in size, ranging from 6 to 20 µm in diameter. The cytoplasm is coarsely granular with many vacuoles, often with included bacteria giving the amoeba a cluttered appearance as in *E. coli* (Plate 49:1, 2). The nucleus is large, often with an indistinct nuclear membrane. The karyosome is large with a central mass usually surrounded by a vesicular material that sometimes extends to the nuclear membrane (Plate 49:1, 2).

The cysts range from 5 to 15 µm in diameter and are usually near round to oval (Plate 49:4, 5), but occasionally are bizarre in shape (Plate 49:6). One or more large glycogen vacuoles are always present and these

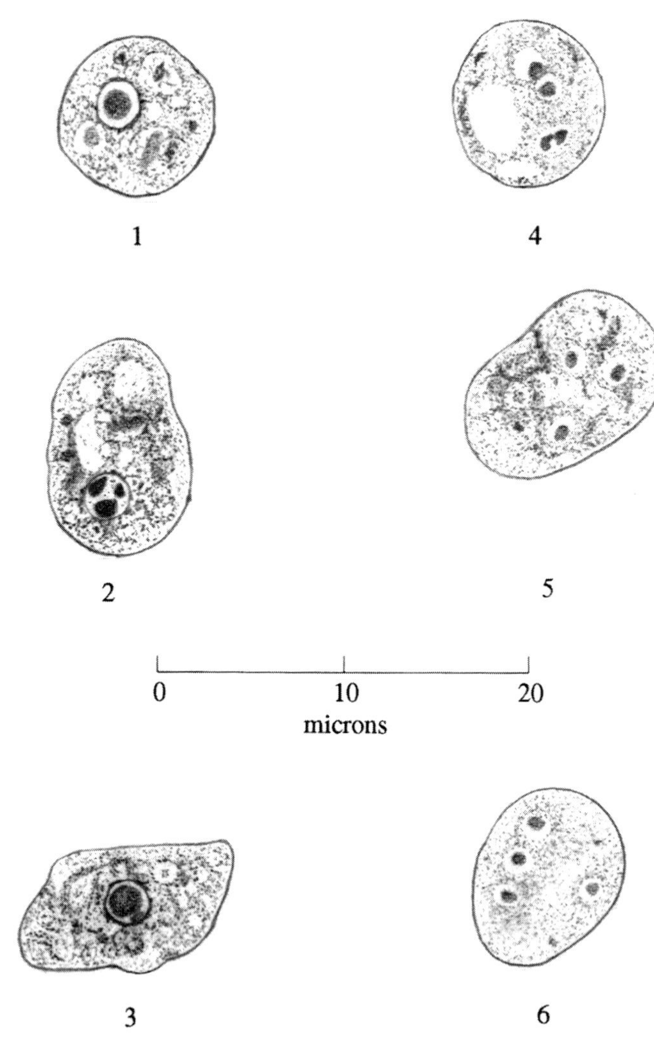

Plate 48. Diagrams of *Endolimax nana*. 1–3: Trophozoites; 4, 5: immature cysts; 6: mature cyst. In the trophozoite of *E. nana*, the cytoplasm is finely granular with many small vacuoles, often with included bacteria (1–3). The nuclear membrane is distinct and easily recognized because of the cytoplasmic granules that appear to adhere to it. The inside of the nuclear membrane is smooth. The karyosome is large and appears flattened (1, 3) but is often made up of several parts as it is in division (2). The early cyst may have one or more large vacuoles (4). The nuclei appear as a punched out holes each with a moderately large, flattened karyosome (4–6). The mature cyst has four small nuclei and a finely granular, very smooth appearing, clean cytoplasm (6).

are surrounded with dispersed large cytoplasmic granules (Plate 49:4, 5). The nuclear membrane is usually indistinct. The karyosome is large and retains the vesicular material at least on one side.

Dientamoeba fragilis

The trophozoites of *Dientamoeba fragilis* (Plate 50, 64, 80) range in size from 4 to 20 µm but are usually 6 to 12 µm in diameter. They appear very delicate in wet preparations with finely granular cytoplasm and many vacuoles (Plate 50:1–3). Usually there are small bacteria included in a few or many of the vacuoles. There may be one or two nuclei, usually two, with a thin and often indistinct nuclear membrane and a relatively cloudy nucleoplasm. The karyosome is made up of several granules, usually 4 to 6, but they are rarely distinctly clear in wet preparations. The nuclei are often not readily distinguished from small vacuoles with inclusions which are about the same size. The cloudy nucleoplasm makes seeing the structure of the karyosome more difficult. This is particularly true in specimens that have not been fixed properly (see Part 1). The cyst stage is not known.

INTRODUCTION TO THE "KEY FOR DIFFERENTIATING SPECIES OF INTESTINAL AMOEBAE IN FECAL SPECIMENS PRESERVED IN PIF OR MIF"

The Key does not follow the usual binomial pattern of asking for a decision between couplets in every case. There are three choices, a, b, and c, in 4 and 5. The trophozoites are very similar at these points in the Key and trying to prepare couplets complicated the Key more than necessary. In 8, there are four choices — a, b, and c for differentiating between mature cysts, and d referring to immature cysts.

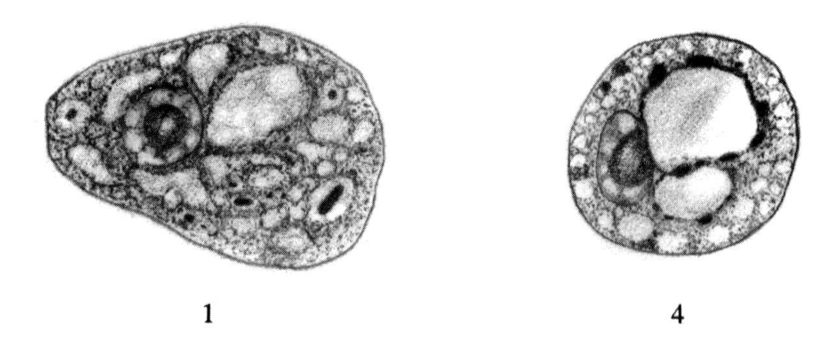

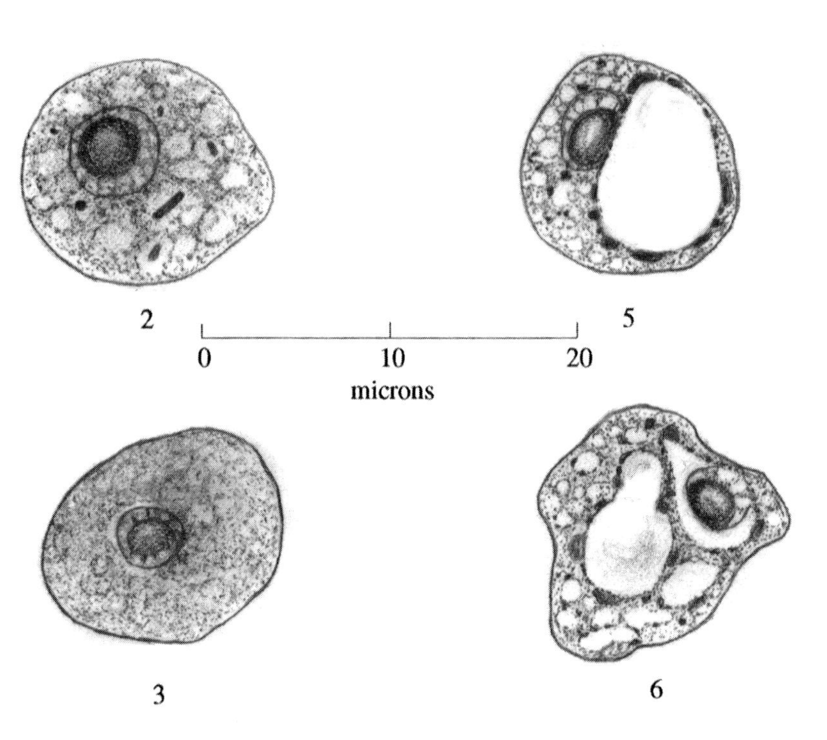

Plate 49. Diagrams of *Iodamoeba butschlii.* 1, 2: Trophozoites; 3: precyst; 4–6: cysts. The trophozoites of *I. butschlii* have coarsely granular cytoplasm with many vacuoles. Vacuoles often have ingested bacteria and are surrounded by particularly large granules (1, 2). The nucleus is large and the nuclear membrane is sometimes indistinct. The karyosome is large, about one third to one half the size of the nucleus, and is often surrounded by a vesicular structure that sometimes appears to extend to the nuclear membrane (1, 2). Like the precyst of other amoebae, vacuoles and inclusions are lost and it has a finer cytoplasmic granulation. The nucleus is usually more distinct (3). The cysts have a single nucleus and one or more large glycogen vacuoles bordered by large cytoplasmic granules (4–6). The nucleus has a large karyosome often surrounded by vesicular material (4, 5). Some cysts are very irregular in shape (6).

Again, this seemed to be a more logical way to approach identifying the cysts.

The Key is based on the process of elimination, i.e., "what can it *not* be?" rather than "what *can* it be?" This arrangement appeared to work better than the usual binomial key. The many individuals who have used the Key have reported that it was easy to use and readily led them to an accurate identification of an amoeba found in fecal specimens they were examining.

The diagrams and photomicrographs used along with the Key should greatly assist individuals in making an identification. A reference manual of color projection photomicrographs along with drawings that show the species of protozoa as seen in MIF-fixed wet preparations can be obtained from Meridian Diagnostics, Inc. (Price, 1979).

Dientamoeba fragilis is included in the Key since it is seen in the amoeboid form in the intestine of man.

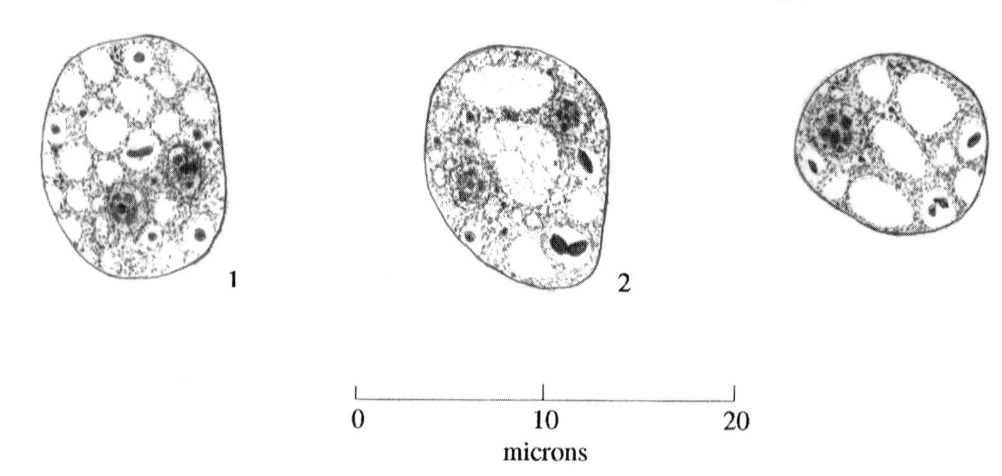

Plate 50. Diagrams of *Dientamoeba fragilis*. 1, 2: Binucleate trophozoites, 3: mononucleate trophozoite. *D. fragilis* is the most delicate appearing of the intestinal amoebae with an extremely fine outer membrane and finely granular cytoplasm with many vacuoles, of which a high percentage have inclusions (1–3). Because the karyosome is made up of several small particles, the nucleus is often difficult to differentiate from a vacuole with an inclusion. Binucleate forms are predominant (1, 2) but some mononucleate forms are usually present (3). Identification can often be made on the basis of the delicate cytoplasm when the nucleus is difficult to define (See the Key at the end of this Chapter and Plates 64, 80). The cyst stage is unknown.

KEY FOR DIFFERENTIATING SPECIES OF INTESTINAL AMOEBAE IN SPECIMENS PRESERVED IN PIF OR MIF

Plate

1. a. Amoeba having cytoplasm with vacuoles, either with or without inclusions; cell membrane smooth; round or irregular with or without pseudopodia 2
   = **trophozoite**

   b. Amoeba with cytoplasm appearing granular either without vacuoles or with large glycogen vacuoles; outer margin smooth, refractile, and without projections or pseudopodia 6
   = **cyst**

2. a. Nucleus with distinct peripheral chromatin; eliminate *Endolimax nana, Iodamoeba butschlii, Dientamoeba fragilis* 3

   b. Nucleus without distinct peripheral chromatin; eliminate *Entamoeba histolytica, E. coli, E. hartmanni, E. polecki* 5

3. a. Parasite ≤10 µm in diameter 11

   b. Parasite >10 µm in diameter 4

4. a. Cytoplasm dense, granular, and appearing cluttered with many vacuoles, some with included bacteria (rarely without inclusions); vacuoles usually only large enough to accommodate inclusions; nucleoplasm light, sometimes with fine granules; nucleus with irregular peripheral chromatin and an eccentric karyosome or one that is ill defined or broken up; if the karyosome is discrete, the diameter of the nucleus is six or more times the diameter of the karyosome; eliminate *Entamoeba histolytica* and *E. polecki*.
   = ***Entamoeba coli***

b. Cytoplasm finely granular and appearing clean, with fewer vacuoles, usually empty but occasionally with one or more ingested erythrocytes in some stage of digestion (erythrocytes being digested do not have the refractivity of bacteria but appear dull and flat), rarely a yeast cell or bacterium may be ingested; nucleoplasm light and often with fine granules; nucleus with evenly granular, peripheral chromatin and a discrete central or nearly central karyosome (if not in division); the diameter of the nucleus is less than six times the diameter of the karyosome; eliminate *Entamoeba coli* and *E. polecki.* **= *Entamoeba histolytica***

c. Cytoplasm appears smooth but varies in density; vacuoles usually relatively small, dispersed, usually some with ingested bacteria or rarely yeast cells; occasionally with numerous vacuoles replacing much of the cytoplasm; nucleus with irregular or regular thin peripheral chromatin, a small, often indistinct, karyosome, and dark, dense nucleoplasm; eliminate *Entamoeba histolytica* and *E. coli.* **= *Entamoeba polecki***

5. a. Nuclear membrane distinct and easily recognized, with cytoplasmic granules adhering to the membrane; karyosome large, flattened, sometimes divided into two or four components; cytoplasm finely granular, with many small vacuoles; eliminate *Iodamoeba butschlii* and *Dientamoeba fragilis.* **= *Endolimax nana***

b. Large karyosome with indistinct nuclear membrane; cytoplasm with many vacuoles, many with ingested bacteria, and with coarse cytoplasmic granules particularly surrounding vacuoles; eliminate *Endolimax nana* and *Dientamoeba fragilis.* **= *Iodamoeba butschlii***

c. Nucleus difficult to define, with an indistinct nuclear membrane, often appearing similar to a vacuole with an inclusion; when discrete, nucleus with a karyosome made up of several distinct granules, binucleate forms predominant; cytoplasm finely granular, with many small vacuoles, some with inclusions, giving the cytoplasm a very delicate appearance, often like clusters of bubbles in a thin clear membrane; eliminate *Endolimax nana* and *Iodamoeba butschlii.* **= *Dientamoeba fragilis***

6. a. Nucleus with distinct peripheral chromatin; eliminate *Endolimax nana, Iodamoeba butschlii,* and *Dientamoeba fragilis* 7

b. Nucleus without distinct peripheral chromatin; eliminate *Entamoeba histolytica, E. coli, E. hartmanni,* and *E. polecki* 10

(***Note***: Immature cysts are often seen, particularly those of *Entamoeba histolytica, E. coli,* and *E. hartmanni.* In immature forms, the nuclei may be in division, irregular, or even different in size. The cytoplasm may have one or more large glycogen vacuoles. In such forms, the presence of distinct chromatoid bars combined with the size may be sufficient to make an identification. If chromatoid bars are not present, the identification of immature forms is usually uncertain.)

7. a. Cyst ≤10 µm in diameter; eliminate large form (race) *Entamoeba histolytica, E. coli,* and *E. polecki* 12

b. Cysts >10 µm in diameter; eliminate *E. hartmanni* and small form of *E. histolytica* 8

8. a. Cysts with more than four nuclei, peripheral chromatin not beaded, irregular, or unevenly distributed; diameter of the cyst is more than six times the diameter of the nucleus; karyosome small and usually eccentric; cytoplasmic granules usually irregular in size but evenly dispersed, rarely forming a pattern of circles (usually in immature cysts); eliminate *Entamoeba histolytica* and *E. polecki.* **= *Entamoeba coli***

b. Cysts with four nuclei, peripheral chromatin dense, granular, and evenly distributed; karyosome distinct and usually central or near central; the diameter of the cyst is less than six times the diameter of the nucleus; cytoplasmic granules usually regular in size, forming a pattern of circles, and never evenly dispersed; eliminate *Entamoeba coli* and *E. polecki.* **= *Entamoeba histolytica***

c. Cysts having one nucleus (rarely two) with dense, dark nucleoplasm; peripheral chromatin usually forming a thin, regular pattern; karyosome small and often indistinct; cytoplasm finely granular usually with large masses that are irregular in shape; sometimes large, glycogen vacuoles present; eliminate *Entamoeba histolytica* and *E. coli*. = ***Entamoeba polecki***

d. Cysts with four or less nuclei and with chromatoid bars 9

9. a. Cysts having chromatoid bars that are smooth and with rounded ends; eliminate *Entamoeba coli* and *E. polecki*. = ***Entamoeba histolytica***

b. Cysts having chromatoid bars that are irregular and with splinter-like ends; eliminate *Entamoeba histolytica* and *E. polecki*. = ***Entamoeba coli***

10. a. Cysts with a single nucleus, one or more large glycogen vacuoles; nucleus with a large karyosome, cytoplasmic granules coarse, dense, and concentrated around vacuoles; eliminate *Endolimax nana*. = ***Iodamoeba butschlii***

b. Cysts with four nuclei appearing as punched out holes in the cytoplasm with a relatively large, flattened karyosome; cytoplasmic granules extremely fine and evenly distributed; occasionally with dense chromatin-like granules present in the cytoplasm; eliminate *Iodamoeba butschlii*. = ***Endolimax nana***

11. a. Trophozoites with a large nucleus in relation to the size of the cell, one cell diameter equal to four or less nuclear diameters; karyosome distinct; peripheral chromatin evenly granular; cytoplasm with few vacuoles; eliminate *Entamoeba hartmanni*. = ***Entamoeba histolytica***, small form

b. Trophozoites with a small nucleus in relation to the size of the cell, one cell diameter equal to five or more nuclear diameters; peripheral chromatin regular or irregular; many small vacuoles often with included bacteria; eliminate small form of *Entamoeba histolytica*. = ***Entamoeba hartmanni***

12. a. Cysts ≤10 µm in diameter, with four nuclei; nuclei large, filling more than half the cell; cytoplasmic granules having a distinct pattern; eliminate *Entamoeba hartmanni*. = ***Entamoeba histolytica***, small form

b. Cysts ≤10 µm in diameter, with four nuclei; nuclei small in relation to the cyst size, filling less than half the cell; cytoplasmic granules irregular, immature forms often with chromatoid bars; eliminate small form of *Entamoeba histolytica*. = ***Emtamoeba hartmanni***

THE INTESTINAL FLAGELLATES

There are a number of flagellates that may be present in the intestinal tract of man. As stated above, three are found more commonly, *Giardia lamblia, Chilomastix mesnili*, and *Trichomonas hominis*.

Giardia lamblia

A typical trophozoite of *G. lamblia* (Plate 37:1) has two nuclei at the anterior end that are connected by axonemes and lie dorsally over a large ventral sucker disk. The trophozoites measure 10 to 20 µm in length by 5 to 15 µm in width (Table 3) and are pear-shaped (see Plates 37, 51, 65, 81). Two sausage-shaped parabasal bodies (Plates 37, 51) lie just posterior to the sucker disk. There are four pairs of flagella: one lateral at the level of the nuclei, one ventral centrally located at the posterior end of the sucker disk, one ventral just below the parabasal bodies, and one at the posterior end of the trophozoite. The two axonemes of the posterior pair of flagella pass through the body functioning as an axostyle and extend beyond the posterior end as flagella.

In lateral view, the trophozoite usually appears long, narrow, and fish-like and often only one of the nuclei is visible. In a semilateral view (Plate 51:3) the trophozoite is thicker, both nuclei can be seen, and it is more easily recognized.

Trophozoites of *G. lamblia* are seen less frequently than cysts and usually only in patients with soft, semiliquid, or liquid stools. They may appear in

Table 3. Size Ranges of Intestinal Protozoa Other than Amoebae Found in Feces

Group	Species	Usual size range in microns: Trophozoite	Cyst
Mastigophora (flagellates)	*Giardia lamblia*	10–20 by 5–15	8–15 by 5–10
	Chilomastix mesnili	6–18 by 5–15	6–10 by 4–6
	Trichomonas hominis	6–15 by 4–10	none
Ciliophora (ciliates)	*Balantidium coli*	50–80 by 30–60	45–65
		Oocyst	**Sporocyst**
Coccidia (sporozoa)	*Isospora belli*	22–30 by 10–18	10–14 by 8–12
	Sarcocystis hominis or *S. suihominis*	18–24 by 16–20 (rare)	12–18 by 8–12
	Cryptosporidium parvum	4–6	No sporocysts
Uncertain classification	*Blastocystis hominis*	(nonmotile, cyst-like structures)	5–30

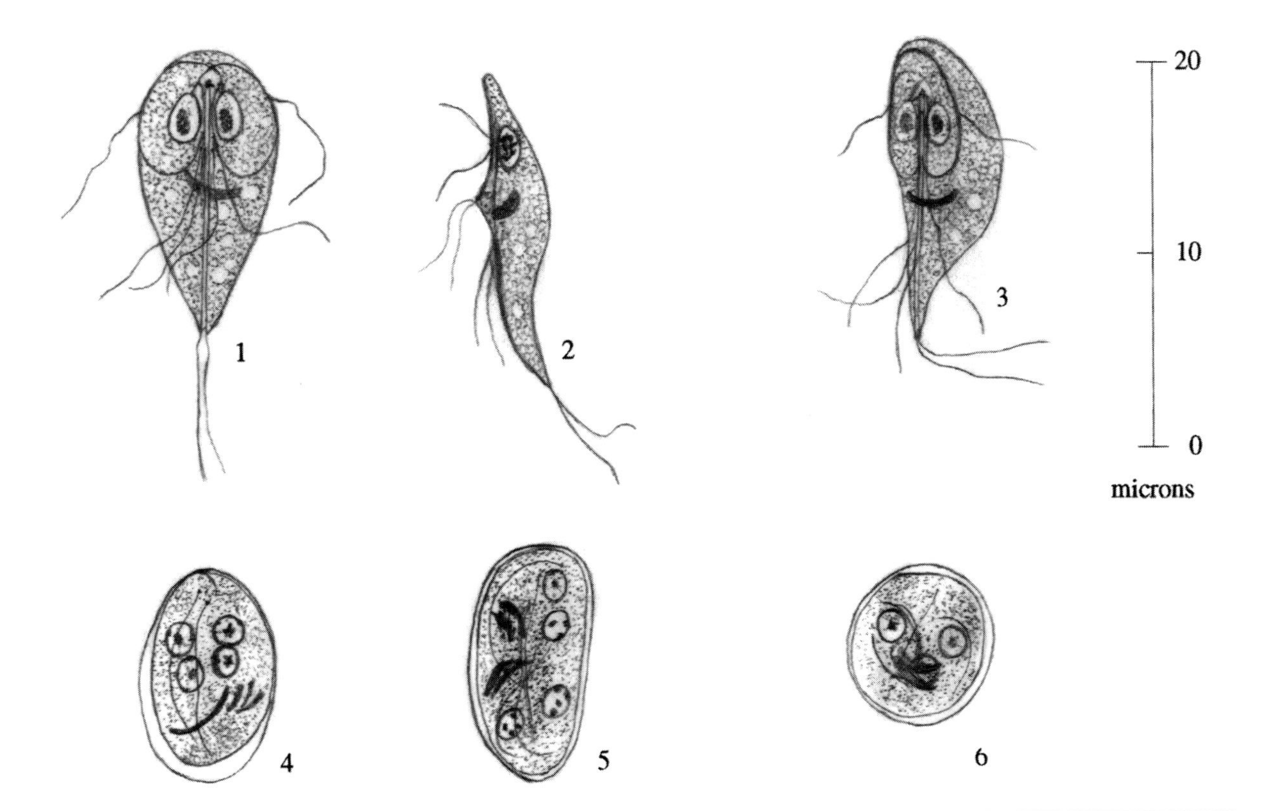

Plate 51. Diagrams of *Giardia lamblia.* 1–3: Trophozoites; 4–6: cysts. Members of the genus *Giardia* are very distinctive. The trophozoite is bilaterally symmetrical with a nucleus on either side and four pairs of flagella for motility (1–3). It has a ventral sucker disk for attachment to the intestinal mucosa which is best seen in a lateral or oblique view (2, 3). Cysts are first seen with two nuclei but division rapidly takes place and most cysts seen have four nuclei and paired organelles (4, 5). The cyst wall is usually clearly separated from the parasite inside (4), at least on one side (5). Since in a fecal preparation, the parasite may lie in a position that obscures some of its structures (6), one must make certain it is *Giardia.*

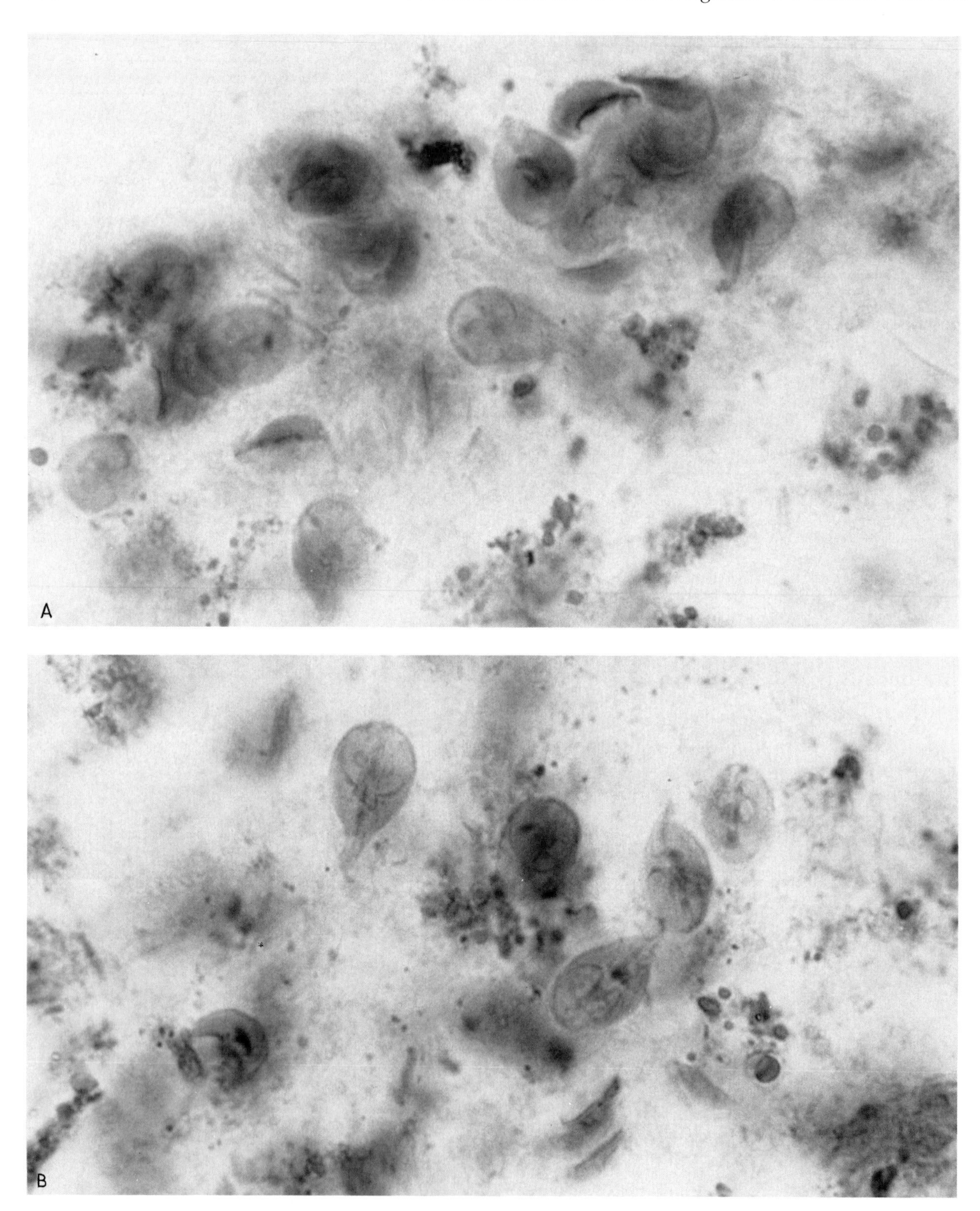

Figure 1. Trophozoites of *Giardia lamblia* clumped together in mucous (a and b: MIF, P—Polychrome IV permanent stain). In a clinically ill patient with diarrhea, it is not unusual for large numbers of *G. lamblia* to be passed in fecal specimens encased in clumps of mucous. If such specimens are strained through a fine screen for concentration, mucous clumps are often held back and do not enter the centrifuge tube. In the cases of such fecal specimens, either direct examination of the specimen or preparation of a permanent slide will often reveal the presence of trophozoites when they are not found after concentration of the specimen.

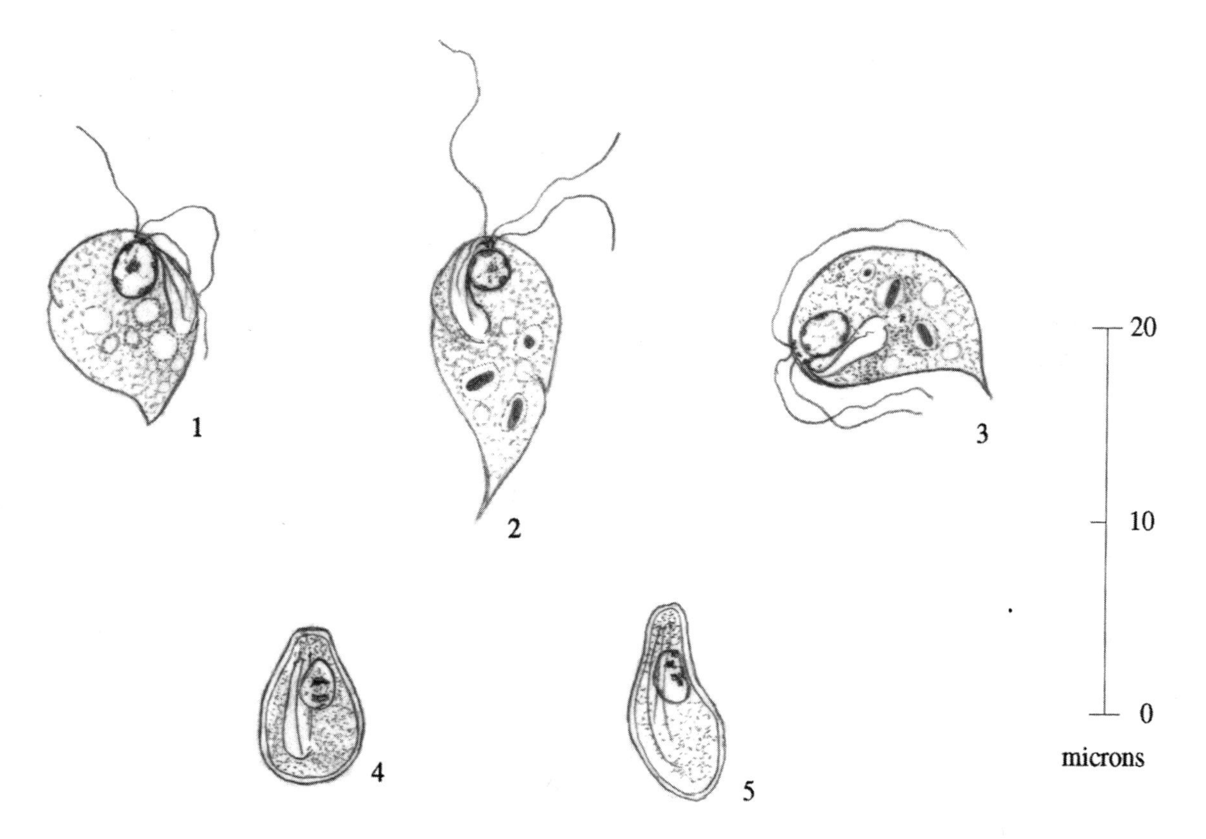

Plate 52. Diagrams of *Chilomastix mesnili.* 1–3: Trophozoites; 4, 5: cysts. The trophozoite stage of *C. mesnili* may appear in various shapes in a slide preparation (1–3) and all structures seen in the diagrams may or may not be discernible. Bacteria are often seen in the vacuoles of active trophozoites (2, 3). Cysts (4, 5) may may be found along with trophozoites or may be the only form found. Cysts may be pear-shaped, with a raised portion at one end, which is typical (4), or may vary in shape (5).

slide preparations in various positions and there is some variation in their general shape from patient to patient. They often appear in clumps of mucus grouped together and piled on top of one another (see Figure 1, page 154). The clumps are usually seen either on direct examination of the specimen in the vial or in permanently stained fecal smears made directly from a fresh or preserved specimen. They are less often seen after concentration because the mucus clumps do not pass through the fine mesh screens that are usually used. The larger mesh screen recommended for the CONSED concentration system does allow small strands of mucus to pass into the centrifuge tube and clumps may be seen when this method is used for concentration.

Apparently, large numbers of trophozoites encyst in the intestine simultaneously every 4 to 7 days and the cysts are passed in the feces. At midpoint between these periods of encystation, there may be relatively few cysts found in the feces.

A typical cyst measures 6 to 10 µm × 4 to 6 µm (see Table 3). After encysting, the parasite quickly divides, showing duplication of all structures, including nuclei. In preserved specimens, the parasite inside the cyst is usually separated from the cyst wall, leaving a space that helps the viewer to find and identify it (Plates 37, 51). The nuclei are usually located at one end of the parasite but they may be scattered (Plate 51:5). When seen on end, the identification of a *Giardia* cyst may be more difficult (Plate 51:6).

In some cysts, many of the structures can be seen. Some of the axonemes and the parabasals are usually seen on close examination of a well-preserved cyst. There appears to be no clear evidence of the sucker disk in the cyst (see also trophozoites and cysts in Plates 65, 66, 81).

Chilomastix mesnili

The trophozoites of *Chilomastix mesnili* vary greatly in size, especially in life where they range from 6 to

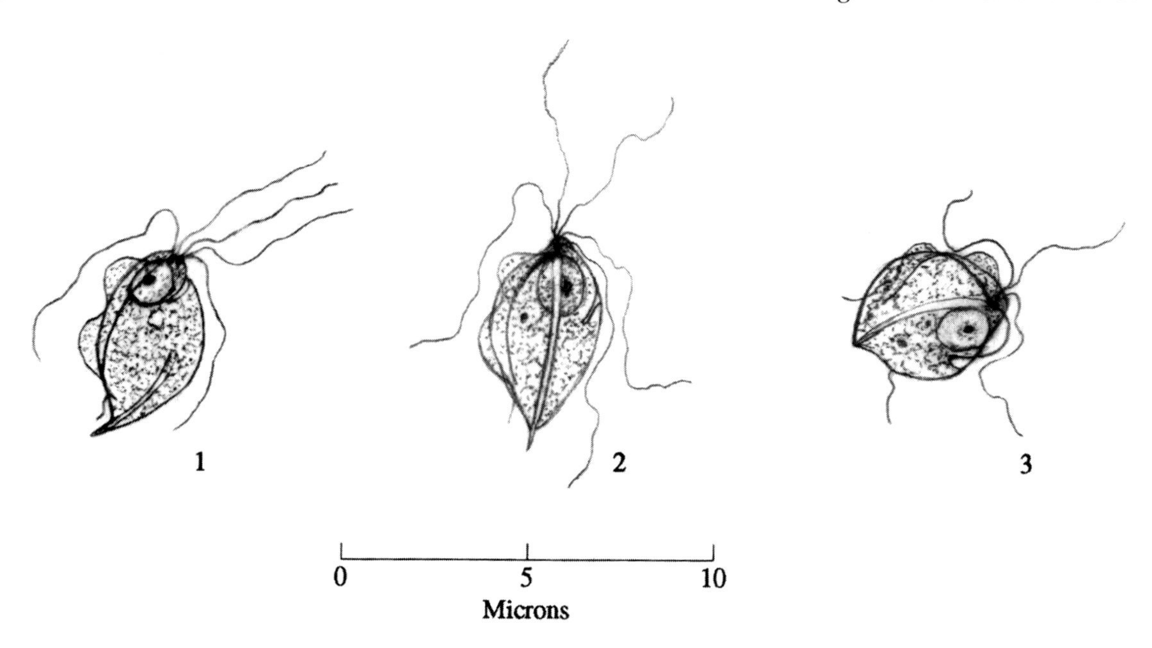

Plate 53. Diagrams of *Trichomonas hominis* 1–3: Trophozoites. The trophozoite is the only known form of *T. hominis.* It can vary from 6 to 15 µm in length, often with trophozoites of more than one size range in the same specimen. Where organisms are very numerous they tend to be clumped together in fixed specimens. The cytoplasm may appear very dense with few vacuoles (1) or thin and vacuolated (3). The axostyle and small cytostome shown here are often not discernible.

18 µm in length (see Table 3). The structures that are usually visible include the nucleus, flagella, cytostome, spiral groove, and food vacuoles. The axonemes do not stain well with the stains usually employed in the diagnostic laboratory. The trophozoites take on various shapes depending on how they lie in the slide preparation.

The fixed trophozoite is usually from 10 to 15 µm in length, depending on how much it is rounded (see Plates 38 and 52). It has a round nucleus usually with a karyosome and some peripheral chromatin. The typical appearance of the trophozoites fixed in MIF or PIF is depicted in Plate 52:1–3. Generally, the trophozoite is asymmetrically pear-shaped, rounded at the anterior end, and pointed at the posterior end (see Plates 38 and 52). A spiral torsion of the body is located near the center of the trophozoite but is not always visible. It has a rounded nucleus lying close to the anterior end, often seen with a small karyosome and irregular peripheral chromatin. The trophozoites have three external flagella, one longer than the others, each originating from a blepharoplast just anterior to the nucleus. A cleft-shaped cytostome with its central cytostomal flagellum lies just below the nucleus (see Plate 38). Food particles ingested through the cytostome, are collected in vacuoles formed at its posterior end, and circulate through the cytoplasm.

The cyst is typically lemon- or pear-shaped with a raised area at one end, but may appear differently depending on its orientation. Rarely the cyst may be oval or irregular. There may be a space between the raised part of the cyst wall and the organism inside. The cyst has a single nucleus similar to that seen in the trophozoite, a cytostome supported by two filaments (axonemes), and retains its central cytostomal flagellum. The cytoplasm is finely granular. Axonemes and blepharoplasts are often more clearly visible in the cyst (see Plates 67 and 82).

Trichomonas hominis

T. hominis may vary greatly in size but is usually 10 to 12 µm in length. It is not uncommon, however, to find organisms of two different size ranges in the same specimen (see Table 3). *T. hominis* appears differently in different fixing media and when stained with different stains (see Plates 68 and 83). In some preparations the axostyle and undulating membrane are not always visible but, in wet preparations (MIF and PIF), the flagella do stain well (see also Plates 38 and 53).

The body is pear-shaped to round. There are three to five flagella, usually five. The nucleus is large, oval to round and lies near but not at the anterior end, has a central karyosome, and a distinct nuclear membrane. The axostyle begins near the anterior end and extends beyond the posterior end of the body. Because of the axostyle, the trophozoite appears more rigid than *C. mesnili.* The undulating membrane is supported by a chromatoid basal rod that begins anterior to the nucleus and extends at an angle over about 60% of the body to near the posterior end. It leaves the body as a short flagellum to give the organism its characteristic appearance. A small cytostome lies just lateral to the nucleus. There is no cyst stage. (See also Plates 68 and 83.)

As in other organisms, it is not possible to get all the structures in focus at the same time. The photomicrographs therefore show only some of the characteristics.

THE INTESTINAL CILIATES

Balantidium coli

Apparently, *Balantidium coli,* the only intestinal ciliate, typically infects hogs, but rats and monkeys may become accidental hosts. Although not scientifically established, anecdotal information suggests that the pig is the reservoir of the infection for man and primates. Man appears to be naturally resistant to infection with *Balantidium coli,* but under certain conditions, such as undernourishment, poor diet, or the presence of other diseases, man may become infected. It follows that the people who are at greatest risk are those who are in lower economic groups, inmates of asylums, and prisoners who are in some way associated with either pigs or pig handlers. Diagnosis is made by finding trophozoites or cysts in fecal specimens.

The trophozoites usually measure 50 to 80 × 30 to 60 μm, are ovoidal with a narrower anterior end, and have a delicate outer pellicle beneath which is a clear zone of ectoplasm (see Plate 39:1). The surface is covered with spiral, longitudinal rows of cilia that lie in grooves and arise from granules in the ectoplasm. At the anterior end there is a mouth opening to a cytostome that is lined with cilia for food gathering. Food vacuoles arise from the base of the cytopharynx and circulate through the body. At the posterior end is a distinct excretory opening, the cytopyge. There are two contractile vacuoles, one in the anterior third and the other in the posterior third of the body. There is a large, usually kidney-shaped macronucleus and a very small micronucleus located adjacent to and near the center of the macronucleus. The micronucleus is not always visible (see Plate 69).

The cyst (Plate 39:2) is round to oval, measures 50 to 65 μm, and is protected by a thick outer and a thin inner cyst wall. The newly encysted organism remains active and, as usually seen in freshly preserved feces, retains its cilia for a time after encystation. Eventually, the cilia are absorbed and the organism becomes quiescent within the cyst walls. The contractile vacuoles and the macronucleus remain visible but the food vacuoles and other structures disappear.

THE INTESTINAL COCCIDIA

Isospora belli

Infection with *Isospora belli* usually originates from sources contaminated with feces from infected individuals but autoinfection may also occur since sporulated oocysts have been found in the duodenum. The number of species of *Isospora* infecting man is questionable. *Isospora natalensis,* and *I. chilensis* have been reported. The species formally referred to as *I. hominis* is now recognized as *Sarcocystis hominis.*

Oocysts of *I. belli* are light-bulb shaped, being nearly oval but narrowing bluntly at one end (see Plates 6, 40). Occasionally, they are oval. A small plug-like spot, the micropyle, lies in the center of the narrower end. After the combination of the two gametes, the zygote forms an outer cyst covering to become an oocyst. Within the oocyst wall, the zygote divides to form two sporoblasts. Each sporoblast secretes a cyst wall to become a sporocyst. Within the sporocysts, four sporozoites develop along with a residual body (see also Plate 70).

Sarcocystis hominis and *S. suihominis*

Man is the definitive host of *S. hominis* and *S. suihominis* and becomes infected by eating inadequately cooked, infected meat, beef or pork, respectively, but man may also become an aberrant, intermediate host by ingesting oocysts or sporocysts from a contaminated source (Beaver et. al., 1979).

Sporocysts, rather than oocysts, are usually seen in the feces of an infected individual. By the time they reach the outside environment, sporozoites have formed (see Plate 6). Infection usually runs a mild clinical course and is usually self-limiting (see also Plate 71).

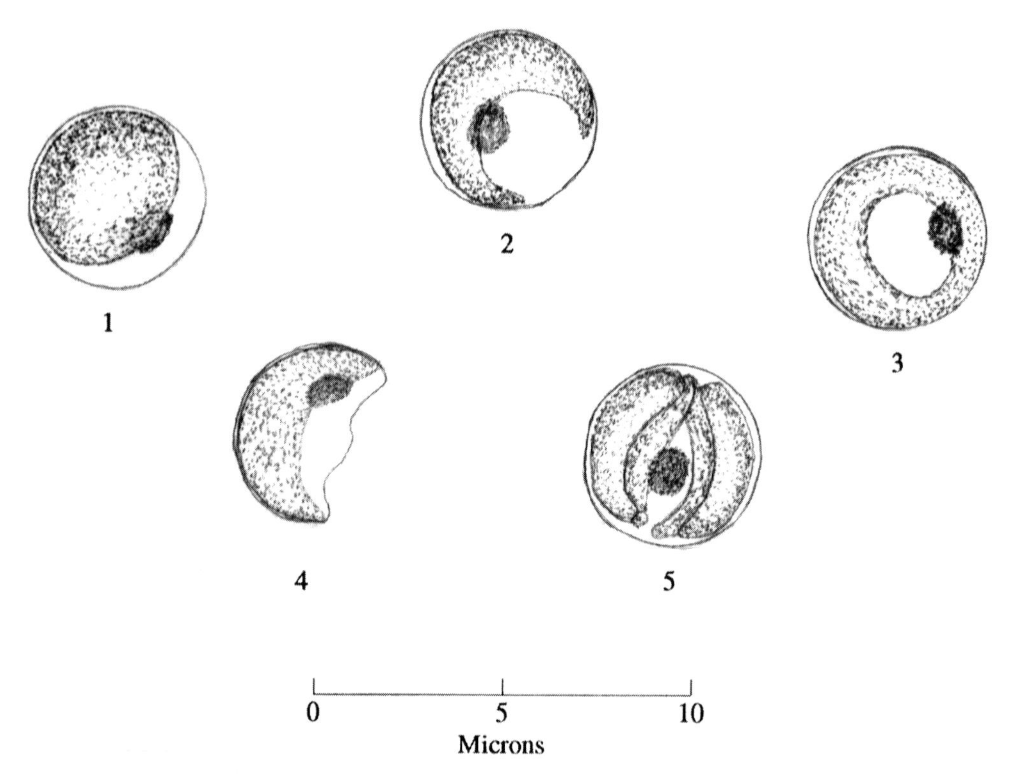

Plate 54. Diagrams of *Cryptosporidium parvum.* 1–3: Immature oocysts; 4: partially collapsed oocyst; 5: mature oocyst (not to scale). The oocysts of cryptosporidia are thin-walled and from 4 to 6 µm in diameter. The sporont appears as a dense oval body (1) with a small, dark staining area lying on one side that appears to persist throughout development (1–5). The sporont within the oocyst sometimes appears sickle-shaped (2, 4) or sometimes donut-shaped (3). Four sporozoites develop free within the oocyst (5) without a sporocyst wall.

Cryptosporidium parvum

The oocyst of *C. parvum* is extremely small and, when seen in fecal specimens, has an appearance similar to some yeast cells (see Plate 54). It is *not* easily identified in direct wet mounts prepared from fresh feces and it is preferable to fix specimens in one of the acceptable collecting/preserving solutions since viable oocysts are directly infective. With careful examination using 1000× magnification, the trained, experienced worker may be able to identify cryptosporidia on wet mounts made from formalin-fixed specimens stained with iodine, or on wet mounts prepared from MIF- or PIF-fixed specimens (see Plate 72). When small cyst-like bodies 4 to 6 µm are seen and *Cryptosporidium* is suspected, the specimen should be concentrated and wet mounts and smears prepared from the concentrate, the latter for staining. Confirmation usually requires acid-fast stains, fluorescent stains, or monoclonal antibody preparations.

If a mature oocyst is ingested by a suitable host, the oocyst wall breaks down and sporozoites are released in the intestine. The freed sporozoites may attach to enterocytes (host cells lying on the surface of the microvilli of the small intestine) to initiate an infection. The intercellular stages of *C. parvum* are apparently restricted to the villus border of the intestinal mucosa where they develop asexually (merogony) in parasitophorous vacuoles. The merozoites that are formed are freed and attach to new cells to repeat merogony.

Eventually, as is usual for coccidia, some of the merozoites that attach to the cells initiate the sexual cycle (gametogony) in which microgametes and macrogametes are formed. The 16 microgametes produced by the microgamont are freed to fertilize the macrogamete. After fertilization, the zygote secretes an oocyst wall and within it the zygote (sporont) begins its development to form the four sporozoites. The oocysts, either those attached to the villi or free in the fecal material, stain acid-fast but other stages do not (see Plate 72:1).

Clinical disease is generally self-limiting, except in patients with suppressed or deficient immune systems, such as patients with AIDS, where it can be a serious clinical disease. At this writing, there is no known effective treatment for cryptosporidiosis in animals or man.

C. parvum appears to have worldwide distribution. Infection is frequently associated with a farm environment or with occupations involving animals, especially cattle. Infections are seen also in children in day care centers and, rarely, in individuals in situations where there seems to be no logical link with the usual sources of infection. Water supplies can become a source when contaminated with wastes from infected animals. Usually, diagnosis is made by finding oocysts in fecal specimens that have been concentrated and stained with a modified acid-fast stain. Because of the low number of laboratory requests made for fecal examinations for *Cryptosporidium*, there is little question that these parasites are not considered in most cases of diarrhea and that the prevalence of infections is underestimated.

PROTOZOA OF UNCERTAIN TAXONOMIC POSITION

Blastocystis hominis

Another organism seen frequently in fecal specimens is *Blastocystis hominis*. It was first named *B. enterocola* by Alesieff in 1911 but the name was changed in 1912 by Brumpt to *B. hominis* (Zierdt, 1991b). The organism was generally accepted as a nonpathogenic yeast by most investigators. More recently, Zierdt et al. (1967) classified it as a protozoan.

B. hominis has been reported as the cause of diarrhea in immunodeficient patients (Zierdt, 1991a,b) but it is also present in the intestine of many apparently healthy individuals. The usual forms seen in fecal preparations are shown in the diagrams (Plate 41:1, 2, 5). They vary greatly in size from about 5 to 30 μm in diameter, have a large central area that has the appearance of a vacuole, and an outer ring of cytoplasm. Several pale-staining nuclei are embedded in the cytoplasm along with darker bodies referred to as volutin. Because their refractive index is similar to that of intestinal amoebae, some forms may be mistaken for *Endolimax nana* or other cysts of amoebae (Plate 41:3). Others are dense, thick walled, and have a darkly staining central area (Plate 41:4). *Blastocystis* divides by binary fission (Plate 41:6).

Zierdt reviewed the information available on *B. hominis* in 1991 (Zierdt, 1991b) and provided strong evidence of its position as a protozoan parasite and further subscribed to its pathogenicity, at least under certain conditions. Markel and Udkow (1986) question the pathogenicity of the organism and suggest the possibility of another organism being present but not found in symptomatic cases. Since *B. hominis* appears in fecal specimens of apparently healthy individuals as well as those with gastro-intestinal symptoms, its role as a pathogenic protozoan parasite remains unclear. Regardless of its effect on its host, the organism is an indicator of fecal contamination and should be reported when found on examination of fecal specimens (see Plates 73 and 84).

PHOTOMICROGRAPHS OF INTESTINAL PROTOZOA

Two series of photomicrographs of intestinal protozoa are presented. The first series is of protozoa as they appear in wet preparations (page 160), and the second series is of protozoa as seen in permanently stained fecal films (page 194).

Photomicrographs Prepared From Wet Preparations

In the photomicrographs prepared from wet preparations, representatives of the amoebae, flagellates, ciliates, coccidia, and *B. hominis* are presented. The stage of the parasite (trophozoite T and cyst C) and the fixing solution (formalin, MIF, or PIF) used in collecting and preserving the specimen are given in the legend for each photomicrograph.

The Plates are arranged by parasite species with a number of photomicrographs representing each species. Separate Plates for trophozoites and cysts are presented for *Entamoeba histolytica*, *E. coli*, and *Giardia lamblia*. For other species, trophozoites and cysts appear on the same Plate. Cryptosporidia are presented as they appear *in situ*, in acid-fast stain, and in wet mounts.

Although many of the diverse variations within a species are not shown, the variations that have been used should offer considerable assistance in identification. It is advisable to use all of the aids provided especially when confronted with a difficult identification. Plates 55 through 73 are of intestinal protozoa in wet preparations.

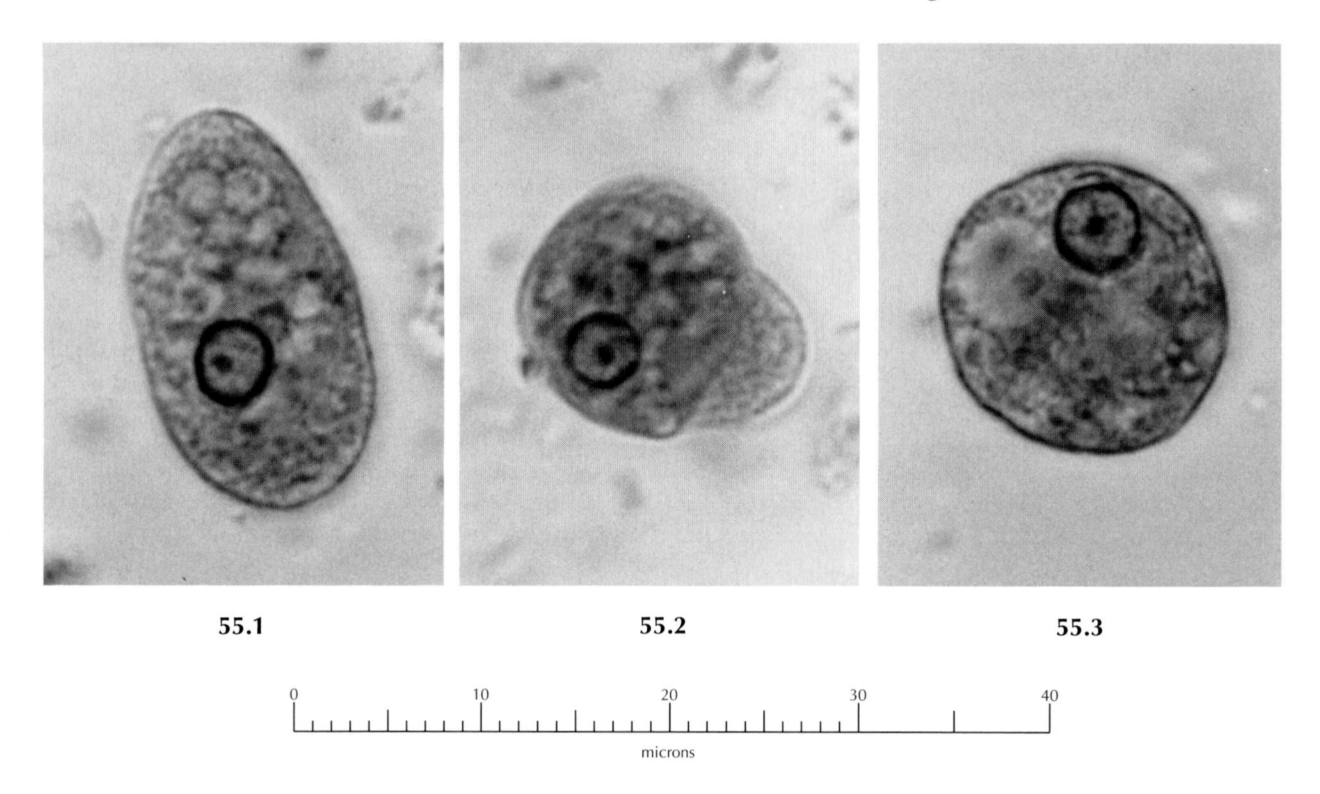

Plate 55. *Entamoeba histolytica* trophozoites (large race). 1–9: MIF. The trophozoite may show a pseudopodium relatively free of the granules seen in the inner cytoplasm (1, 2). Some trophozoites have a few larger vacuoles (3, 5) while others may appear with many small vacuoles (1). Vacuoles are usually empty but they may contain recently ingested red blood cells (not shown) or remnants following partial digestion (3). The cytoplasmic granulation in properly fixed trophozoites is generally smooth (1–8). When there is a delay in fixation after the specimen is passed, the cytoplasm becomes highly vacuolated and the smooth granulation of the cytoplasm is lost (9). In the nucleus, the peripheral chromatin is relatively even (1–9) and may appear slightly beaded (2, 3, 7, 8). The karyosome is fairly large and quite distinct (1–9). It is most often located near the center of the nucleus but in 1, 6, and 8 it is eccentric. *E. histolytica* is no doubt the most important intestinal protozoan infecting man. It is cosmopolitan and occurs at all elevations, sea level to over 15,000 ft. It infects people of all races and economic levels but is more common in those living in areas less affluent. There appear to be racial and geographic variations in clinical disease. For example, in South Africa, infection produces more serious disease in blacks than in Asians or whites; in Colombia, infections usually result in more serious disease in Medellin than in Bogota. Whenever found, *E. histolytica* should always be considered a potential pathogen.

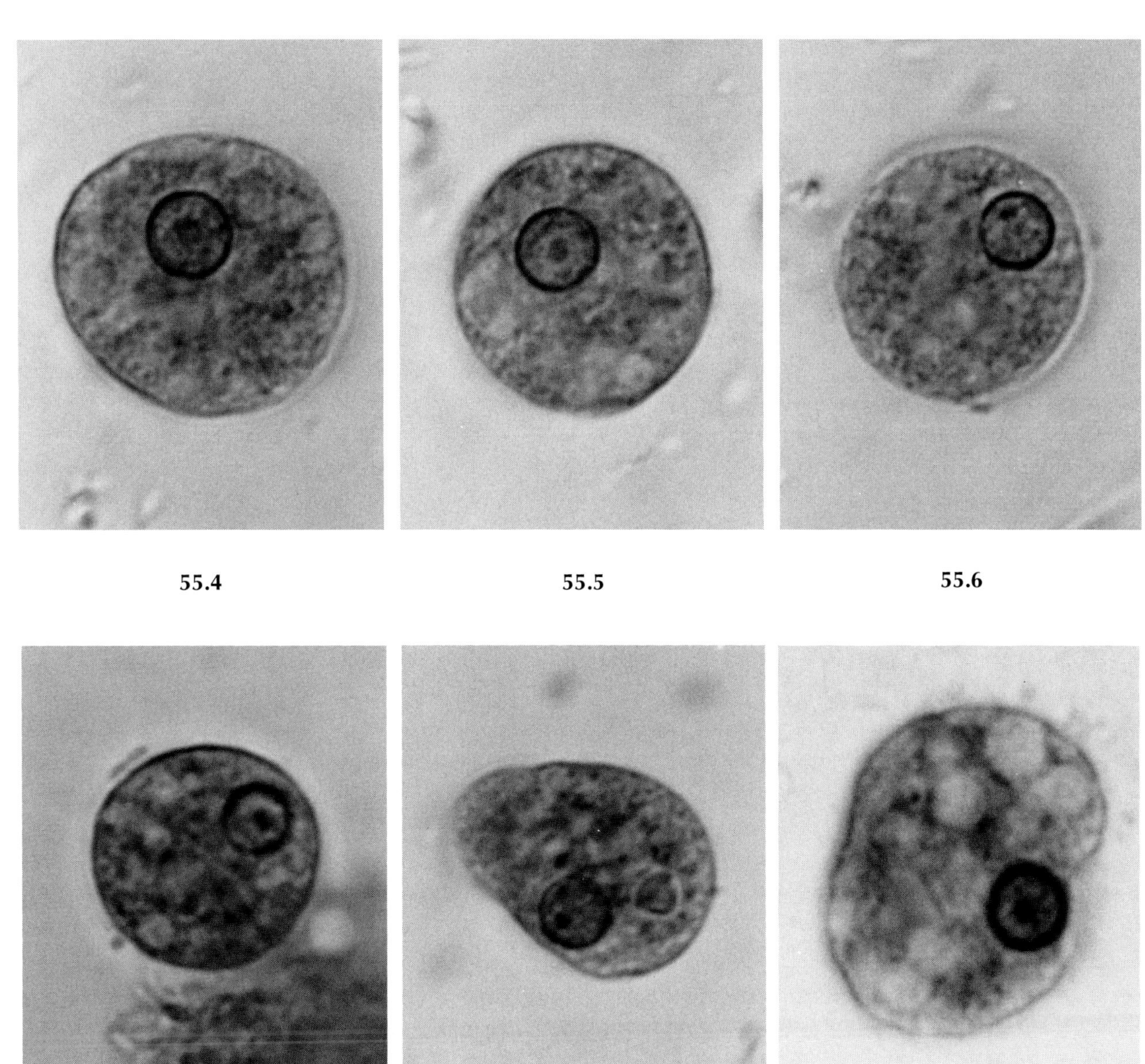

55.4 55.5 55.6

55.7 55.8 55.9

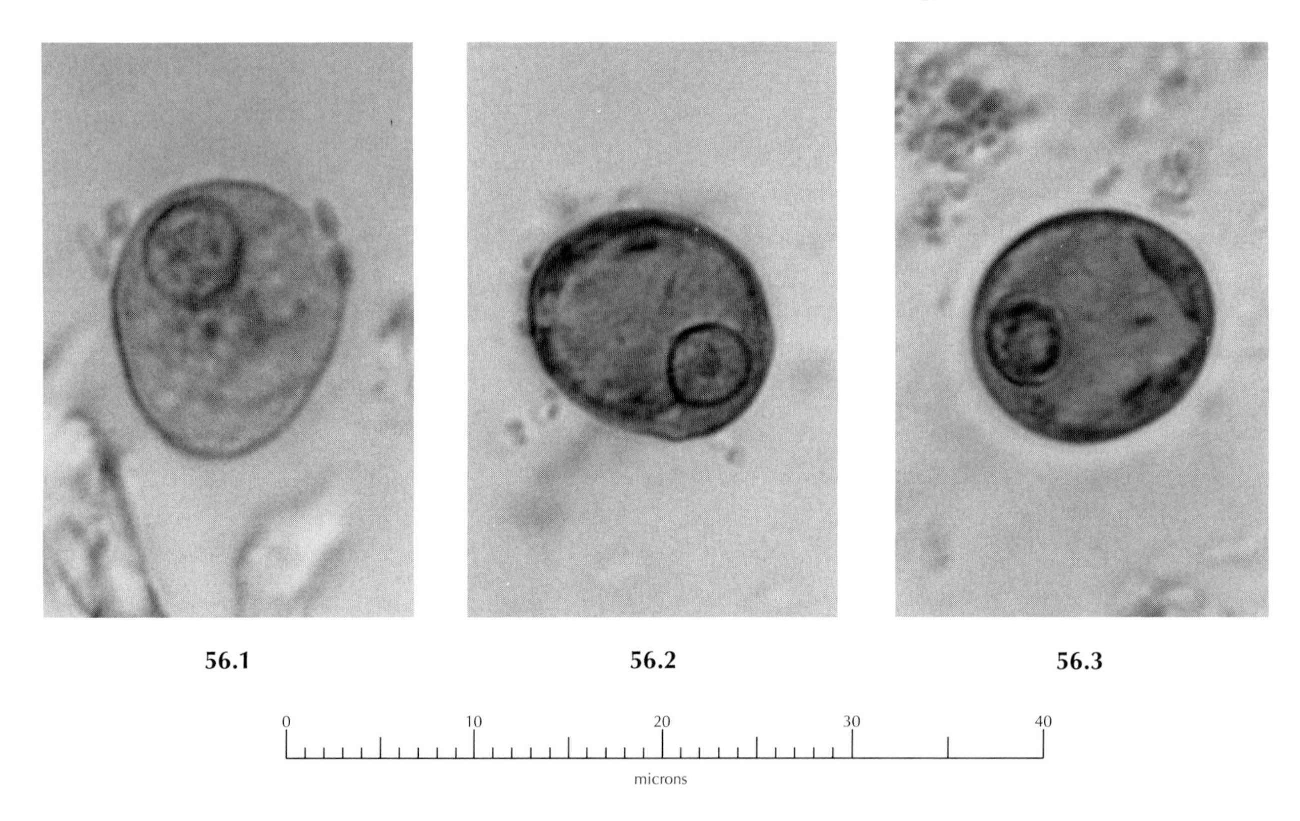

Plate 56. *Entamoeba histolytica* cysts (large race). 1–9: MIF. Almost immediately after the cyst wall is formed, the nucleus may begin division (1). In this form, there are no cytoplasmic structures or inclusions. Alternately, a glycogen vacuole that is quite large may form (2, 3). Nuclear division continues to form two nuclei (4, 5), then four nuclei (6–9). Peripheral chromatin is evenly beaded and the karyosome is near central (9). While nuclear division is taking place, changes are also taking place in the cytoplasm. Cytoplasmic materials may concentrate into small masses (4), coarse granules (6), or chromatoid bars (7). After the four nuclei are formed, vacuoles and structures in the cytoplasm disappear. The cytoplasmic granules begin to be organized (8), become equal size, and are arranged in a pattern of small circles. Only then is the cyst mature and infective (9). All the changes in the cyst occur while it is in the intestine. Cysts that are not mature when evacuated, do not continue their development outside of the host. Cysts are usually passed to a new host from a contaminated source, often water (including ice), dirty eating utensils, or dishes and, less often, food. Food handlers most often are responsible for contaminating the source of infection through unclean work habits. Farm workers may be responsible for contaminating produce but, unless the produce goes directly from garden to kitchen or table, cysts probably will not survive long enough to initiate an infection.

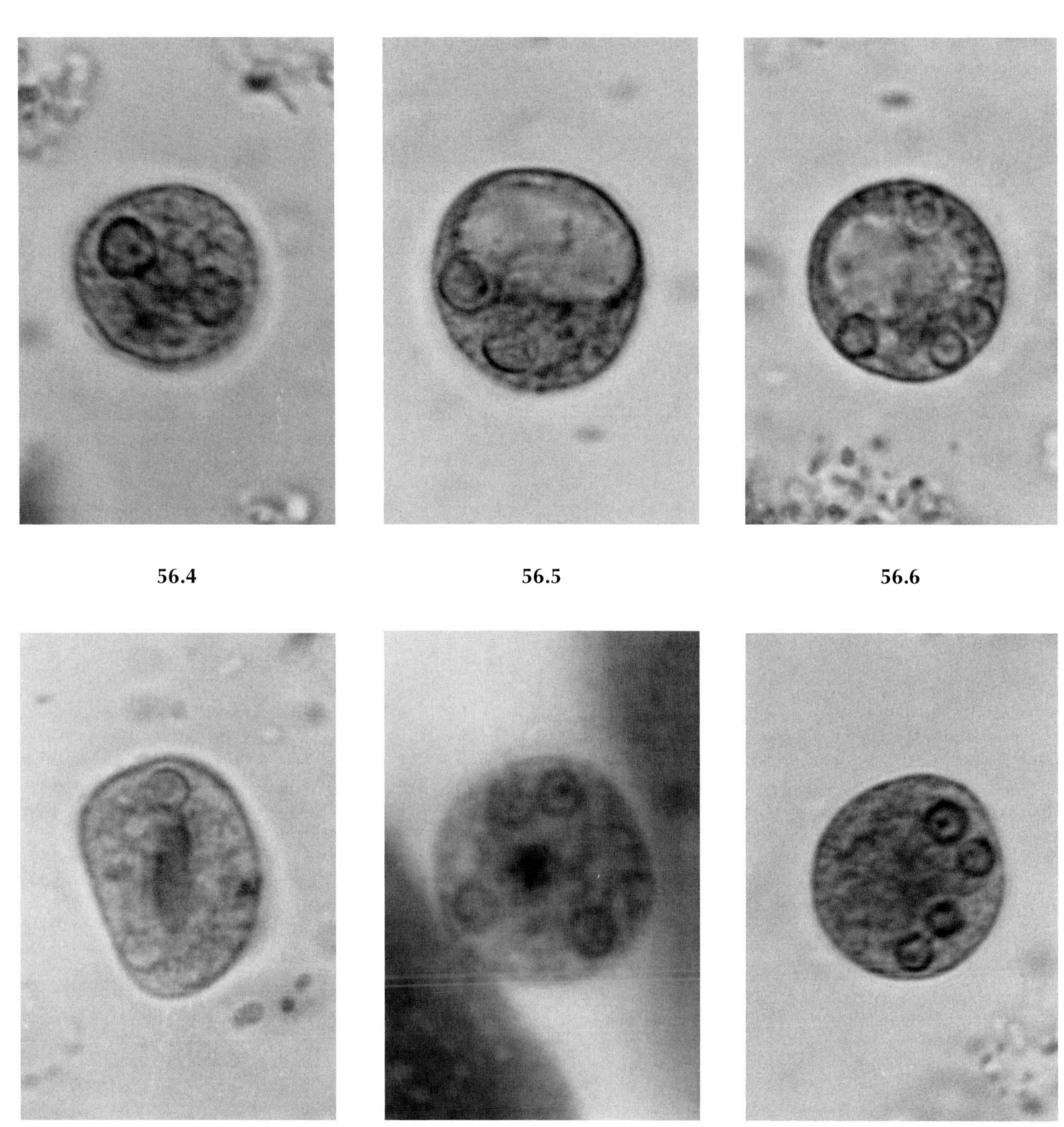

56.4 56.5 56.6

56.7 56.8 56.9

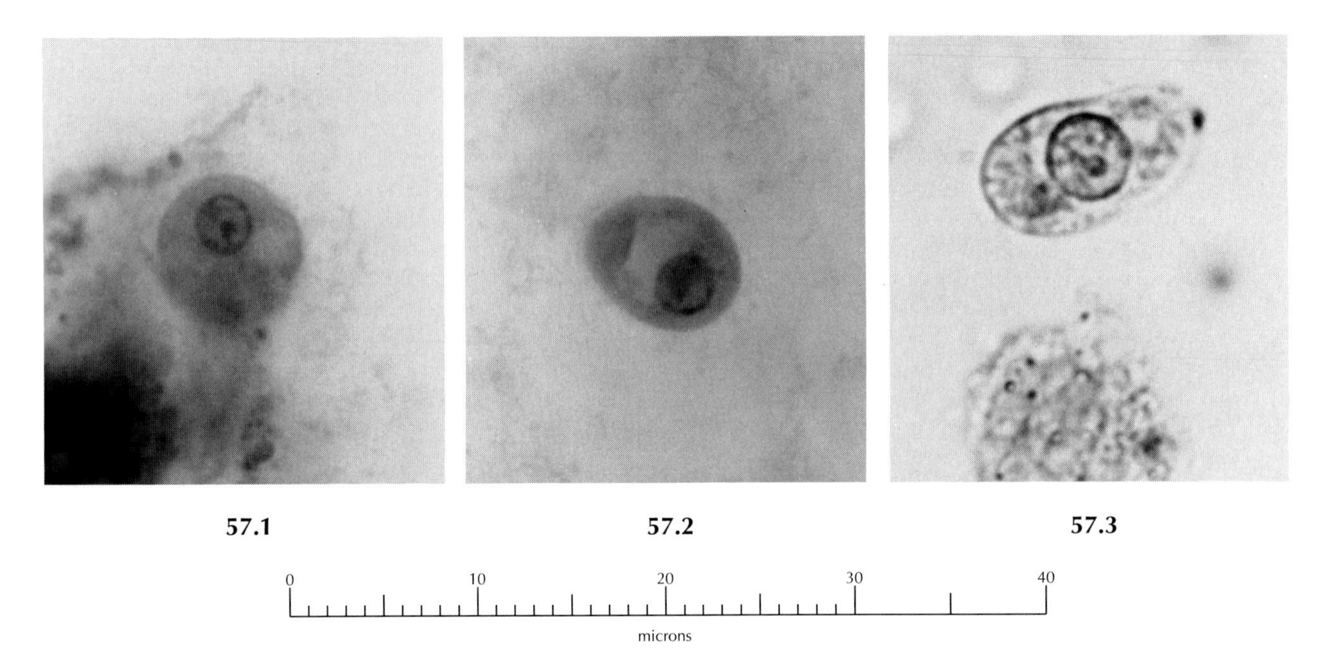

Plate 57. *Entamoeba histolytica* trophozoites (1–5) and cysts (small form) (6–9). 1: T, PIF; 2–5: T, MIF; 6–9: C, MIF. Trophozoites and cysts are usually less than 10 μm in diameter. When there is a single nucleus, it usually measures more than one fourth the diameter of the trophozoite or cyst (1 to 6) and is approximately the size of the nucleus seen in the large form. Rarely, it is nearly half the diameter of the amoeba (3). The trophozoite may have a few large vacuoles without inclusions (1, 2). Occasionally some type of inclusion may be present (4). The precyst, like that of the large form, loses all vacuoles before encysting (5). The nuclei of the cyst lose their structural appearance when in division (7). The early cyst may have a large glycogen vacuole (6) or initiate the formation of chromatoid bars (8, 9). Eventually, four nuclei are formed. Chromatoid bars may be present until the cyst is fully mature (9). When four nuclei are present, their diameter is greater than or equal to one fifth the diameter of the cyst, whereas the diameter of the nuclei of the mature cyst (four nucleate) of *E. hartmanni* is *less* than one fifth the diameter of the cyst (see Plate 58:7–9). The existence of a small form (race) of *E. histolytica* is not generally accepted. The small form is occasionally seen in patients where the large form is present. In patients where it occurs alone, there usually are no overt clinical symptoms. Untreated patients in which the small form was originally found were followed over several months. In some patients who developed intercurrent infections, the large form rather than the small form appeared in their stools.

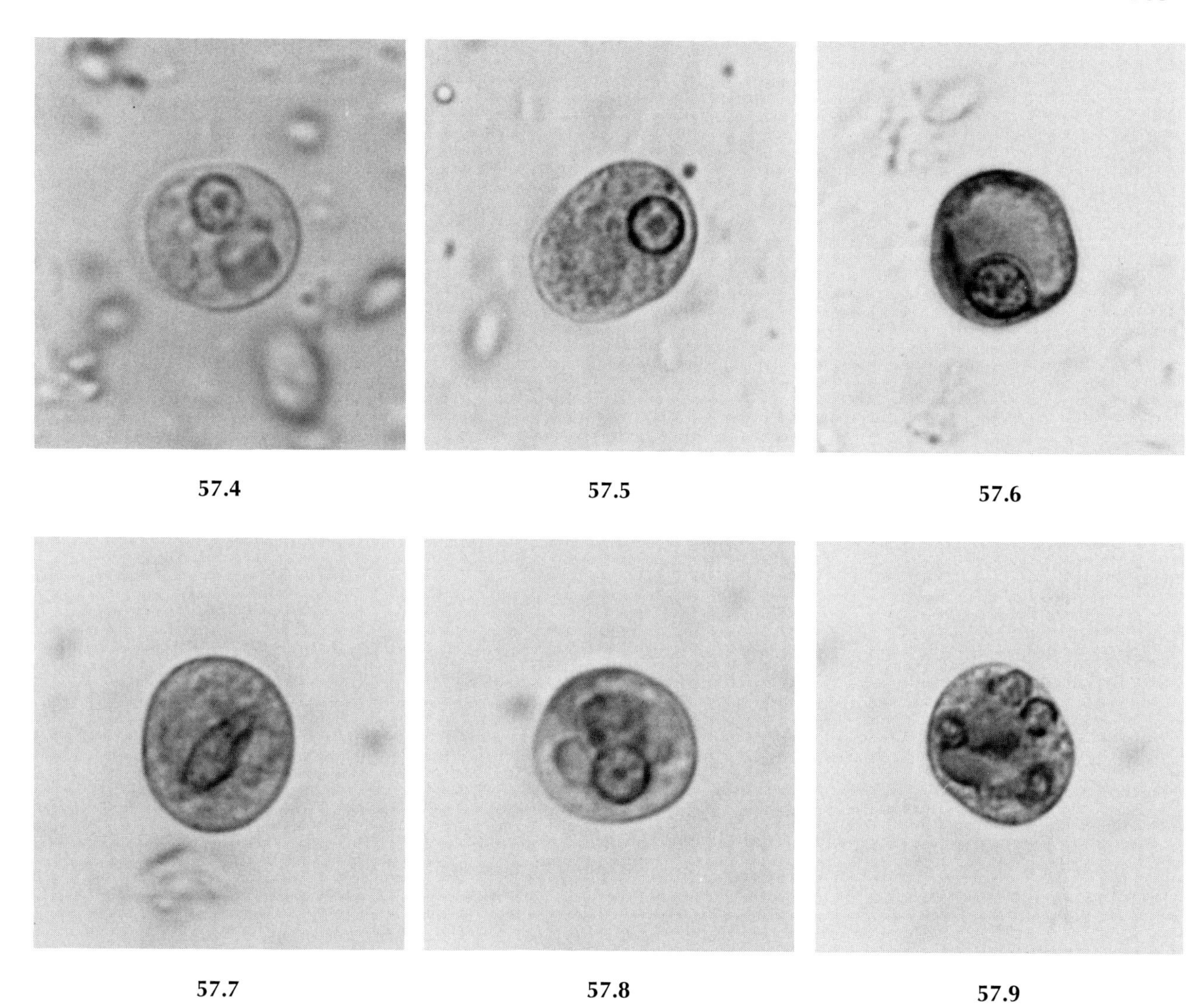

57.4 57.5 57.6

57.7 57.8 57.9

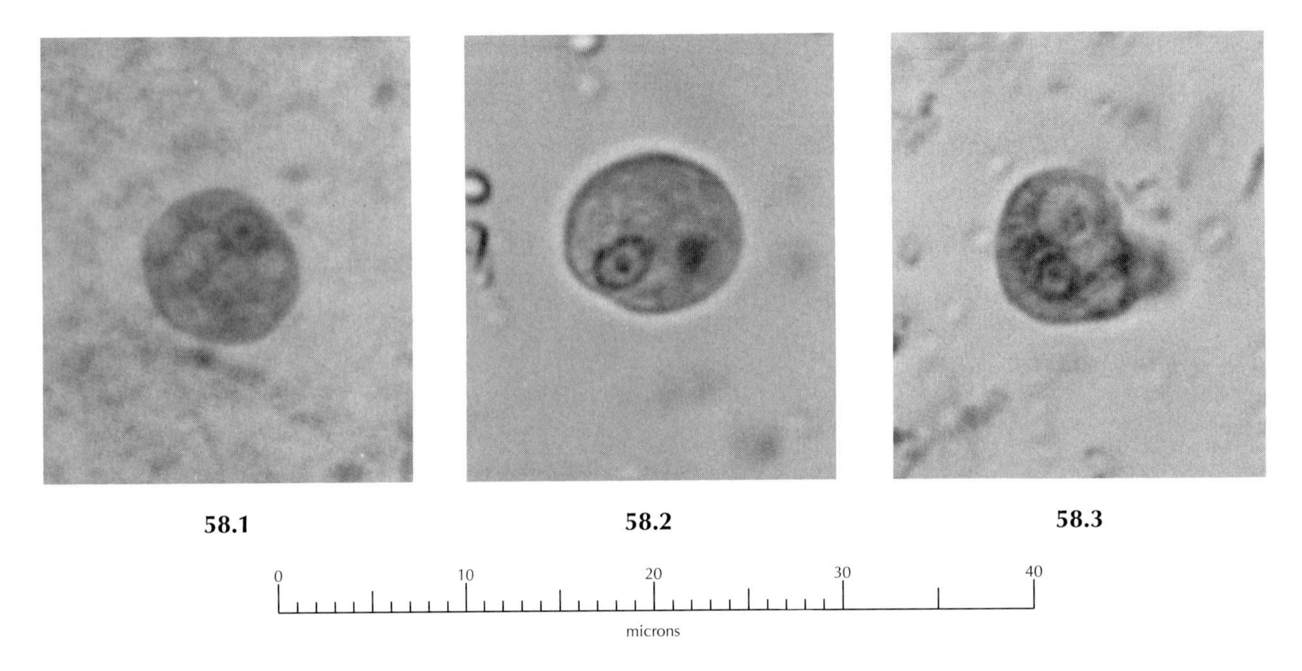

Plate 58. *Entamoeba hartmanni* trophozoites and cysts. 1: T, MIF; 2, 3: T, PIF; 4–8: C, MIF; 9: C, PIF. Trophozoites usually have small vacuoles (1–3) often with included bacteria (1) which helps distinguish them from the small form of *E. histolytica*. Often they are round but some may appear with pseudopodia extended (3). The nucleus usually measures less than one third of the diameter of the trophozoite. It has evenly distributed, often beaded peripheral chromatin and a distinct, usually central karyosome (1–3). After encysting, the cytoplasm becomes condensed (4) and large glycogen vacuoles may form (5). Division yields first two nuclei (6) then four (7–9). As they divide the nuclei become much smaller and take up only about one fourth of the volume of the cyst. Round end chromatoid bars may form in the immature cyst (8, 9) but will be absorbed when the cyst reaches maturity. *E. hartmanni* is distributed worldwide. It is not considered to be pathogenic but is an indicator of fecal contamination. It should be differentiated from the small form of *E. histolytica* especially in patients with gastrointestinal symptoms.

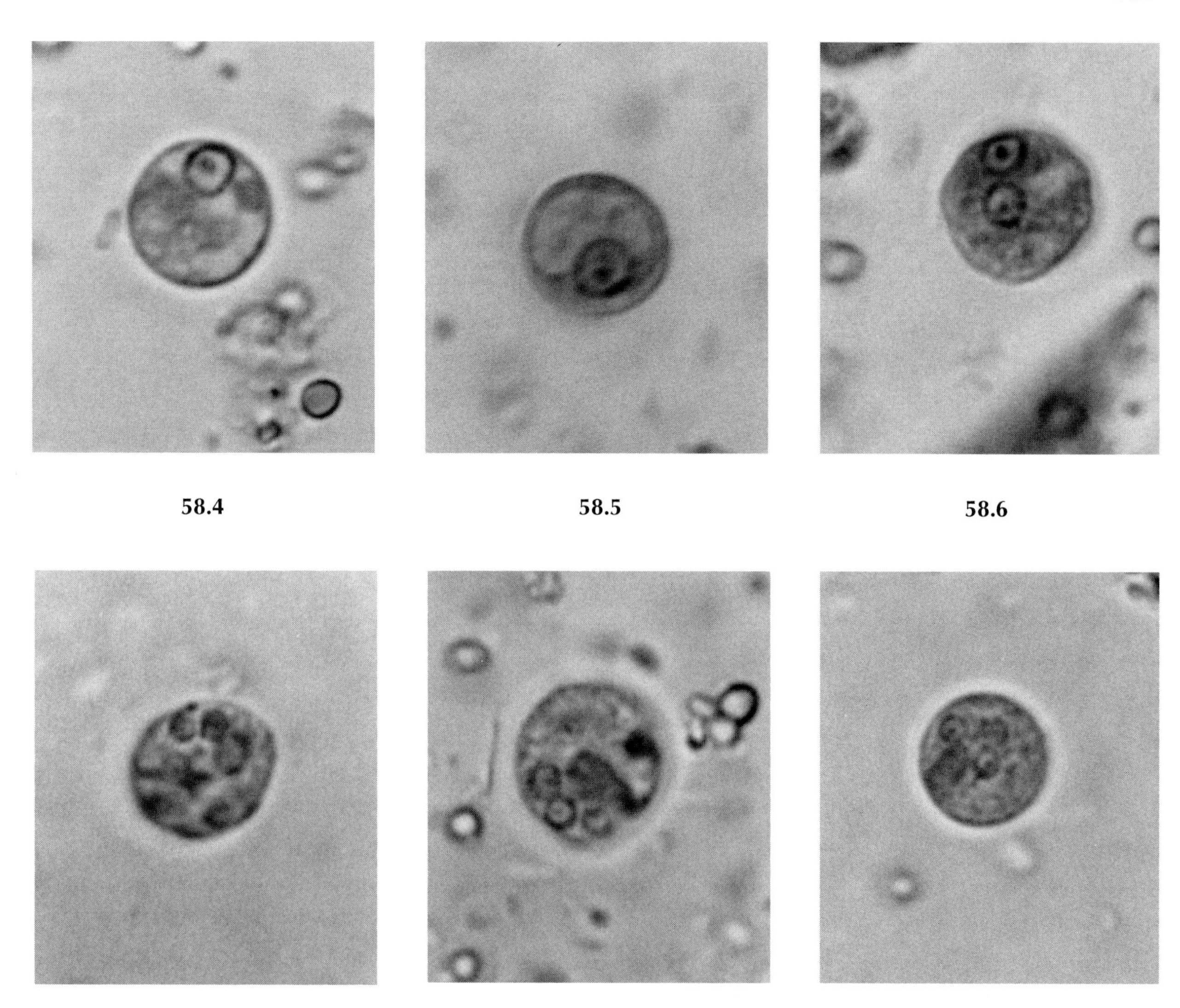

58.4 58.5 58.6

58.7 58.8 58.9

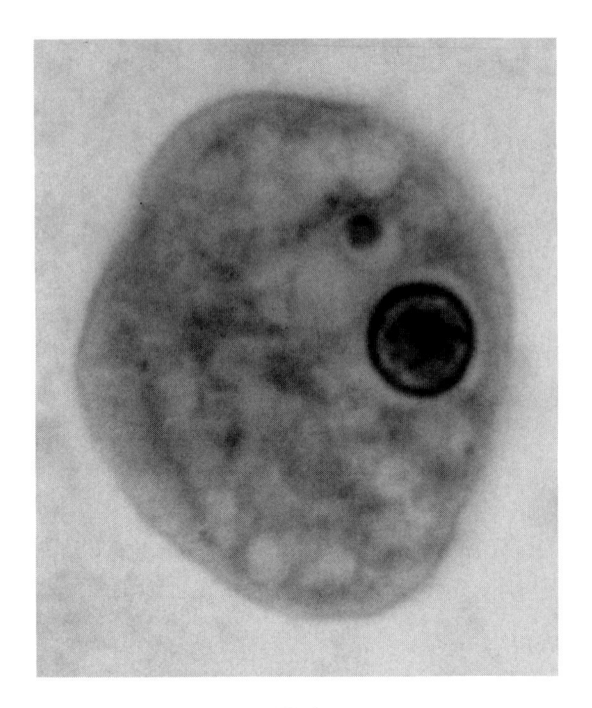

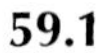

59.1

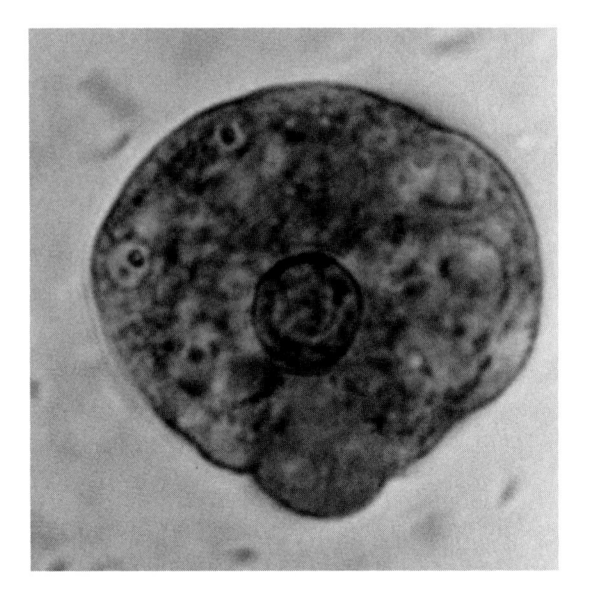

59.2

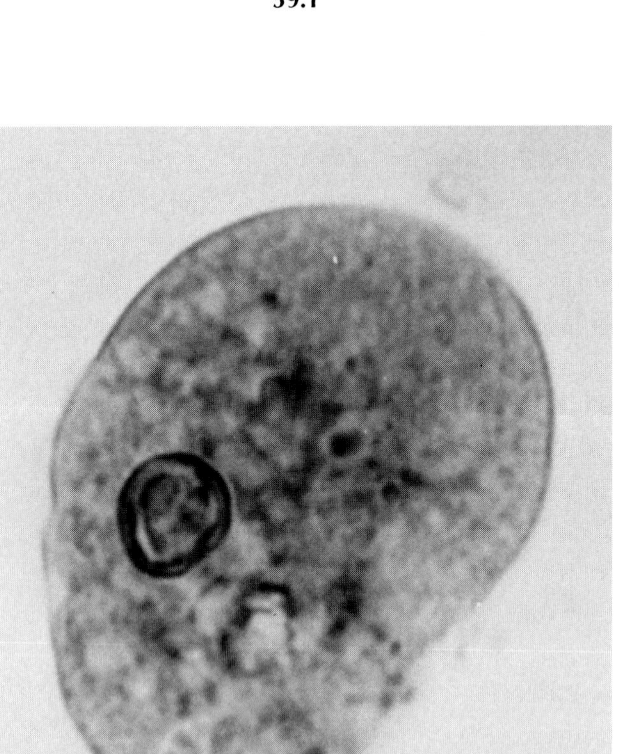

59.3

Plate 59. *Entamoeba coli* trophozoites. 1: PIF; 2–4: MIF; 5: PIF; 6: MIF. The cytoplasm of *E. coli* trophozoites is usually filled with vacuoles (1 to 6) most of which are small. Some individual trophozoites have only a few inclusions (1) but most active trophozoites have many inclusions (2–6), usually bacteria. Because of the many vacuoles and the included bacteria, the cytoplasm often has a cluttered appearance (2, 3, 5, 6). Usually, the trophozoite stays in one place and projects small pseudopodia (2). Occasionally, it will show progressive motility (6). The pseudopodia are not free of granules as in *E. histolytica* (2, 6). The peripheral chromatin of the nucleus is most often irregular in the trophozoite (2, 3, 5) but may appear relatively smooth and evenly dispersed (1, 4). The karyosome is often difficult to detect in the trophozoite because of other materials in the nucleus that stain (1–5). The karyosome is eccentric when seen. When the precyst is formed (not shown), *E. coli* appears more similar to *E. histolytica* (see Table 2 and Plate 47). Since *E. coli* is transmitted from contaminated sources as is *E. histolytica*, both are often present in specimens from the same individual. At such times, it is especially important to identify both species. *E. coli* is a commensal; however, when found, it indicates contact with raw feces. It is sometimes seen in cases of intestinal upset where pathogenic bacteria or viruses are the cause. Again, differentiation from pathogenic *E. histolytica* is important.

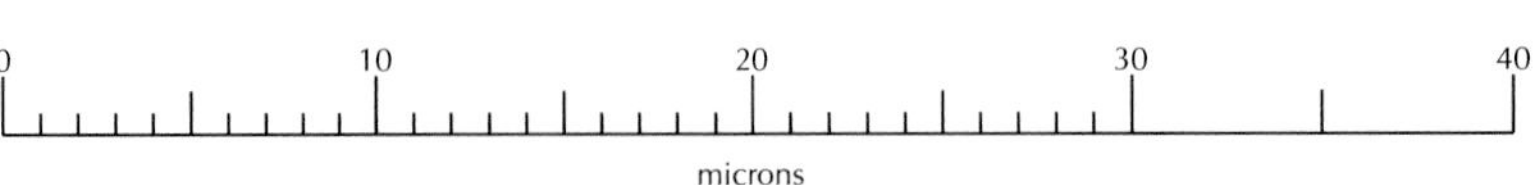

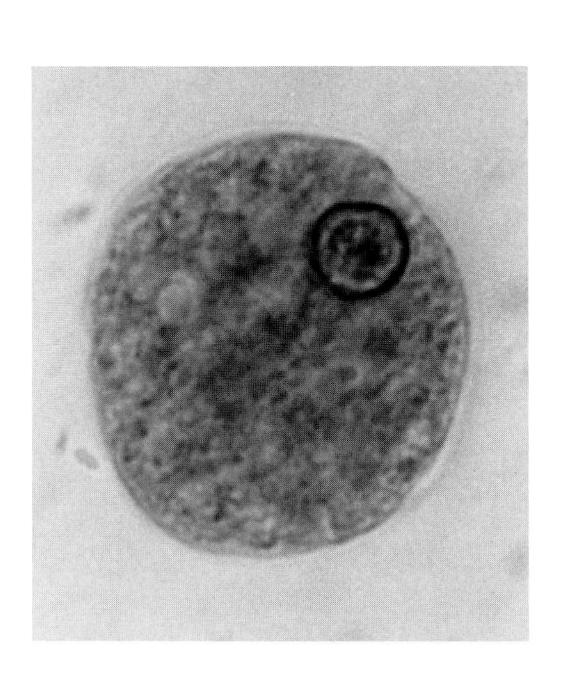

59.4

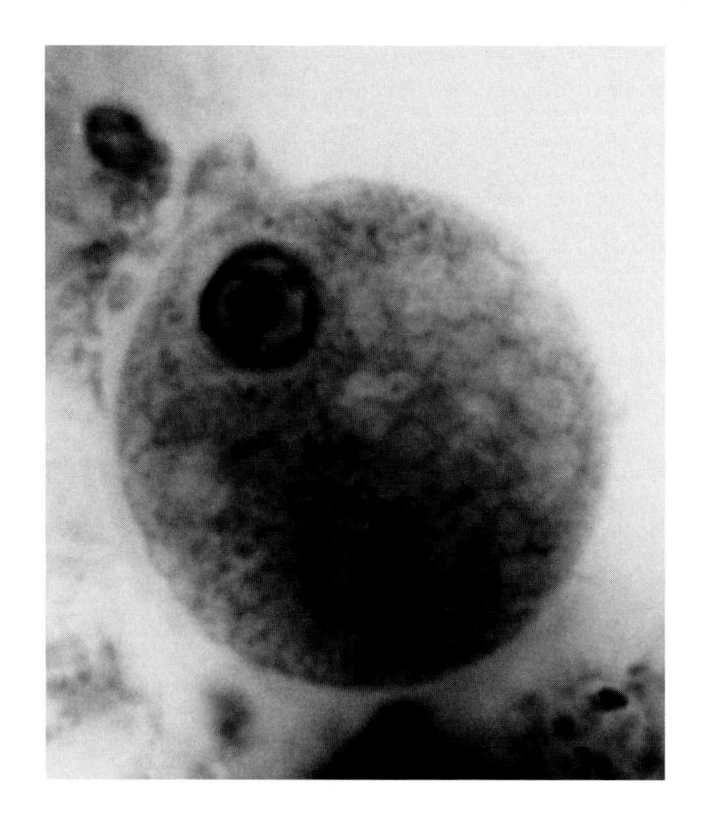

59.5

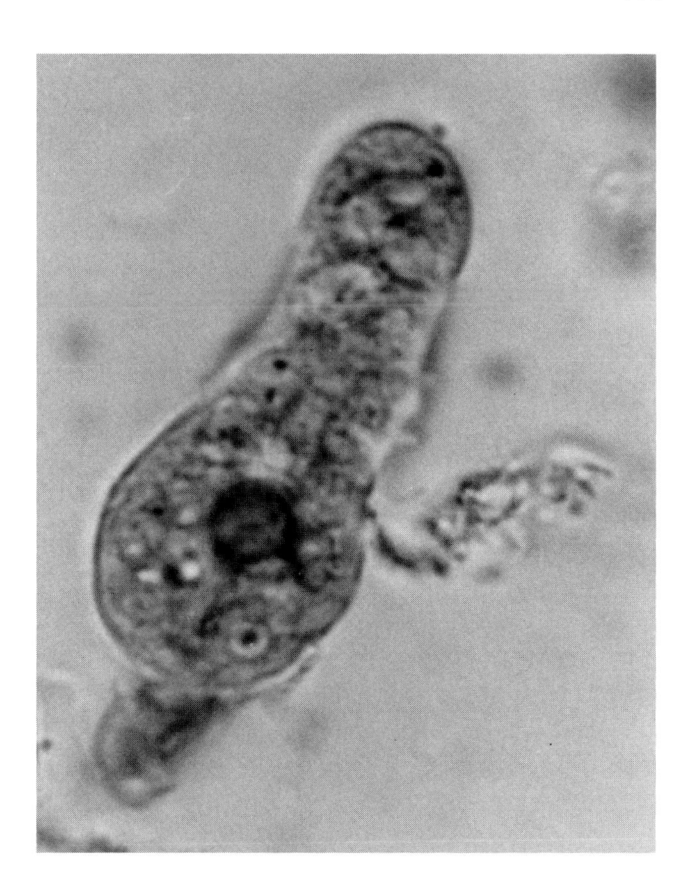

59.6

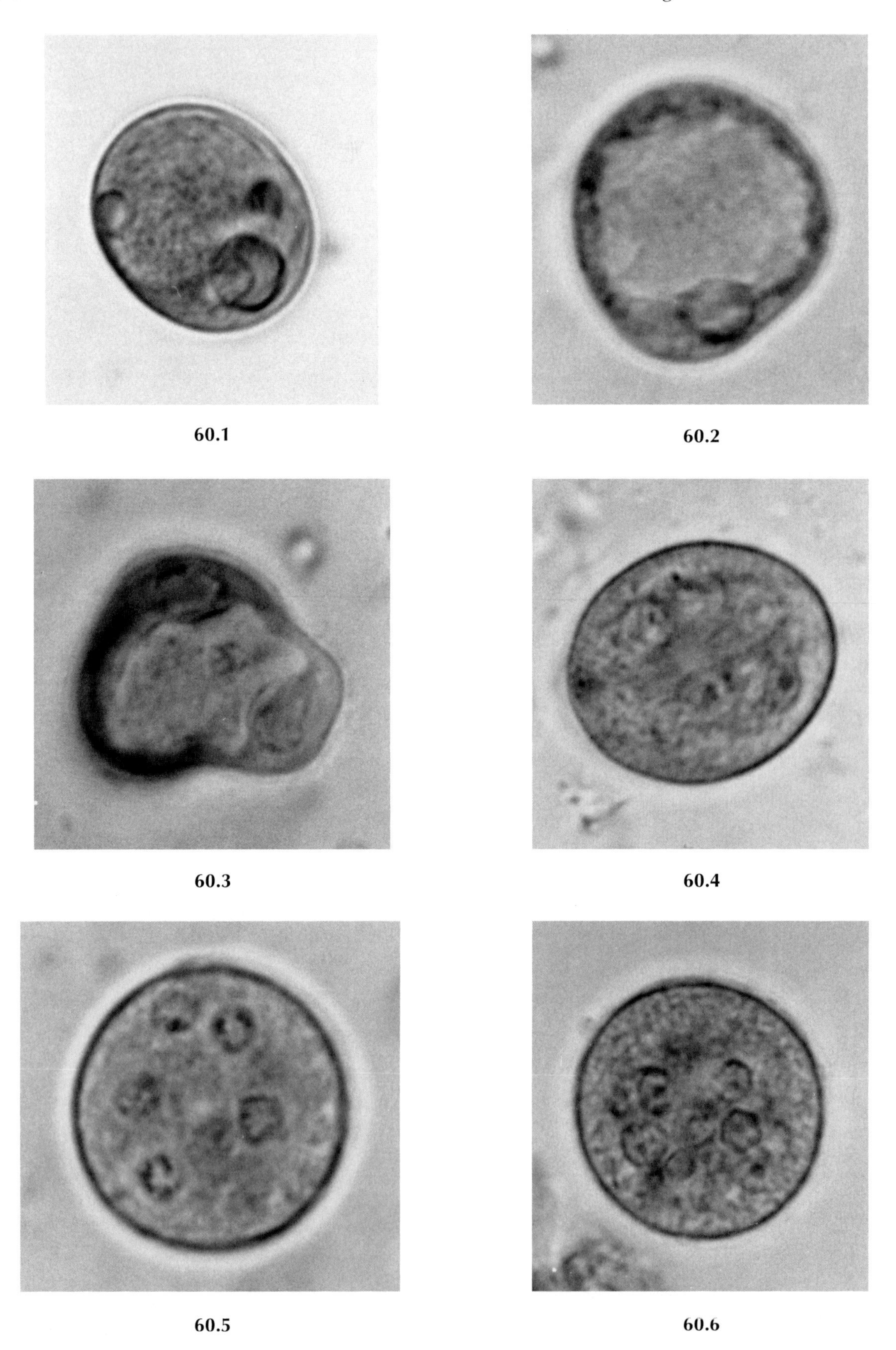

60.1

60.2

60.3

60.4

60.5

60.6

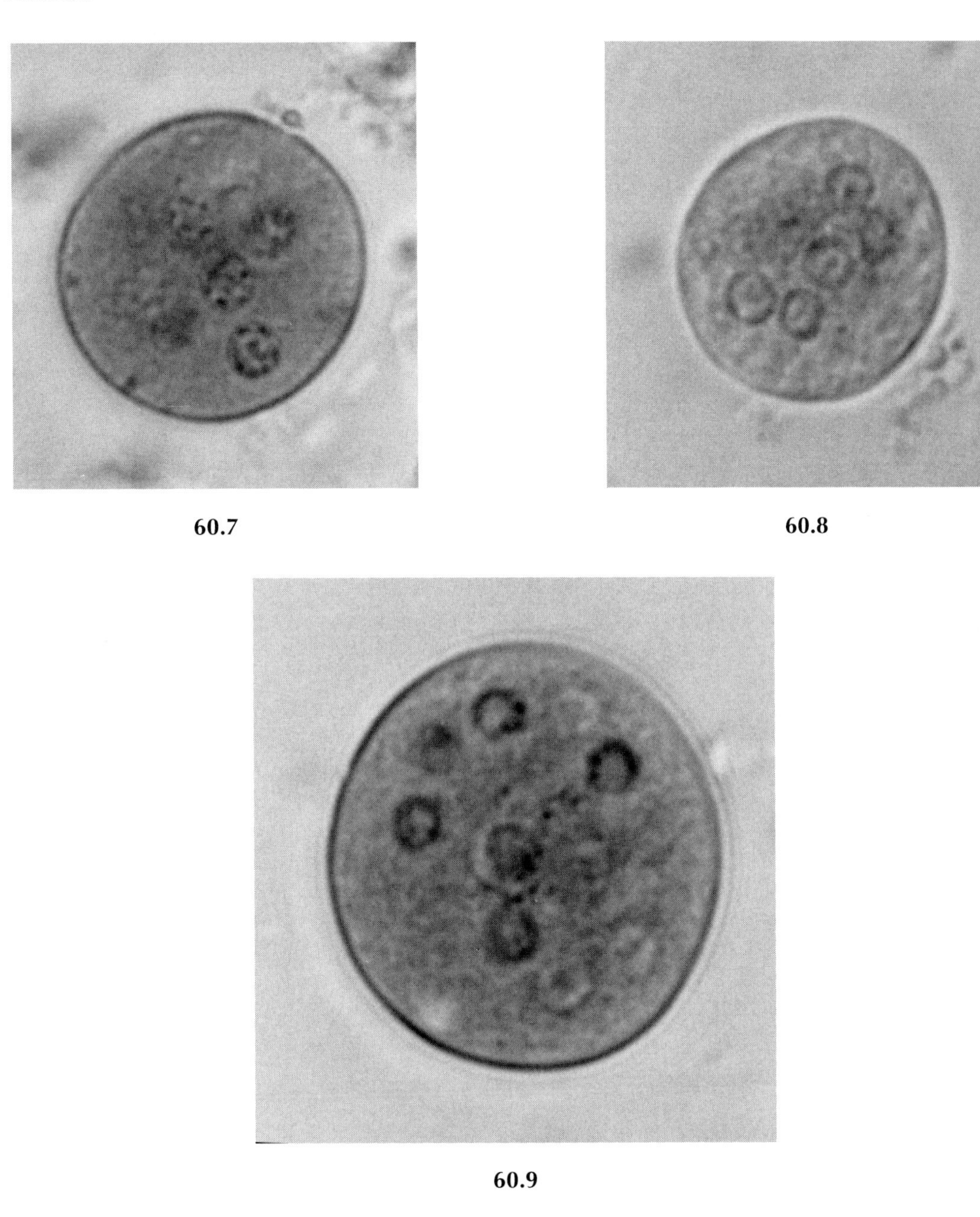

60.7

60.8

60.9

0 10 20 30 40

microns

Plate 60. *Entamoeba coli* cysts. 1–5: MIF; 6: PIF; 7–9: MIF. The cysts of *E. coli* vary greatly in size from 15 to 30 μm in diameter but they are usually between 15 and 20 μm (1–8). The early cyst almost always has a large glycogen vacuole (1–3). The mononucleate cyst (1) often appears similar to that of *E. histolytica* (see Plate 56). The vacuole of the binucleate cyst often is very large and appears to press the cytoplasm and nuclei against the cyst wall (2, 3). As the cyst matures, nuclear division continues, the glycogen vacuole usually disappears, and occasionally bundles of splinter-like chromatoid bodies appear (4). When nuclear division has produced eight nuclei, the cytoplasm begins changing. Granules first appear irregular in size and arrangement (5), then they begin to form a pattern of circles (6), and finally granules become evenly dispersed (7–9). The nuclei are usually round, the peripheral chromatin is irregularly beaded, and the karyosome is distinct (6–8). Characteristically, there are eight nuclei but nuclear division may continue. The cyst in 9 had 16 nuclei. The mature cyst is the infective form that can be transmitted to a new host. Immature cysts that are passed with the feces apparently do not develop further. *E. coli* is one of the most commonly found intestinal parasites.

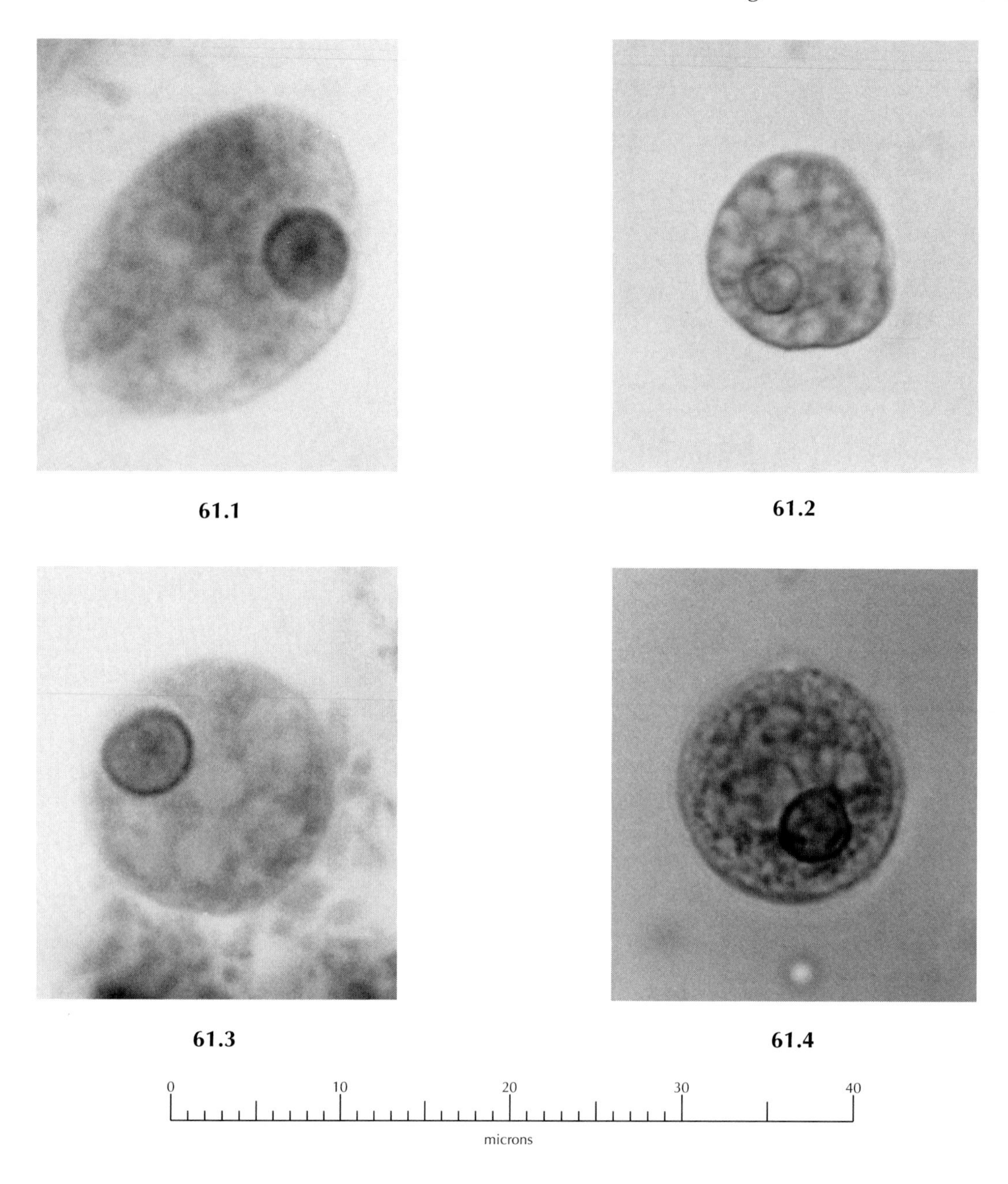

Plate 61. *Entamoeba polecki* trophozoites and cysts. 1: T, MIF; 2–5: T, PIF; 6: C, MIF; 7–8: C, PIF; 9: C, MIF. The trophozoites of *E. polecki* may vary greatly in size (1, 2) and its cytoplasm may have few relatively large vacuoles as does *E. histolytica* (1, 3) or have many small vacuoles as does *E. coli* (2, 4). Like other amoebae, the precyst loses its inclusions and vacuoles and becomes relatively clean-appearing (5). After the cyst wall is formed, bizarre structures appear in the cytoplasm that are somewhat similar to chromatoid bodies in *E. coli* and *E. histolytica* (6–8). These bodies may be numerous and occupy most of the cytoplasm along with some vacuoles (9). The nucleus of both the trophozoite and the cyst usually has relatively thin and evenly dispersed peripheral chromatin (2–6) but it may appear more coarse (1) especially in cysts (7, 9). The karyosome usually appears as a pinpoint (2, 4–7). The very dense nucleoplasm, which often obscures the karyosome, is a characteristic feature of *E. polecki*. The nuclear membrane is often misshaped (6, 7, 9). Cysts usually have a single nucleus but rarely are seen with two nuclei. *E. polecki* is generally considered to be an intestinal parasite of pigs but is often seen in monkeys. Although it is not usually considered to be a common intestinal parasite of man, it is being seen more frequently in the U.S. in refugees and international travelers.

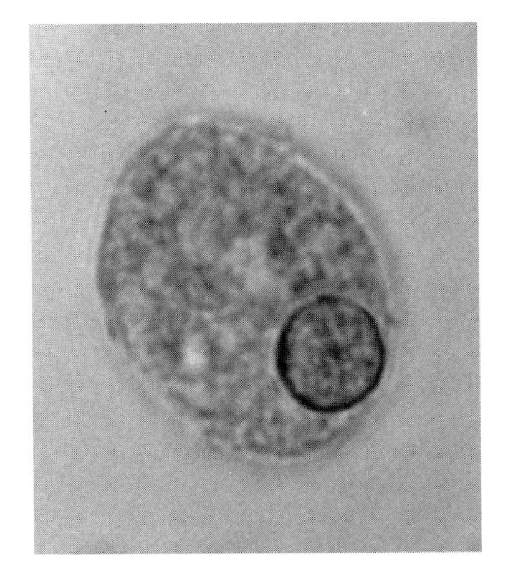

61.5

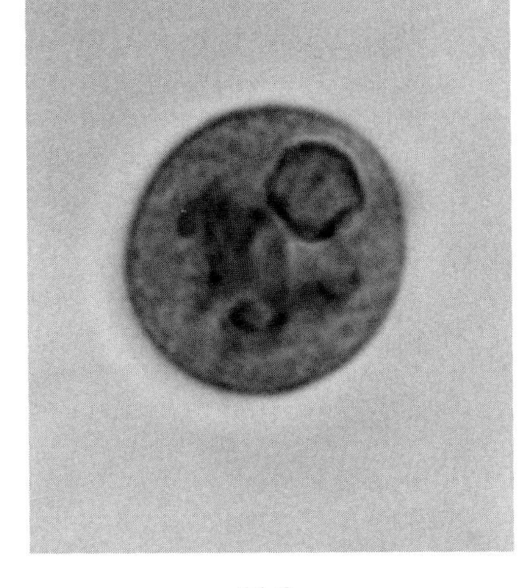

61.6

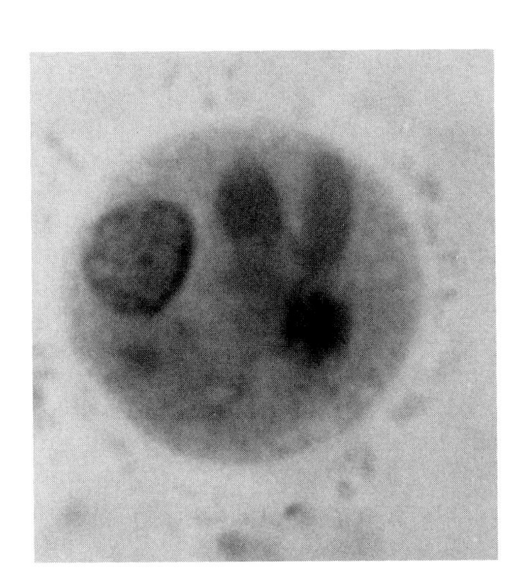

61.7

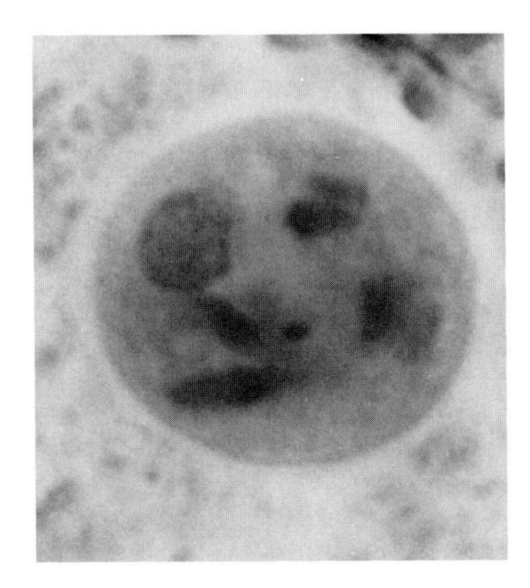

61.8

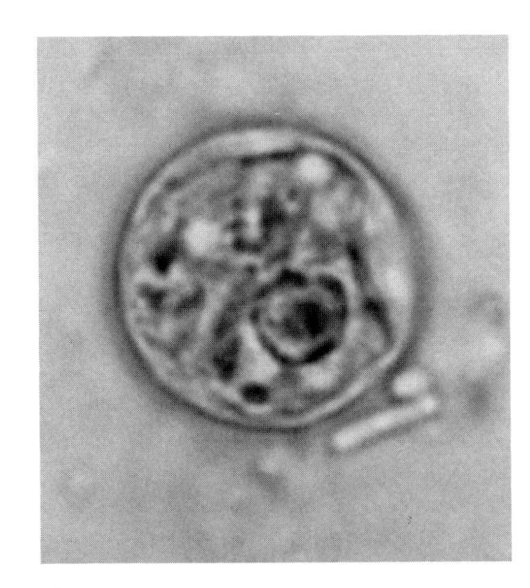

61.9

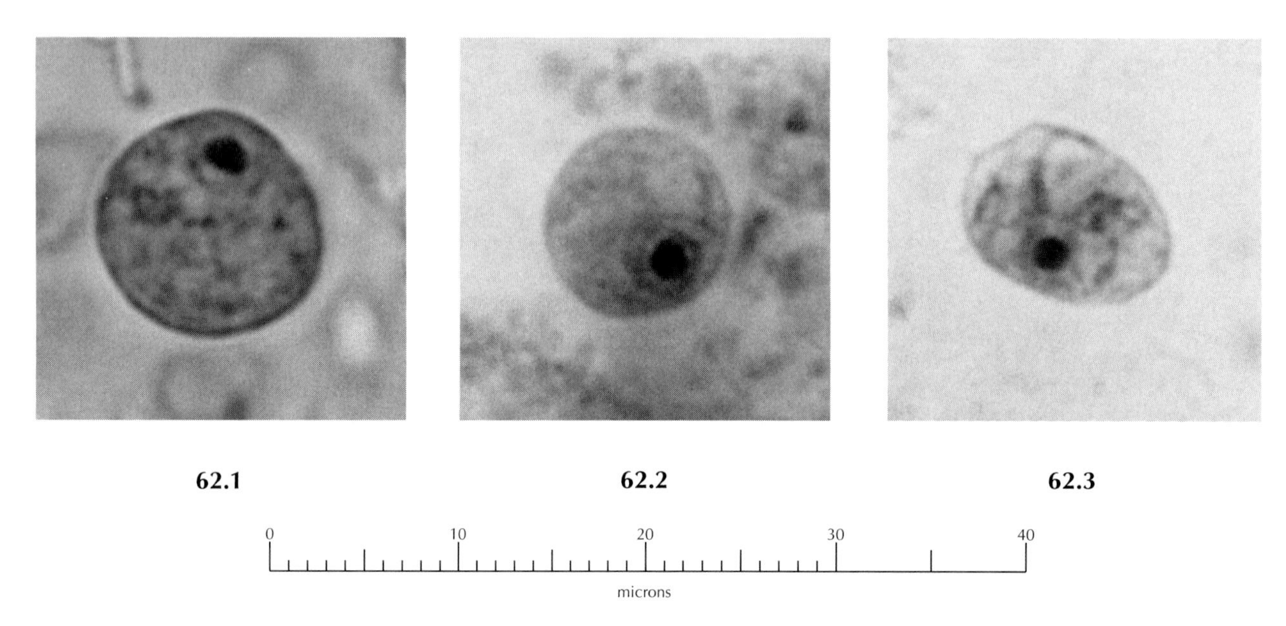

Plate 62. *Endolimax nana* trophozoites and cysts. 1: T, MIF; 2–5: T, PIF; 6: C, PIF; 7–9: T, MIF. The cytoplasm of *E. nana* usually includes a number of small vacuoles (1–5). Because of their many vacuoles, the very active trophozoites (3, 4) may appear more delicate and similar to *Dientamoeba fragilis.* Usually, the trophozoites are rounded up and appear inactive (1, 2, 5). The nucleus is distinguished by the large flat karyosome (1–3) and the cytoplasmic granules appear to adhere to the nuclear membrane (1, 2, 4). In some trophozoites the karyosome is divided into several parts as if in division (5). Cysts, when first formed, sometimes have glycogen vacuoles that may persist until nuclear division is completed (6). The four nuclei appear to be in punched holes in the cytoplasm with no sign of a nuclear membrane (7–9). Many of the cysts appear to have bodies in the cytoplasm that stain darkly and are clearly not nuclei (8). The cytoplasm of the mature cyst becomes very smooth and almost nongranular (9). *E. nana* is a commensal that inhabits the large intestine of man. Since it is transmitted from a contaminated source, it is often seen in the presence of other parasites in fecal specimens. A small cyst may be no greater in diameter than 5 or 6 μm and can readily be overlooked. Cysts have a high refractive index and are often seen more readily using the principle of the Becke line (see Appendix).

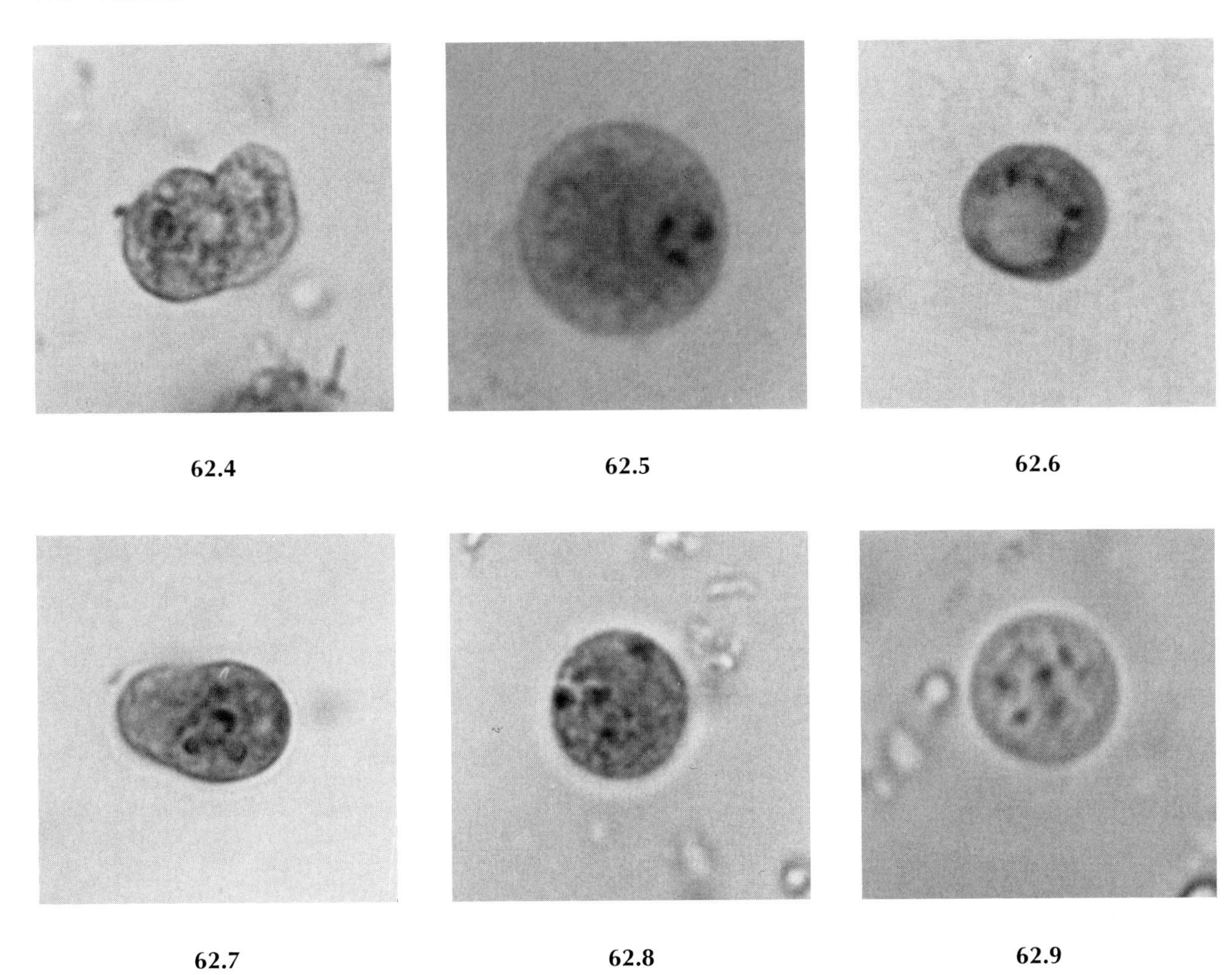

62.4 62.5 62.6

62.7 62.8 62.9

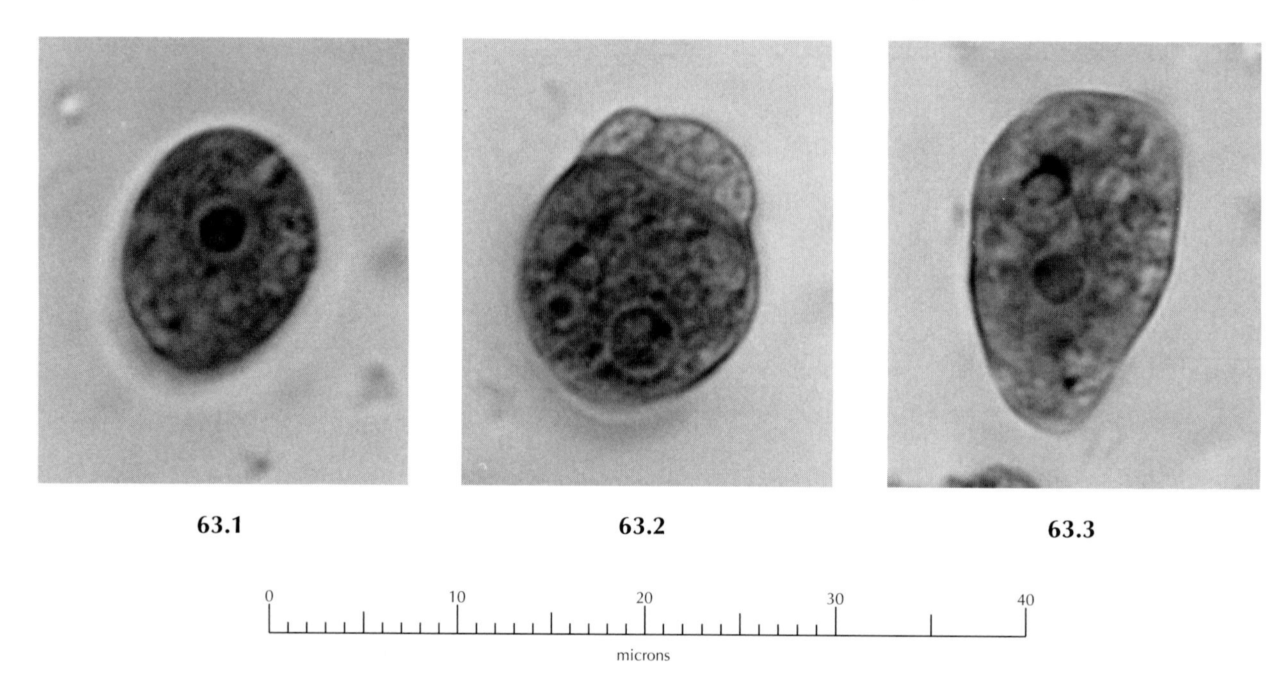

63.1 63.2 63.3

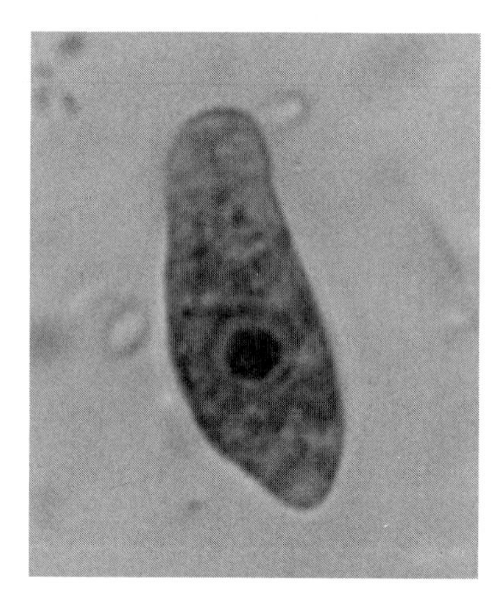

63.4

Plate 63. *Iodamoeba butschlii* trophozoites and cysts. 1: T, MIF; 2: T, PIF; 3–6: T, MIF; 7–9: C, MIF. Trophozoites of *I. butschlii* vary greatly in size (4, 6). The active trophozoite has many vacuoles and ingests bacteria, yeasts, and other material present in the fecal mass (1–6). Pseudopodia vary from short blunt (2) to long narrow projections (6). The nucleus is unusually round with an indistinct membrane and a large karyosome (1–6). In wet preparations prepared from MIF- or PIF-fixed specimens, the karyosome in the trophozoite appears to be more evenly stained than in the cyst and the nuclear membrane of the cyst is usually irregular in shape (compare 1–6 with 7–9). The cyst has a single nucleus and one or more large glycogen vacuoles (7–9). The cytoplasm of the cyst also contains exceptionally large granules, especially on the periphery of vacuoles (7–9). Like most other intestinal amoebae, *I. butschlii* is cosmopolitan. It is considered to be nonpathogenic but is an indicator of fecal contamination. It must be differentiated from *E. histolytica* and *Dientamoeba fragilis.*

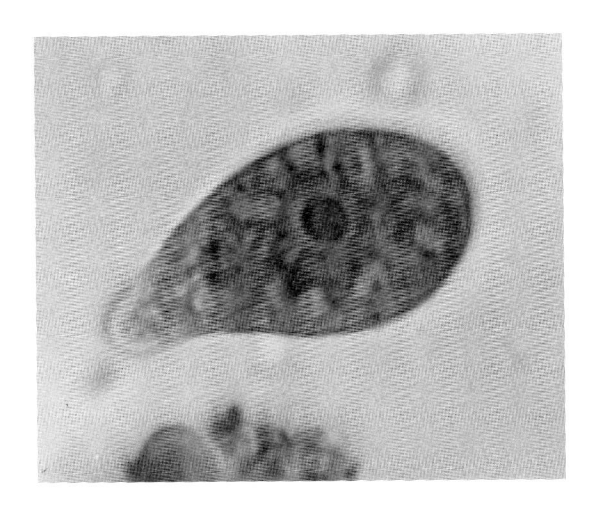

63.5

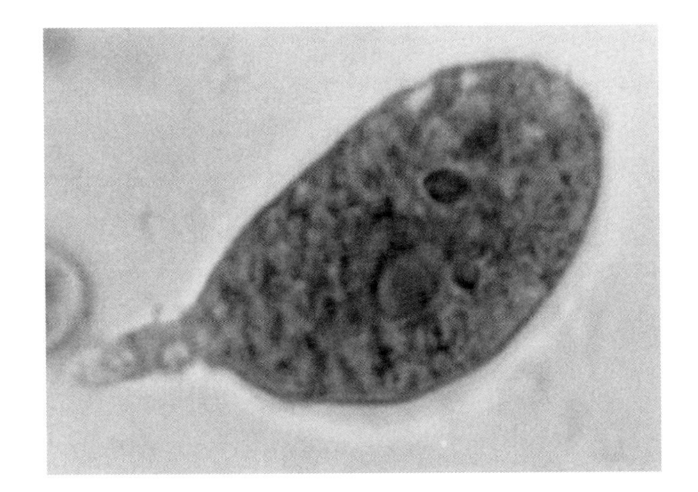

63.6

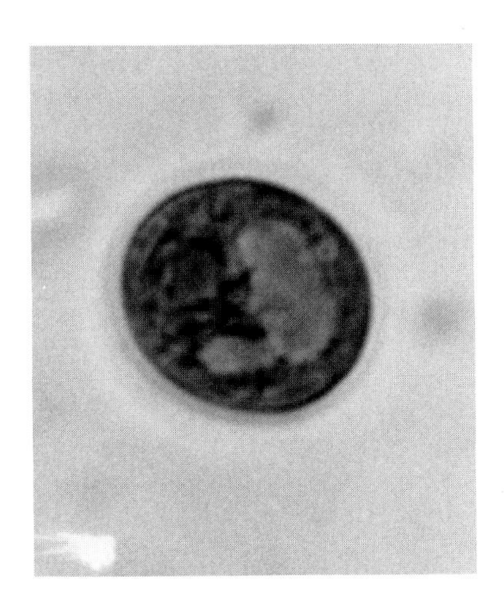

63.7

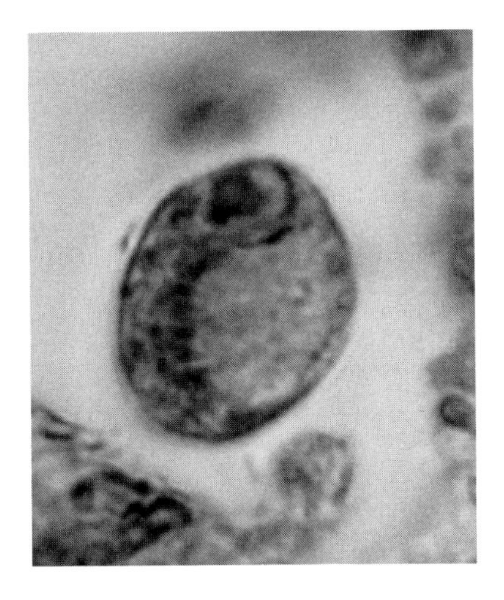

63.8

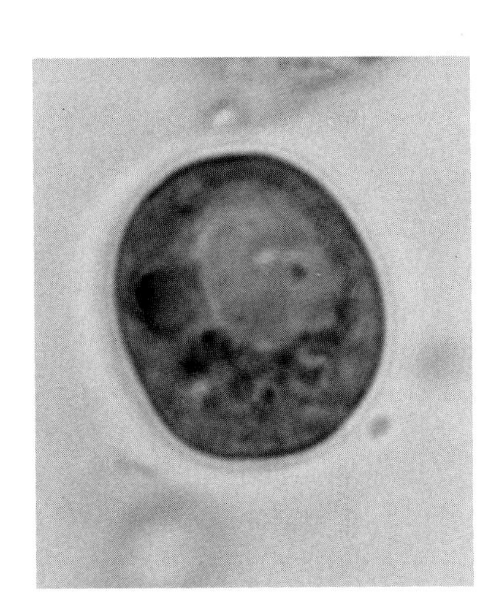

63.9

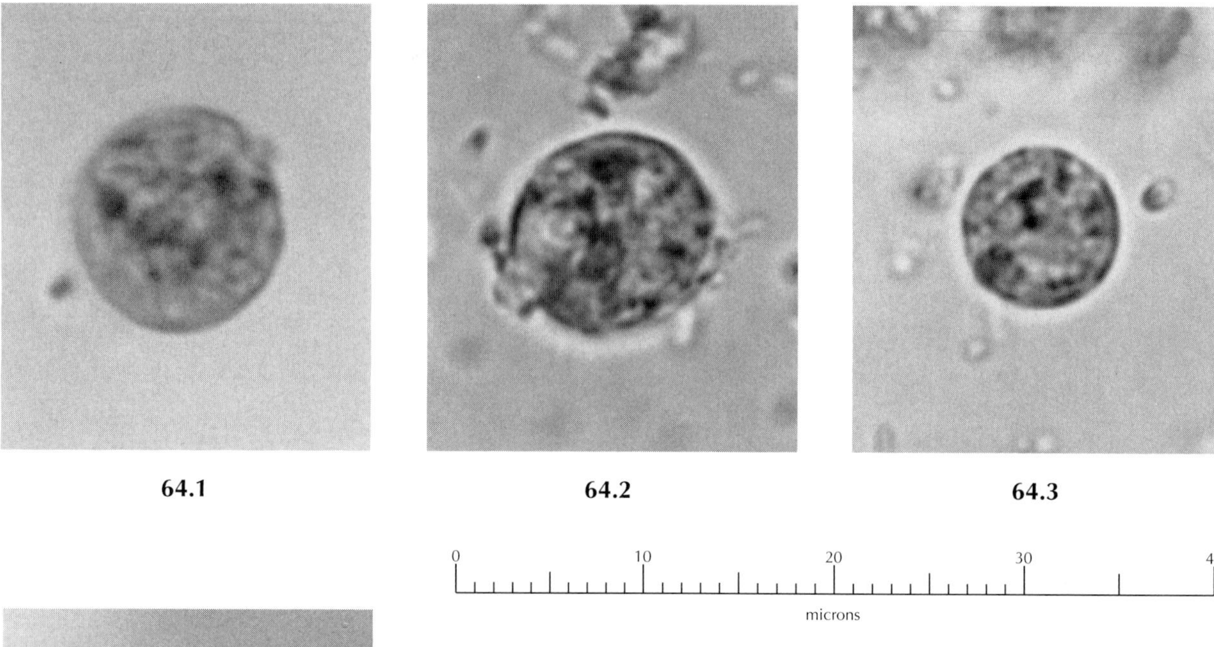

64.1 64.2 64.3

64.4

Plate 64. *Dientamoeba fragilis* trophozoites. 1: PIF; 2–9: MIF. The amoeboid trophozoites of *D. fragilis* may vary greatly in size (compare 3 and 8). The cytoplasm appears to be more uneven than in other intestinal amoebae. Vacuoles are very numerous with many vacuoles having ingested material. The trophozoites usually have two nuclei (1–3) but these are less well defined in wet preparations than in permanently stained fecal films (4–9). Nuclei are difficult to distinguish from some vacuoles with included material (4, 5, 8). In wet preparations the structure and texture of the cytoplasm is distinctive, which allows identification when the nuclei are not clearly discernible (see the "Key"). Trophozoites with a single nucleus are seen less frequently (9). Although no specific pathologic changes in the intestine have been associated with infections with *D. fragilis,* anecdotal information has incriminated it in some cases of intestinal upset; when the organism is eliminated with treatment, the symptoms disappear. It should be considered as a possible cause of illness and reported as such.

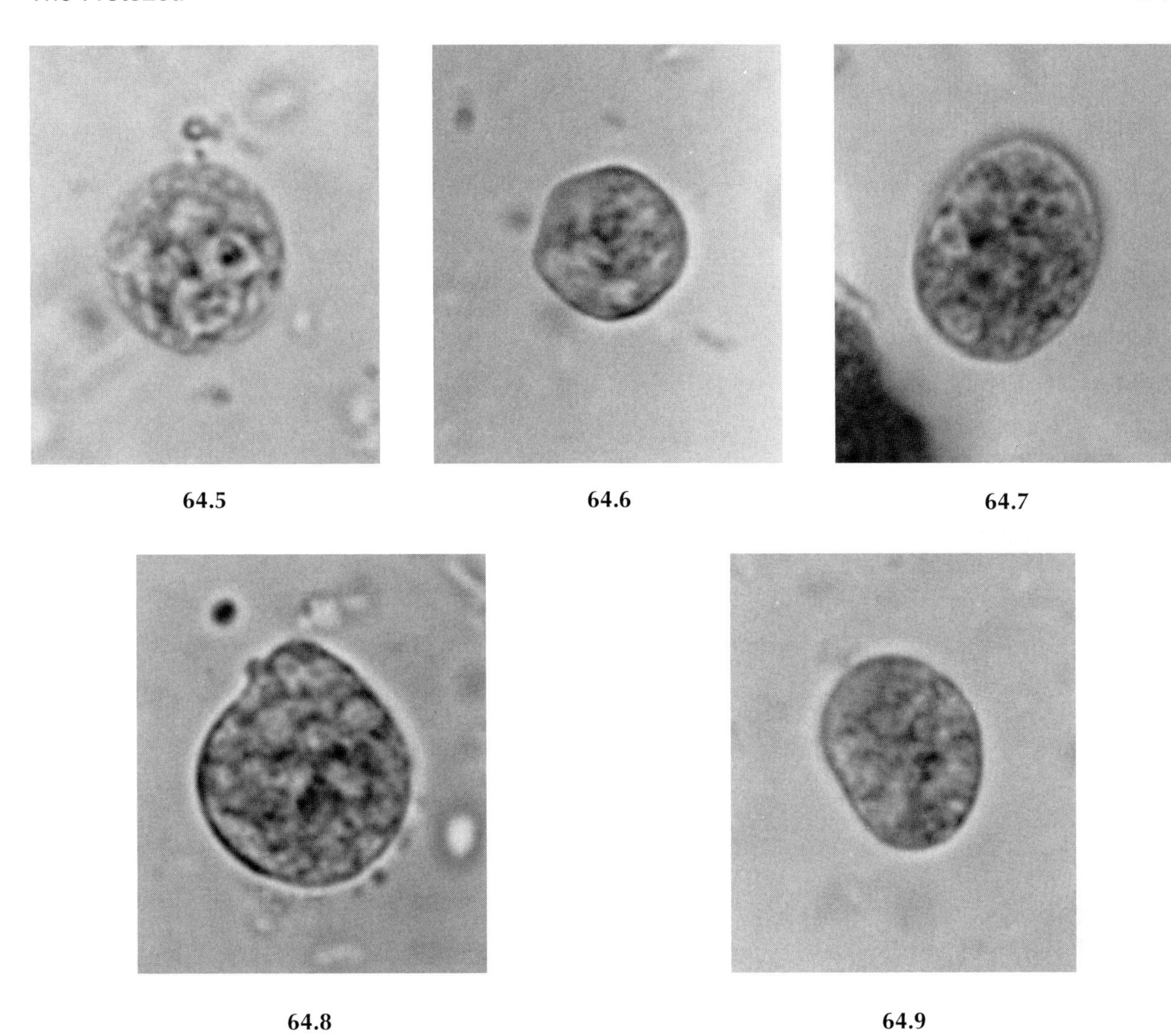

64.5

64.6

64.7

64.8

64.9

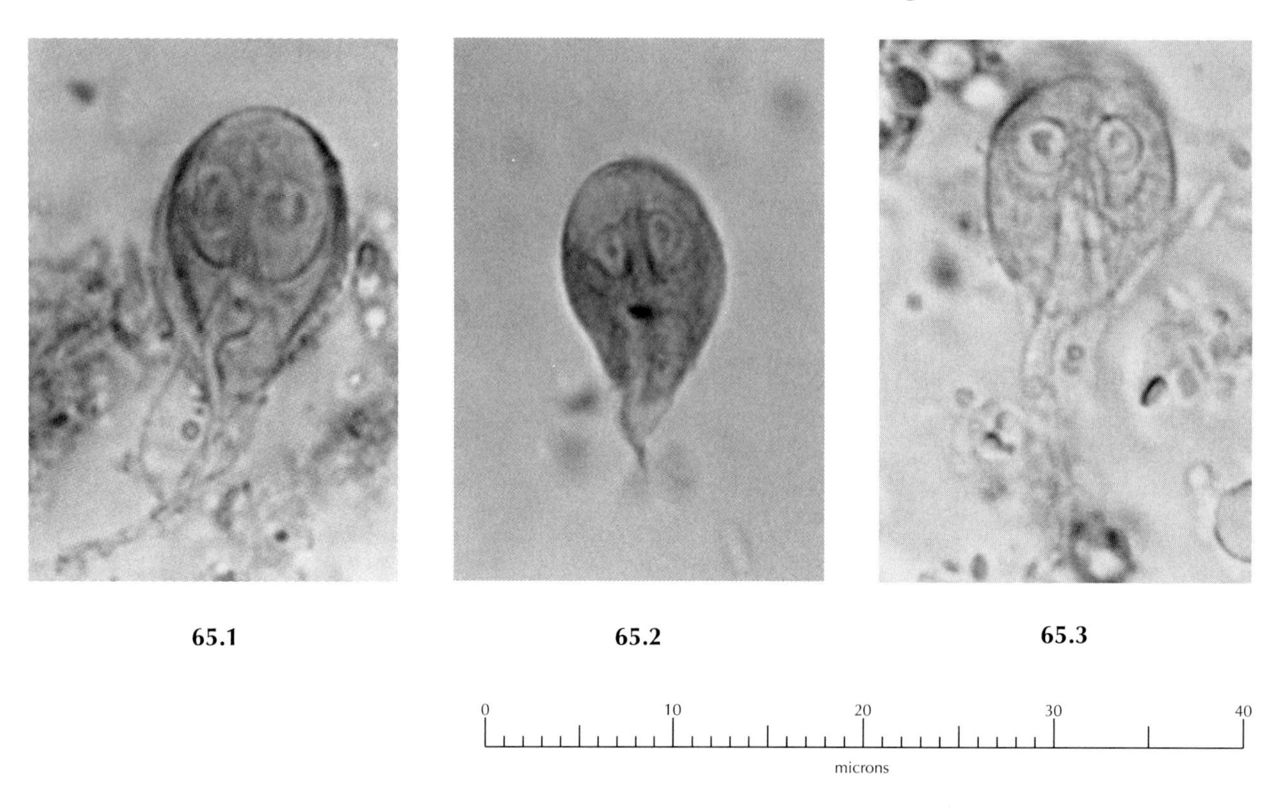

65.1 65.2 65.3

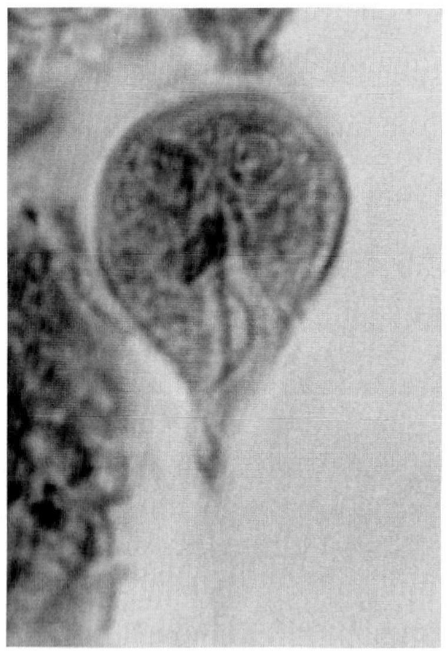

65.4

Plate 65. *Giardia lamblia* trophozoites. 1: PIF; 2 and 3: MIF; 4: PIF; 5 and 6: MIF; 7: PIF; 8 and 9: MIF. The trophozoites of *G. lamblia* are distinctive and should not be confused with other species of intestinal protozoa. They are bilaterally symmetrical, with paired nuclei and flagella. When seen from the dorsal or ventral view (1–5, 9) the paired organelles can be readily seen (see Plate 37). Some trophozoites are oddly shaped (4) and others may be in a lateral or semilateral position (7, 8) which makes them more difficult to recognize. If fixation is delayed, they may become distorted (5). When they are numerous, they may attach to one another (6) or appear in clusters (see Figure 1, page 154). *G. lamblia* is cosmopolitan and is as common in cold climates as in tropical regions. The trophozoites inhabit the duodenum of the small intestine where they are found in the lumen and attached to the mucosa. Although they do not invade the tissues, they are considered to cause malabsorption which can lead to relatively complicated symptoms. Elimination of the infection without chemotherapy does occur and is suspected to be related to immune response to the infection. More severe symptoms occur in immunodeficient patients. Chemotherapeutic treatment is not always successful and may have to be repeated.

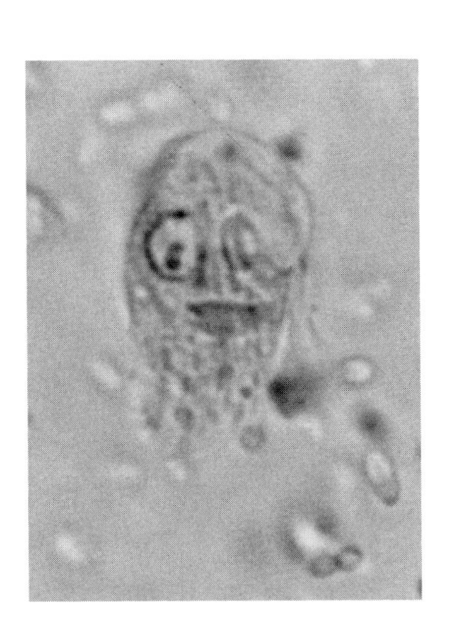

65.5

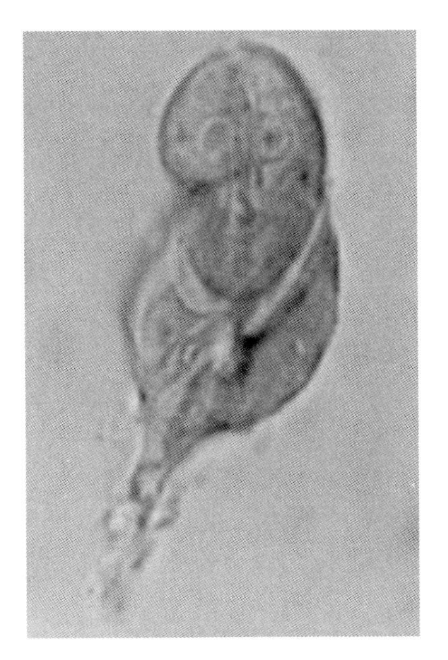

65.6

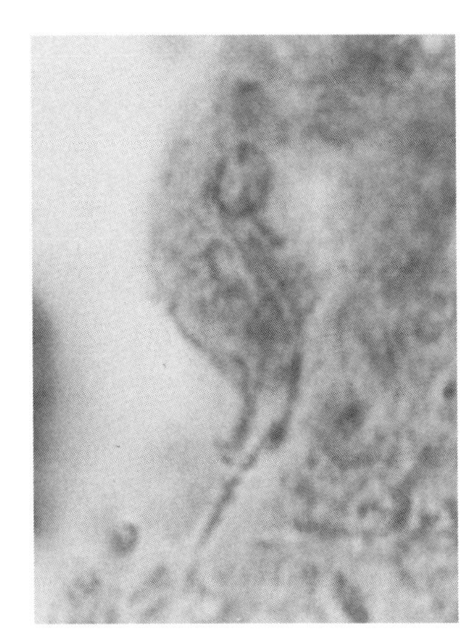

65.7

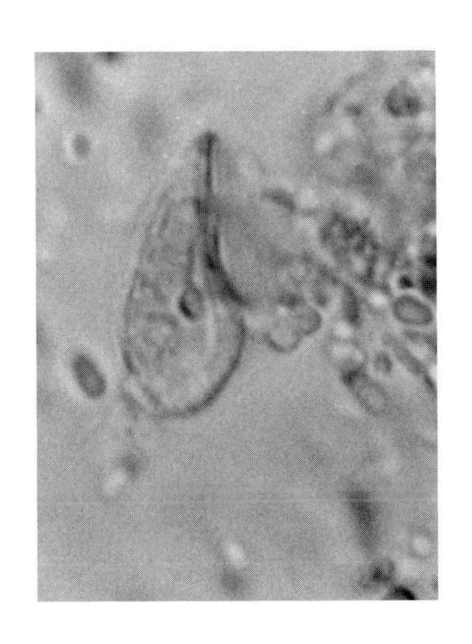

65.8

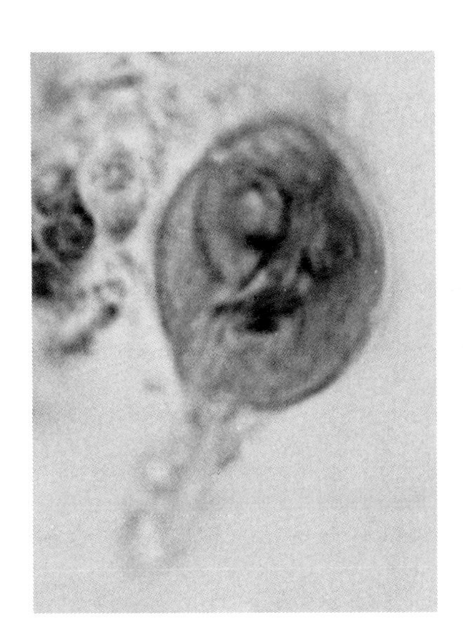

65.9

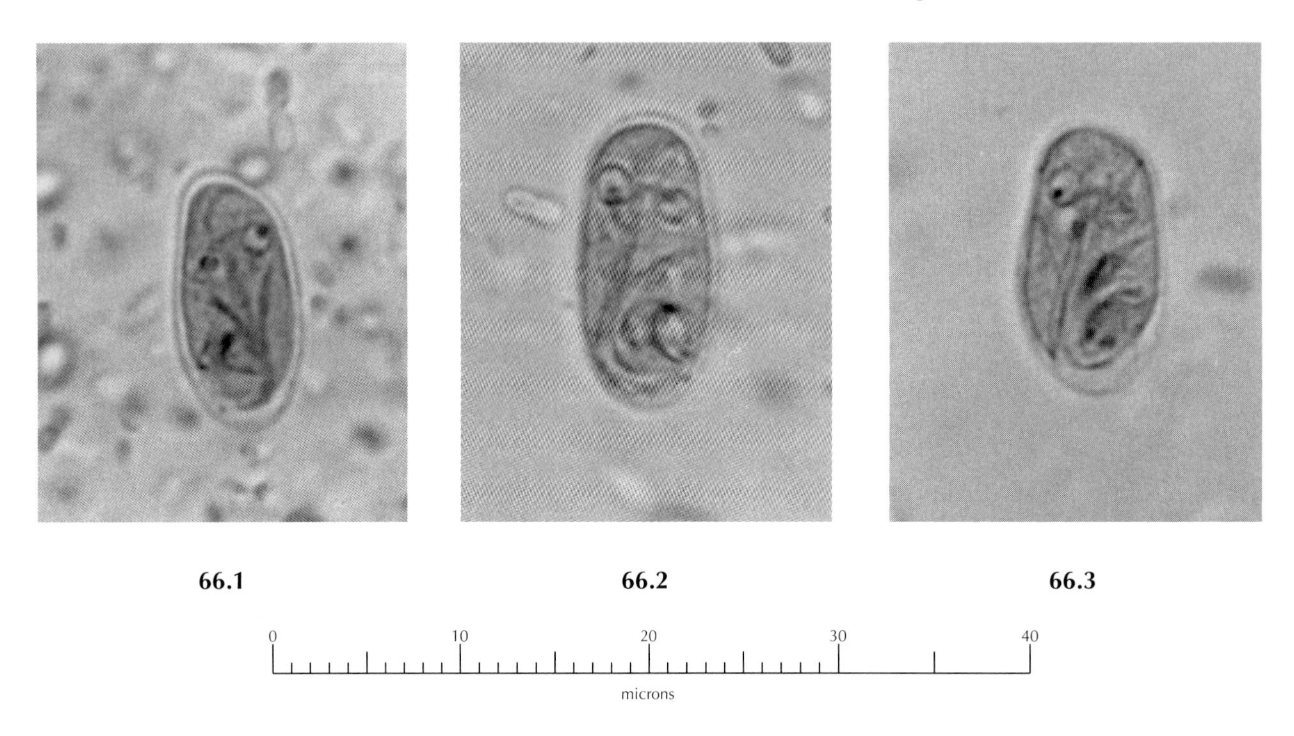

Plate 66. *Giardia lamblia* cysts. 1: PIF; 2 and 3: MIF; 4 and 5: PIF; 6 and 7: MIF; 8: PIF; 9: MIF. Like the trophozoites, the cysts of *G. lamblia* are distinctive and are usually easy to recognize. The protozoan is usually separated from the cyst wall leaving a space that may help in finding and recognizing the cyst (1–5). In some cysts, especially those more recently formed, there is no apparent separation (6). Occasionally, cysts may display the bilateral symmetry seen in the trophozoite (1), but usually the organelles have no pattern and are randomly dispersed (2–6). Some cysts may be distorted (7) or are lying on end (8) and are more difficult to recognize. If fixation is delayed, the protozoan within the cyst dies and becomes unrecognizable (9). Such cysts are referred to as ghost cysts. The mature cyst with four nuclei is the infective stage. The cysts are passed in the feces irregularly with maximum numbers present at 5- to 7-day intervals. Very few cysts may be present in fecal specimens during part of the interim periods so several specimens should be collected 2 or 3 days apart to better ensure finding cysts. Water is most often the source of the infective stage for man and waterborne outbreaks occur frequently in some geographic areas. Some domestic and wild mammals may act as reservoirs for infection in man.

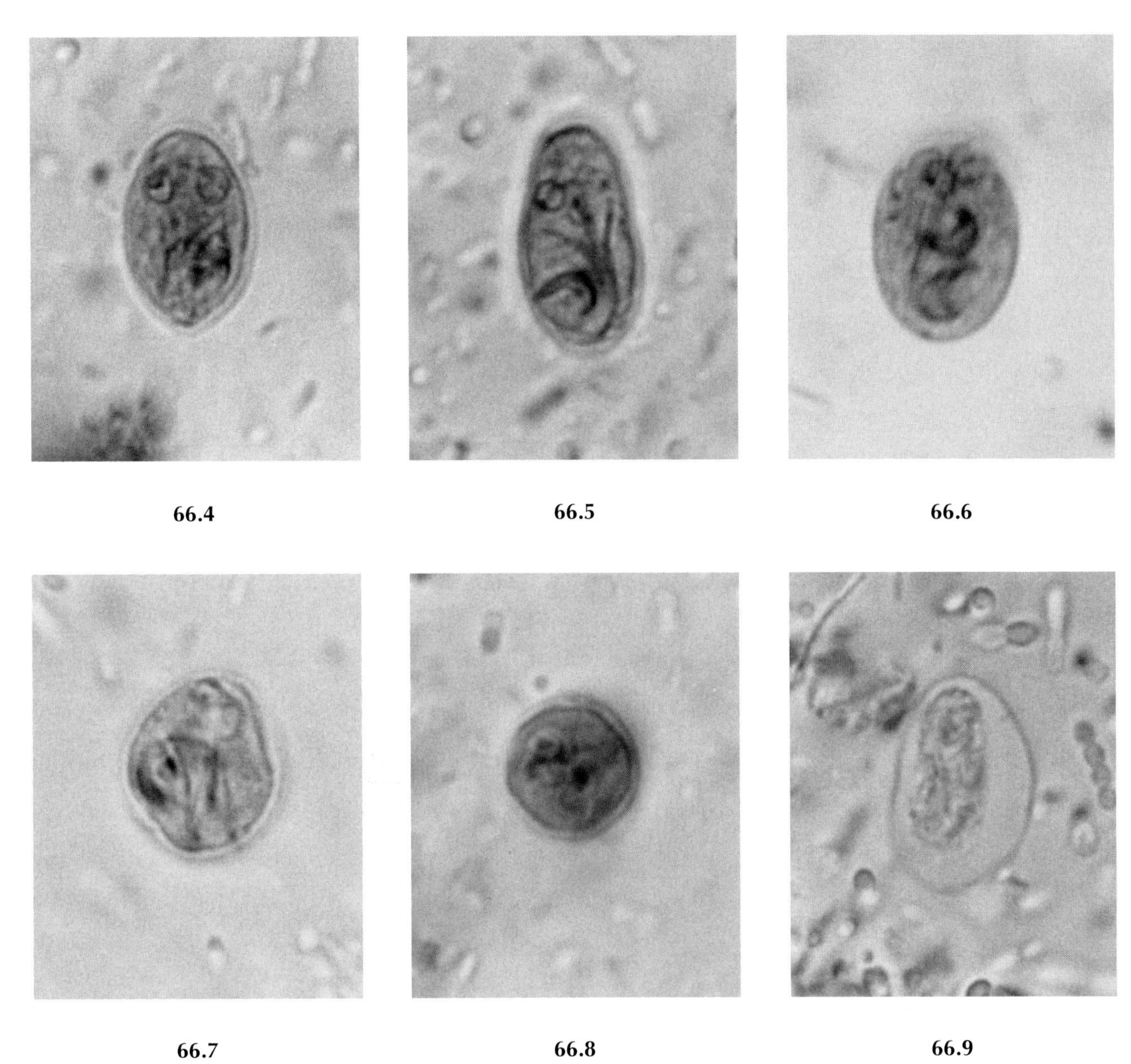

66.4 66.5 66.6

66.7 66.8 66.9

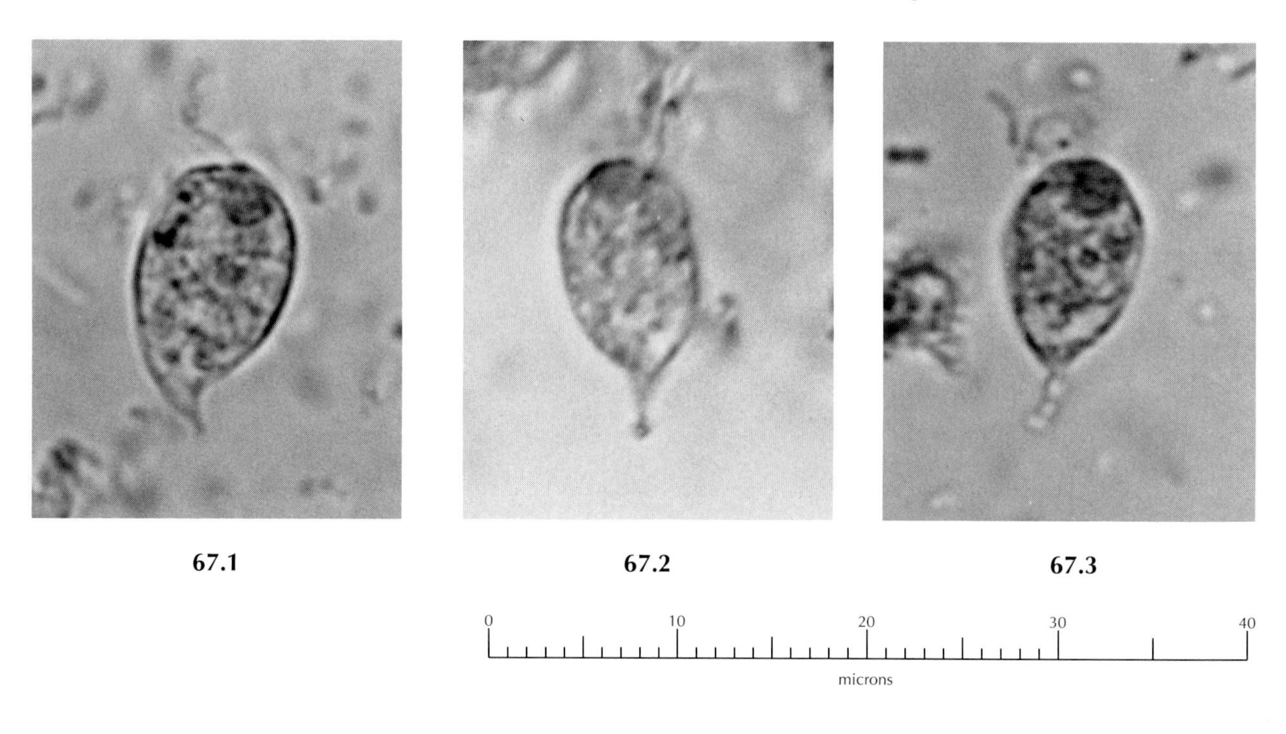

67.1 67.2 67.3

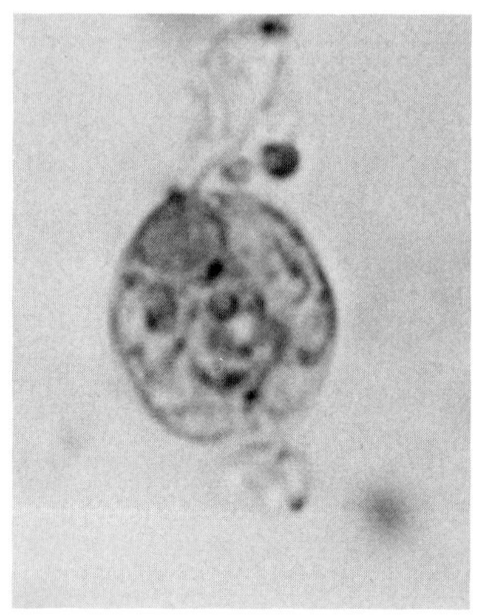

67.4

Plate 67. *Chilomastix mesnili* trophozoites and cysts. 1: T, PIF; 2: T, MIF; 3: T, PIF; 4–6: T, MIF; 7–9: C, MIF. Trophozoites of *C. mesnili* may appear elongated with a blunt anterior end and tapering to a pointed posterior end (1–3) or they may appear round with no distinctive shape (4–6). The nucleus lies close to the anterior end and three flagella arise just anterior to the nucleus (1–5). The blepharoplasts from which the flagella arise may be seen on the upper right side of the nucleus in Figure 5. The cytostome is sometimes visible when the trophozoite is lying in a lateral position (1 left side, 2 right side). Small food vacuoles, many with inclusions, are dispersed throughout the cytoplasm (1–3). Cysts are often seen in the presence of trophozoites or may be present when trophozoites are rare. The cyst is usually pear-shaped with a raised area at one end separating the cyst wall from the protozoan inside (7, barely visible in 8). Two cysts are seen in 8 with the one at the top in better focus. The nucleus and cytostome with its supporting axonemes may be seen (7, 8 left of nucleus). Cysts may be irregular in shape and more difficult to recognize (9). *C. mesnili* is a nonpathogenic intestinal protozoan and is distributed worldwide. Like other nonpathogenic protozoa, it is an indicator of fecal contamination.

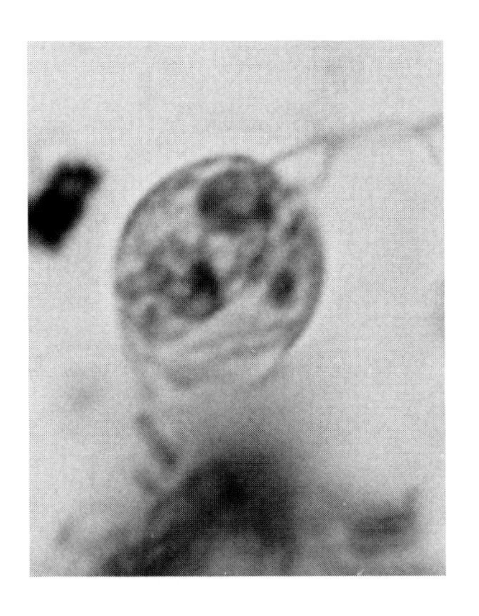

67.5

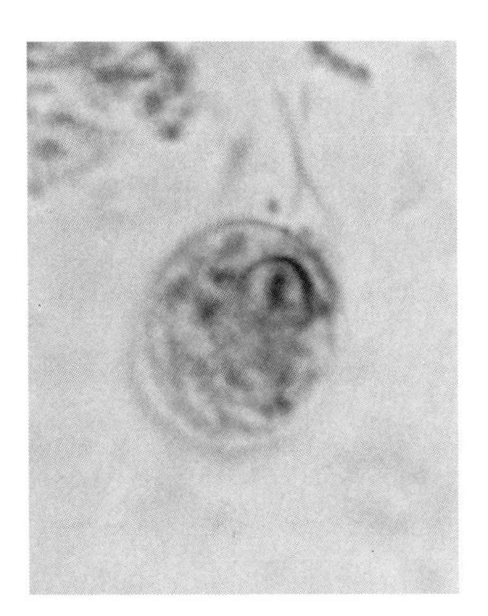

67.6

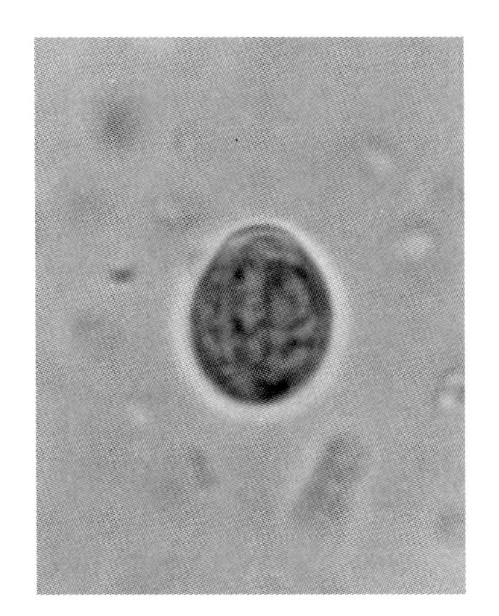

67.7

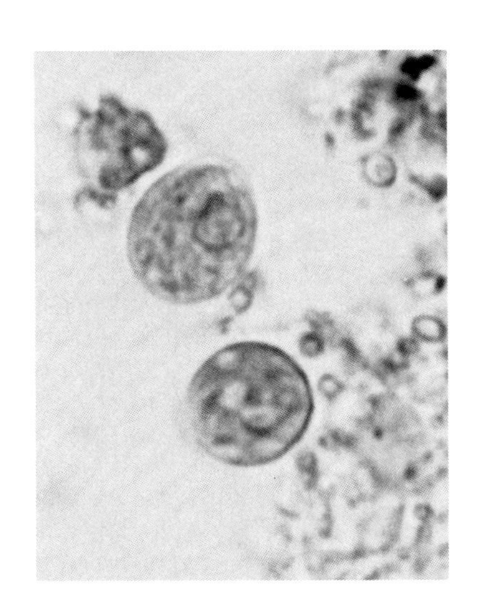

67.8

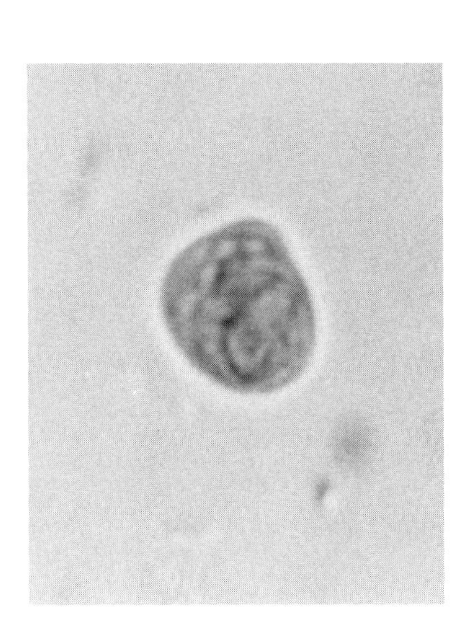

67.9

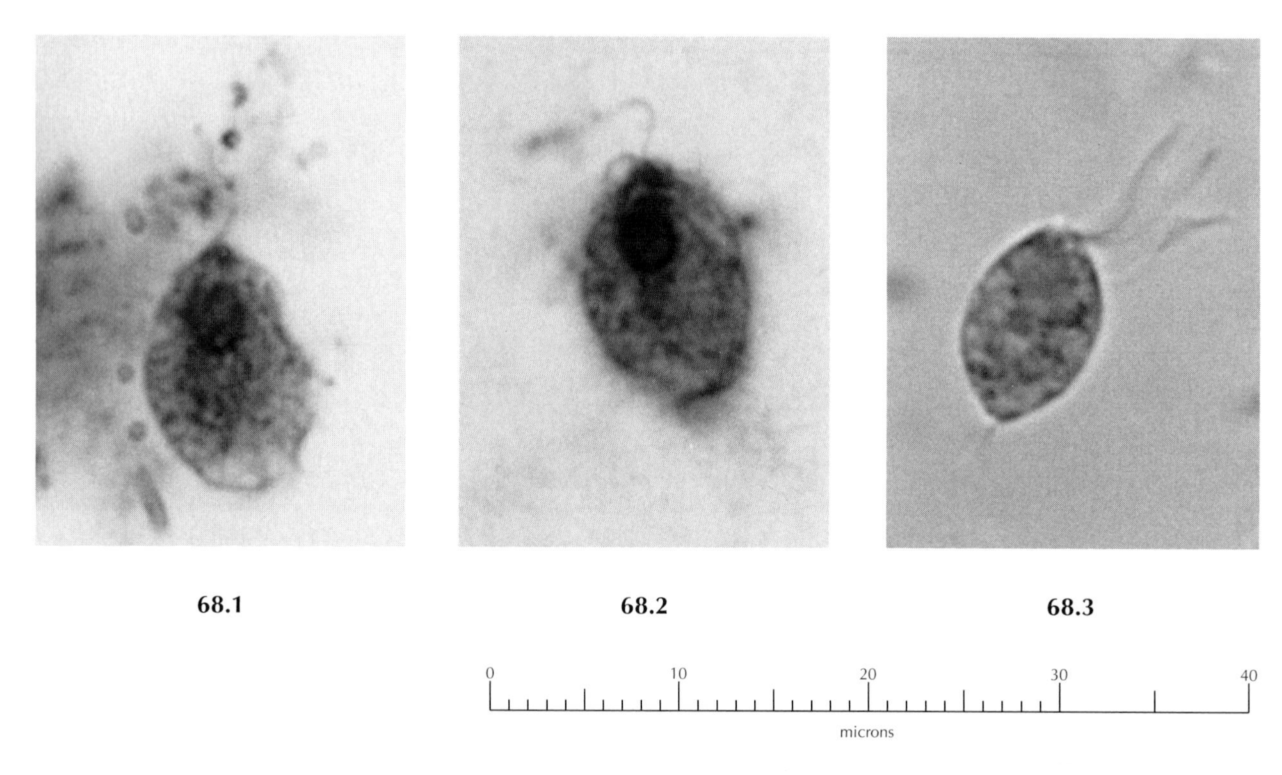

68.1 **68.2** **68.3**

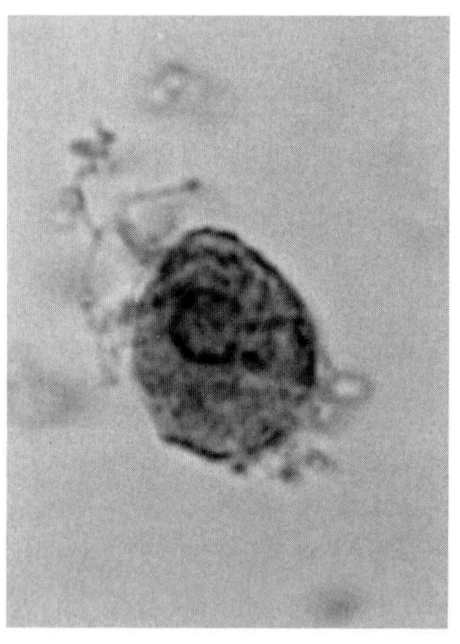

68.4

Plate 68. *Trichomonas hominis* trophozoites. 1–4: MIF; 5–6: PIF; 7: MIF; 8: PIF; 9: MIF. *T. hominis* trophozoites may vary in length from 8 to 20 μm, but are usually 10 to 15 μm. The nucleus may have a dark outer membrane with lighter nucleoplasm (1, 2, 4–7) and the karyosome is often faintly visible (1, 2, 4, 5). The nucleus is sometimes distinct (3). In some views, the undulating membrane can be seen (1 right, 8 left). When the organisms are turned correctly, the small cytostome may be discernible as a light area beginning beside the nucleus (2, right of the nucleus). The flagella can usually be seen in wet preparations (1–3, 7, 8) but are often entangled in debris (5, 6). The axostyle is sometimes clearly visible (7). Trophozoites in division are sometimes seen (9); here two nuclei and two sets of flagella are present. Photomicrographs 7 and 8 were taken of trophozoites in a wet preparation made following concentration by the CONSED sedimentation method. *T. hominis* is cosmopolitan in man and is an indicator of fecal contamination. Since there is no cyst form, the mode of transmission is not clear.

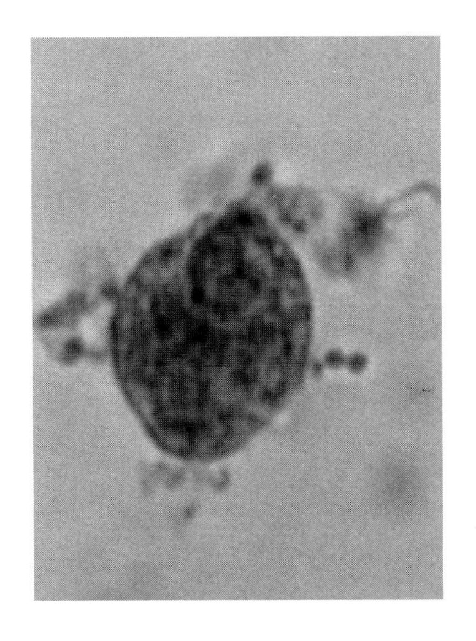

68.5

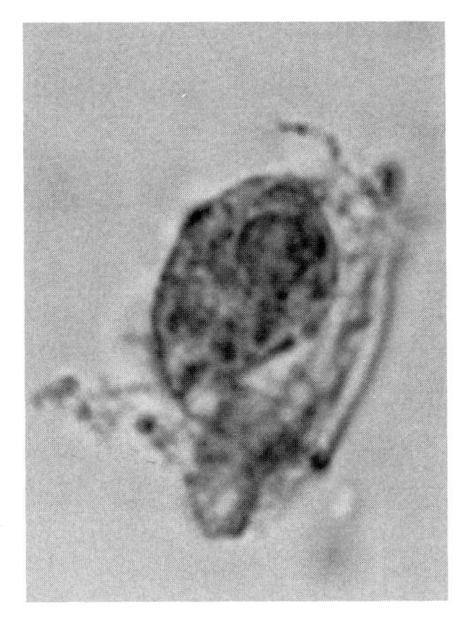

68.6

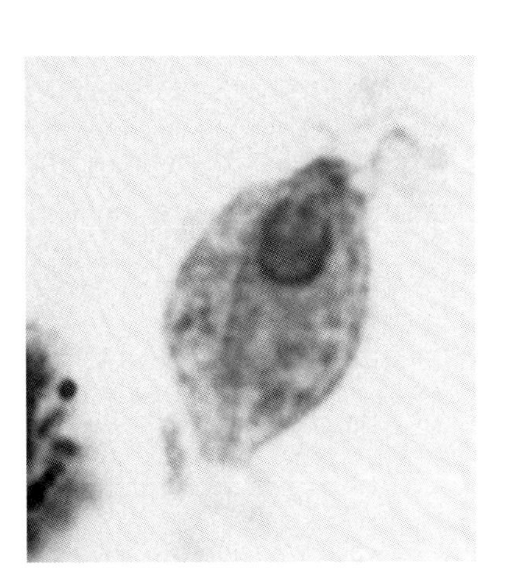

68.7

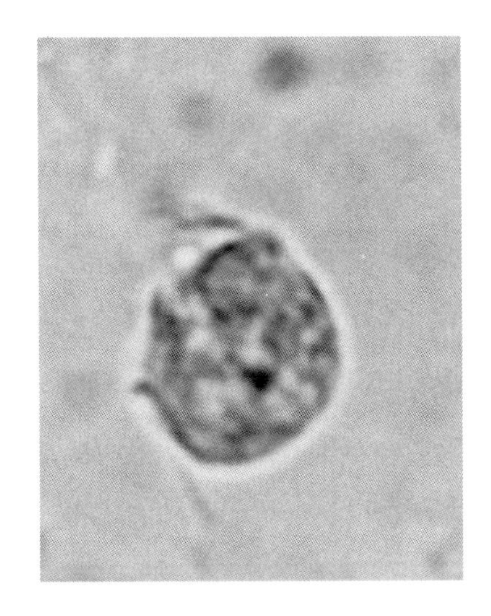

68.8

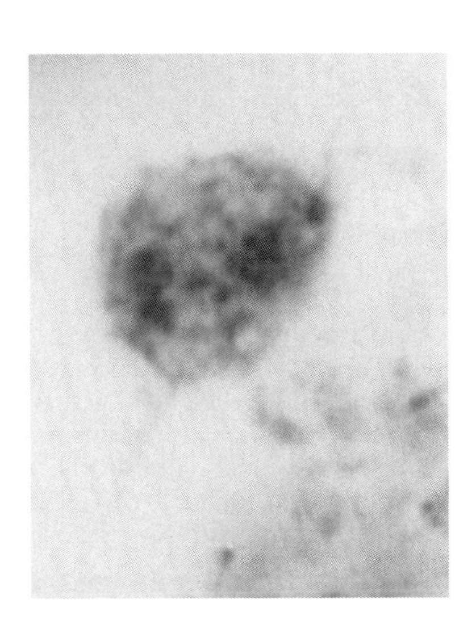

68.9

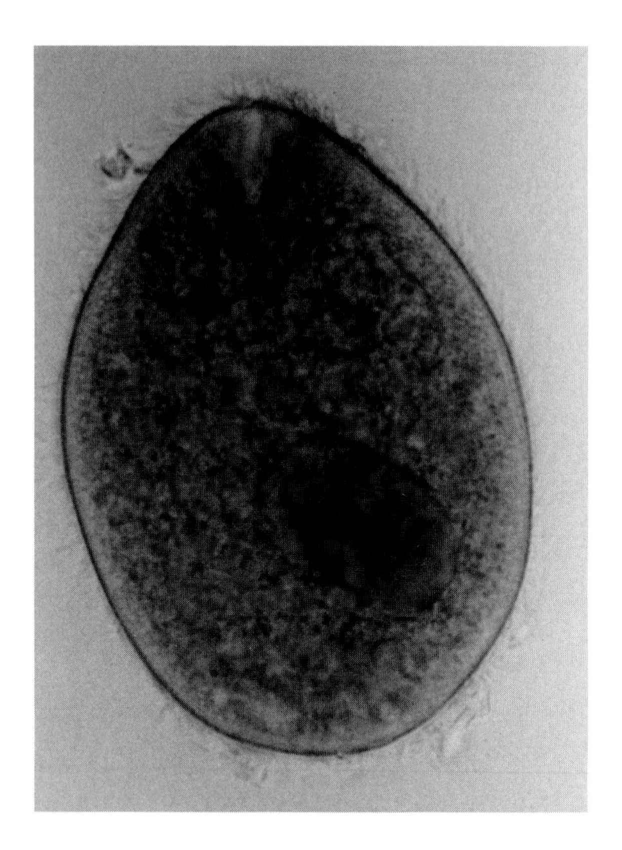

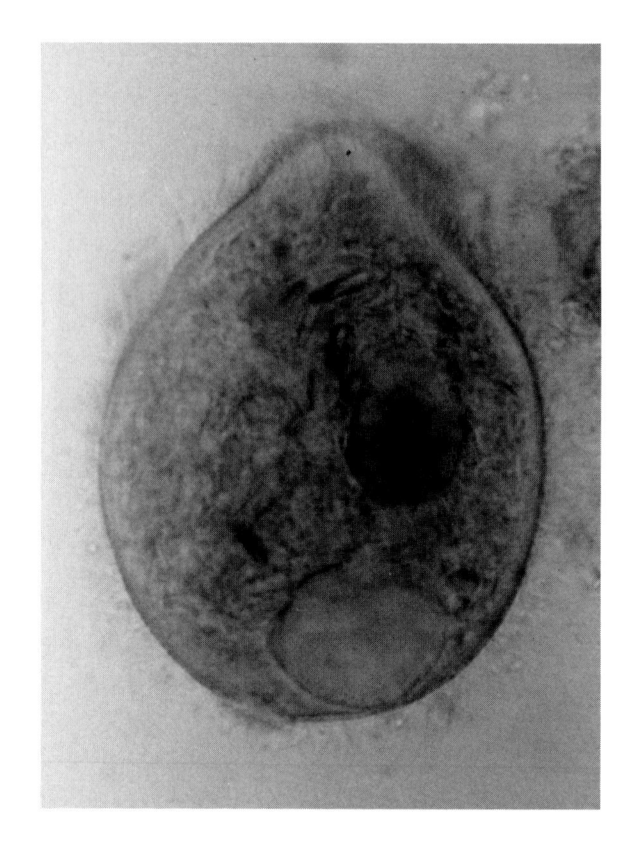

69.1 69.2

Plate 69. *Balantidium coli* trophozoites and cysts. 1–2: T, formalin; 3–4: C, formalin (size of parasites and cysts are about 45% of the standard scale). Although the original specimens were fixed in formalin, portions of the specimens were post-fixed in MIF and the photomicrographs were taken of organisms in preparations made from the specimens post-fixed in MIF. The body of the trophozoite is generally ovoidal but slightly narrowed at the anterior end (1, 2). The body is covered with many slightly oblique, longitudinal rows of cilia. The cilia can be seen extending outwardly at the periphery of the trophozoite (1, 2). Each cilium arises from a basal granule that functions as does the blepharoplast in flagellates. The cytostome is at the anterior end slightly to one side of center (1). Food vacuoles that form at the base of the cytostome often contain ingested bacteria (2). A large, discrete, bean-shaped macronucleus is usually visible as a dark mass (1, 2). The small micronucleus is often adjacent to and may appear embedded in the macronucleus (not shown). A contractile vacuole is clearly seen at the posterior end of the trophozoite (2). The round to oval cysts of *B. coli* have a two-layer cyst wall, an outer dark wall and an inner hyaline wall or membrane (3). In fresh preparations, the cilia remain active in the early cyst and can be seen between the cyst wall and the body of the protozoan (4). *B. coli* infections may be asymptomatic. Symptomatic cases are usually characterized by intermittent diarrhea and constipation, and more severe cases may lead to ulceration of the colon and caecum. See also Plate 39, page 135.

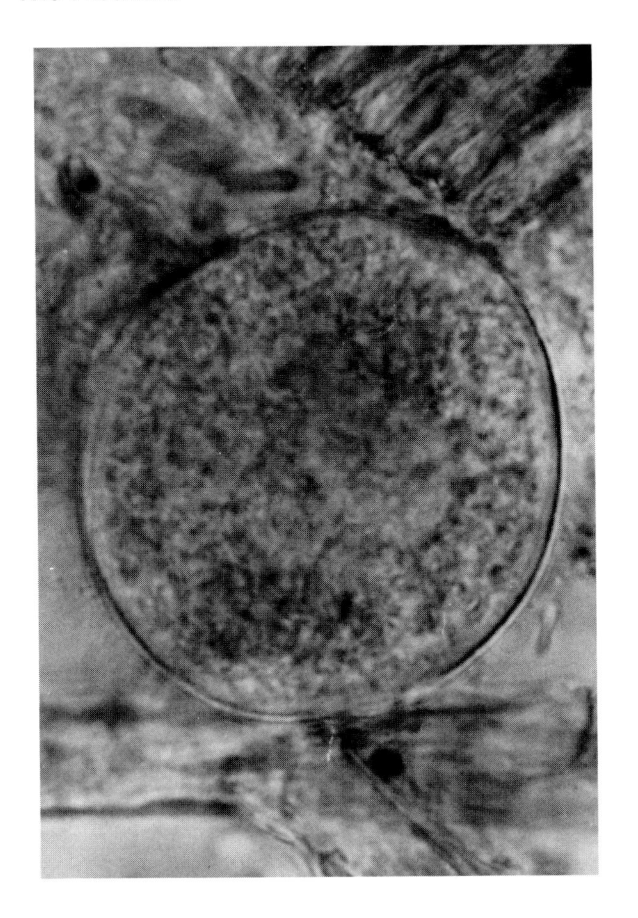

69.3

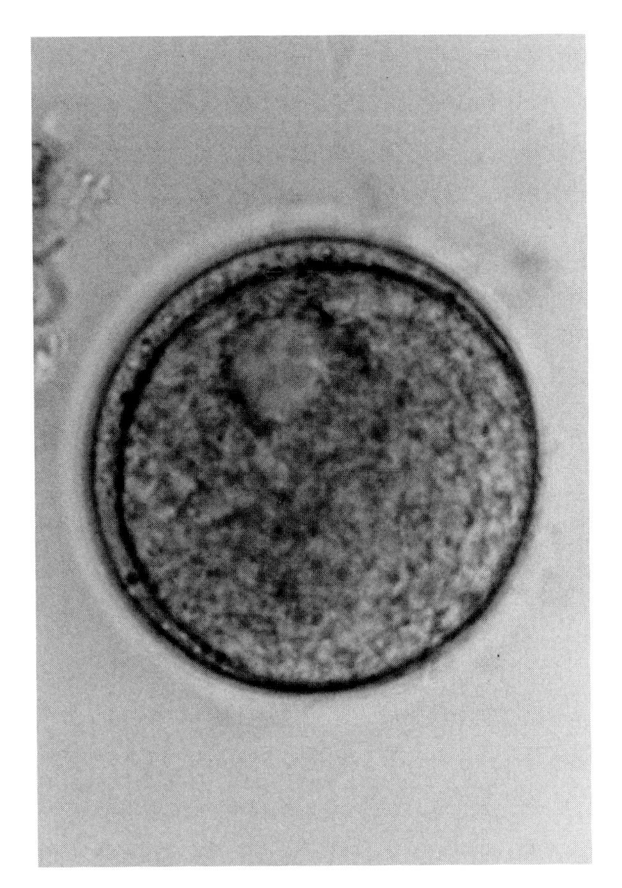

69.4

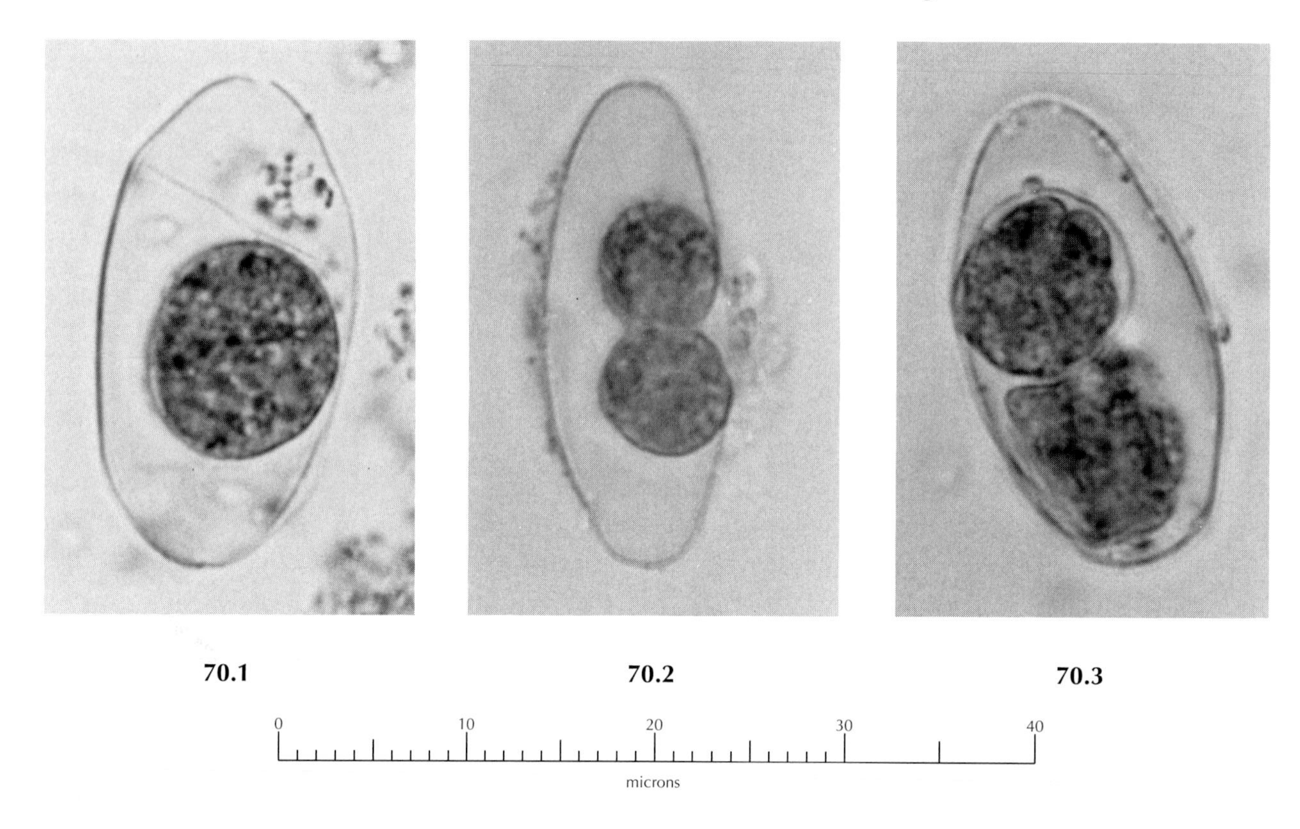

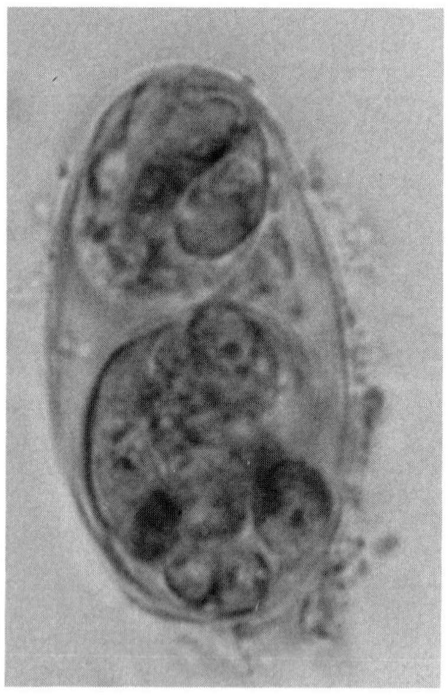

70.4

Plate 70. Oocysts of *Isospora belli.* 1–4: formalin. Although the specimens were fixed initially in formalin, they were post-fixed in MIF and the photomicrographs were taken of oocysts in preparations made from specimens post-fixed in MIF. The oocysts of *I. belli* are usually seen in the unsegmented, sporont stage (1) in fecal specimens from infected individuals. If passage of the fecal specimen is delayed, oocysts may continue to develop to the sporoblast (2), sporocyst (3), or sporozoite (4) stage. The sporozoite stage is the infective stage. *I. belli* infection is world wide but within isolated endemic areas. Infection occurs on ingestion of an infective oocyst. The sporozoites are freed from the oocyst in the small intestine and invade the mucosal epithelium where sexual and asexual reproduction occur. The asexual cycle (schizogony) begins with trophozoites that develop into schizonts. The schizonts produce merozoites that invade new cells to continue the cycle. Merozoites from some schizonts develop into male and female gametocytes to begin the sexual cycle. The microgametocyte (male) combines with the macrogametocyte (female) to form a zygote which in turn lays down a cyst wall to form an oocyst. Oocysts produced in the sexual cycle are passed in the feces. Infection is often self-limiting except in individuals with a compromised immune system. The mode of transmission is by fecal contamination.

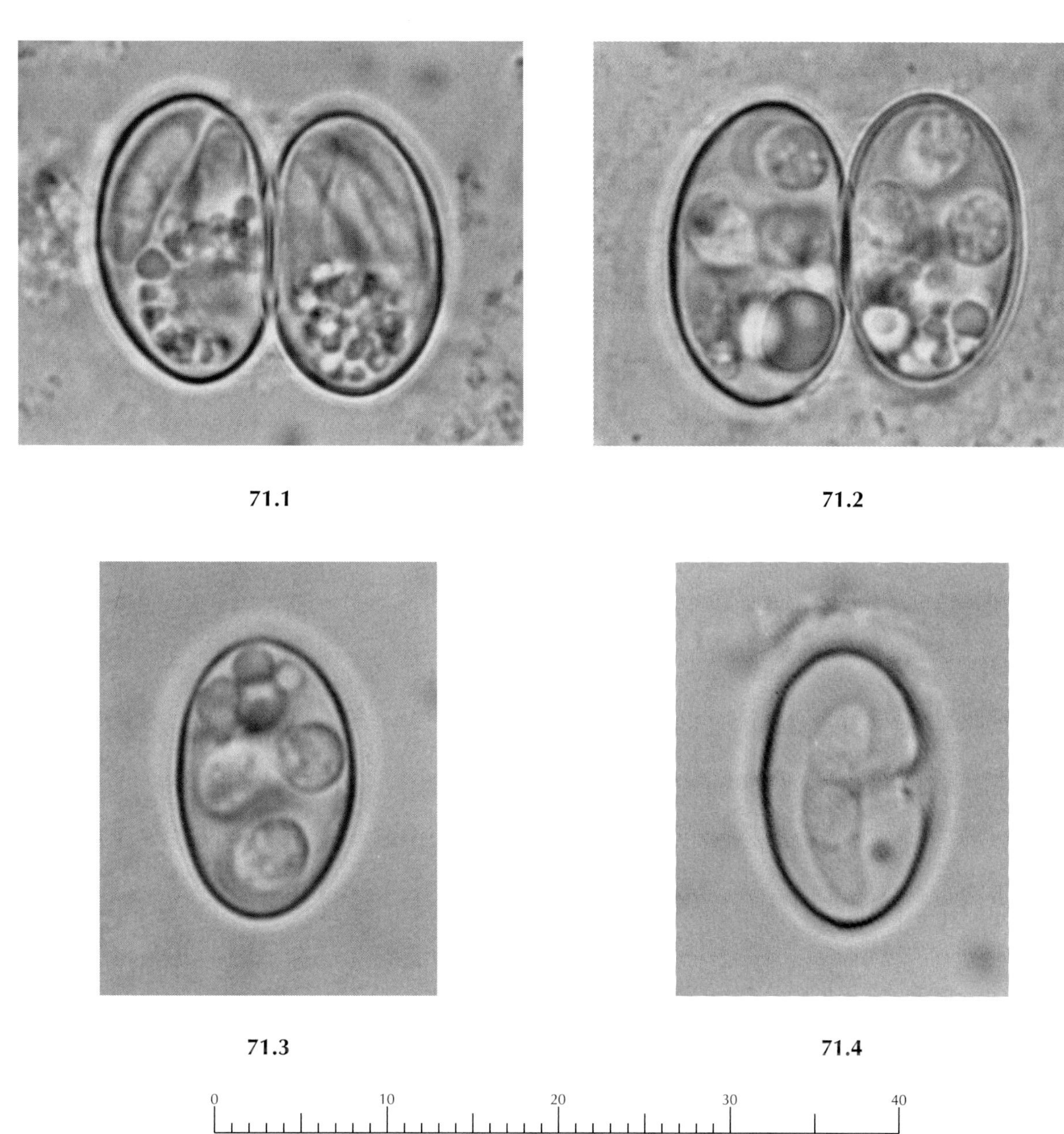

Plate 71. Sporocysts of *Sarcocystis* species. 1–4: formalin. Although the specimens were fixed initially in formalin, they were post-fixed in MIF and the photomicrographs were taken of oocysts in preparations made from specimens post-fixed in MIF. (The specimen from which the photomicrographs were taken was provided by Ronald Fayer, Ph.D., Zoonotic Diseases Laboratory, Agriculture Research Service, United States Department of Agriculture (USDA), Beltsville, Maryland 20705.) The oocyst wall of species of *Sarcocystis* is usually lost in the intestine and sporocysts are found in fecal specimens. They are usually in pairs (1, 2) each sporocyst containing four sporozoites and a residual body. Sporocysts may separate and be found singly (3) and occasionally they will rupture and the sporozoites escape (4). Only one sporozoite remains in the sporocyst (shown in 4). For many years, the sporocysts of *S. hominis* passed in the feces of man were identified as *Isospora hominis* and were not associated with the intermediate stages found in cattle or as accidental aberrant infections in man (Beaver et. al., 1979). Now, two species are recognized to infect the small intestine of man, *S. hominis* with cattle as the intermediate host and *S. suihominis* with pigs as the intermediate host. The cycle in man is similar to that with *I. belli*. Infection in man is usually self-limiting and may last only a few weeks and symptoms are mild. Transmission is by ingestion of inadequately cooked meat from an infected source.

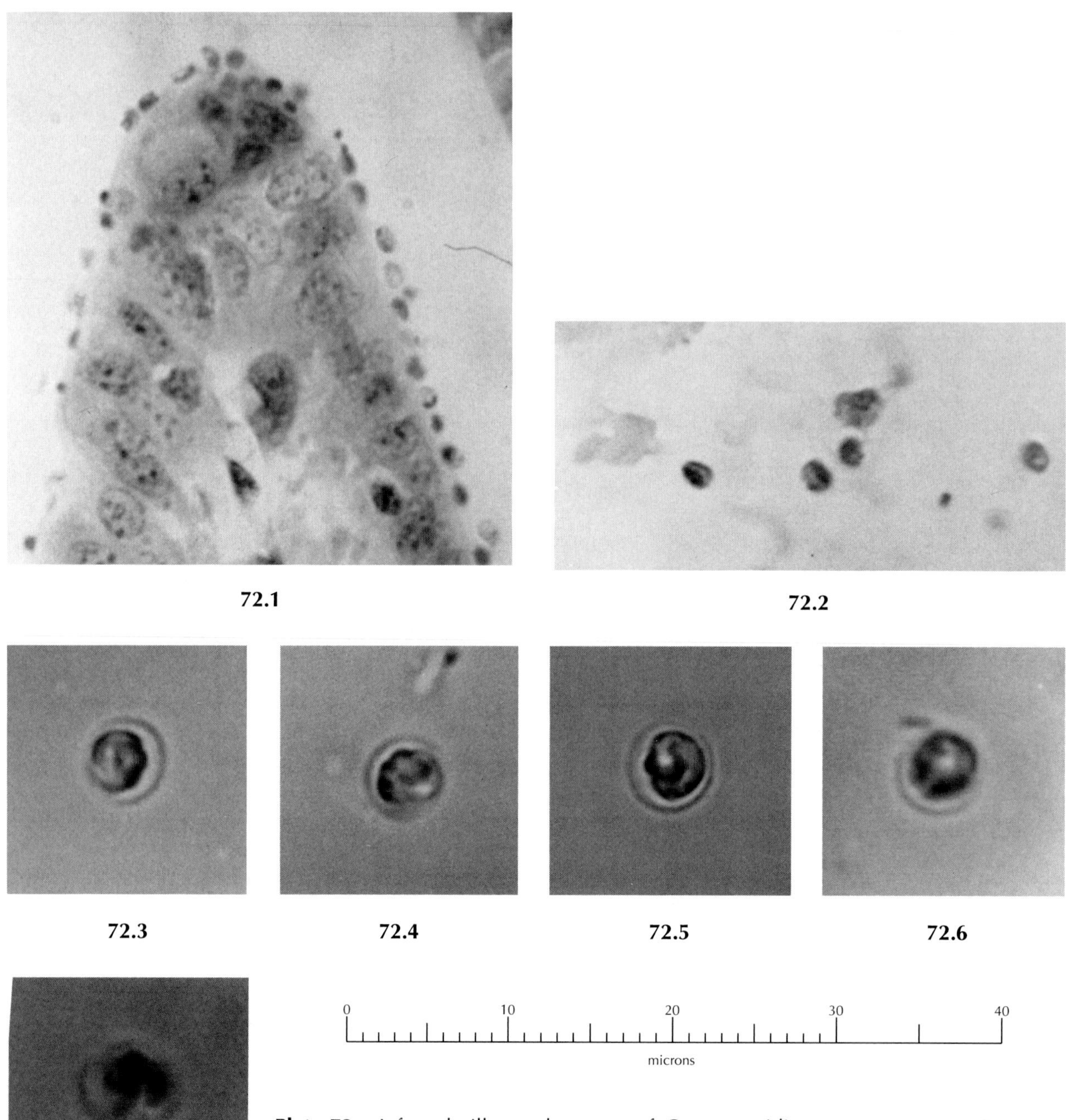

Plate 72. Infected villus and oocysts of *Cryptosporidium parvum.* 1–2: A-F (not to scale); 3–6: MIF. A histologic section of the tip of a villus, with a relatively heavy *C. parvum* infection is shown (1). The stain is Price's acid-fast stain with a methylene blue counterstain. The schizonts and developing gametes stain blue to lavender, but the oocysts are acid-fast and stain bright red. Oocysts are shown in a fecal specimen at the same magnification, stained by the same method. These oocysts are also red and acid-fast. A higher magnification (the standard scale) is used to show oocysts in wet preparations prepared from MIF-fixed specimens. They are at various stages of development: 3, an unsegmented sporont; 4, early development of the sporozoites; 5, sporozoites well developed; 6, early development of sporozoites; and 7, fully developed sporozoites. The parasites do not invade the tissues, but live attached to the cells of the villi. As in other coccidia, schizogony takes place producing merozoites that continue the asexual cycle. Some of the merozoites develop into gametocytes. The male gametocyte produces microgametes that fertilize the macrogamete, producing a zygote that secretes a cyst wall to form the oocyst. Reese et al. (1982) published an excellent article on the infection in a patient with a normal immune system. Infection of suckling mice with oocysts from human and cattle origin produced indistinguishable infections. In individuals with a compromised immune system, *C. parvum* may produce a serious infection for which there is no suitable treatment.

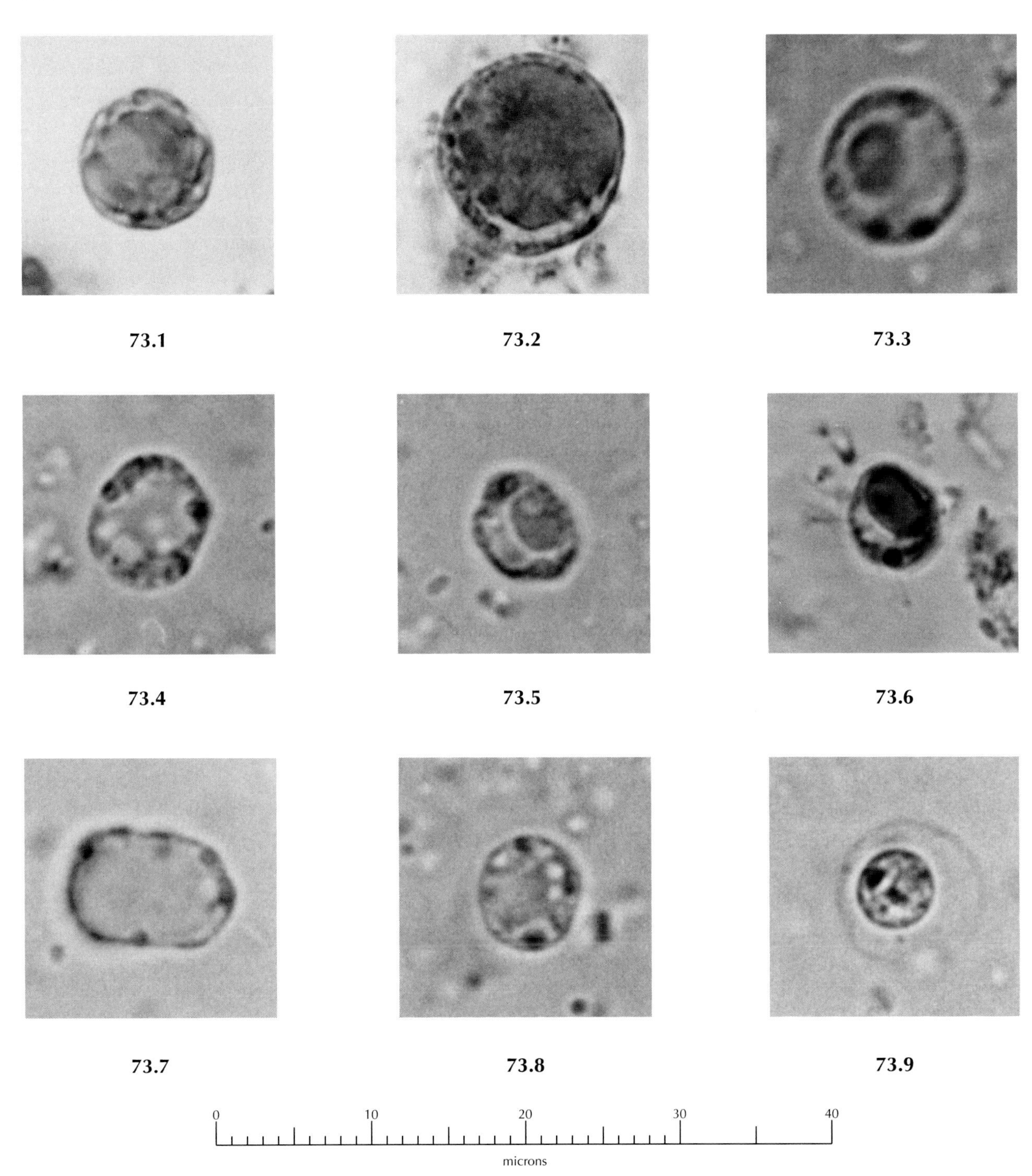

Plate 73. Some forms of *Blastocystis hominis* in fecal specimens. 1–3: MIF; 4–5: PIF; 6: MIF; 7–9: PIF. The cyst-like forms of *B. hominis* that appear in fecal specimens vary greatly and the various forms should be recognized. A variety of forms are shown in the Plate, some which are readily recognized as *B. hominis* (1, 2, 4, 7) while others may not be. Some organisms show the typical pale central mass surrounded by a ring of cytoplasm containing pale bodies (nuclei) and dark volutin granules (1, 4, 7). In some there is a dark central mass (3, 5, 6). In some, the distinct outer ring of cytoplasm is not shown (5, 6). Organisms beginning division are often seen (7). The central portion of the organism may begin to condense (8) and sometimes it has completely condensed, eliminating the central mass, and has pulled away from its outer cell membrane (9). That portion of the life cycle of the organism in the intestinal tract has not been described in the literature sufficiently well to provide good descriptive details of each form. Each photomicrograph depicts a recognizable form of *B. hominis* seen in fecal preparations.

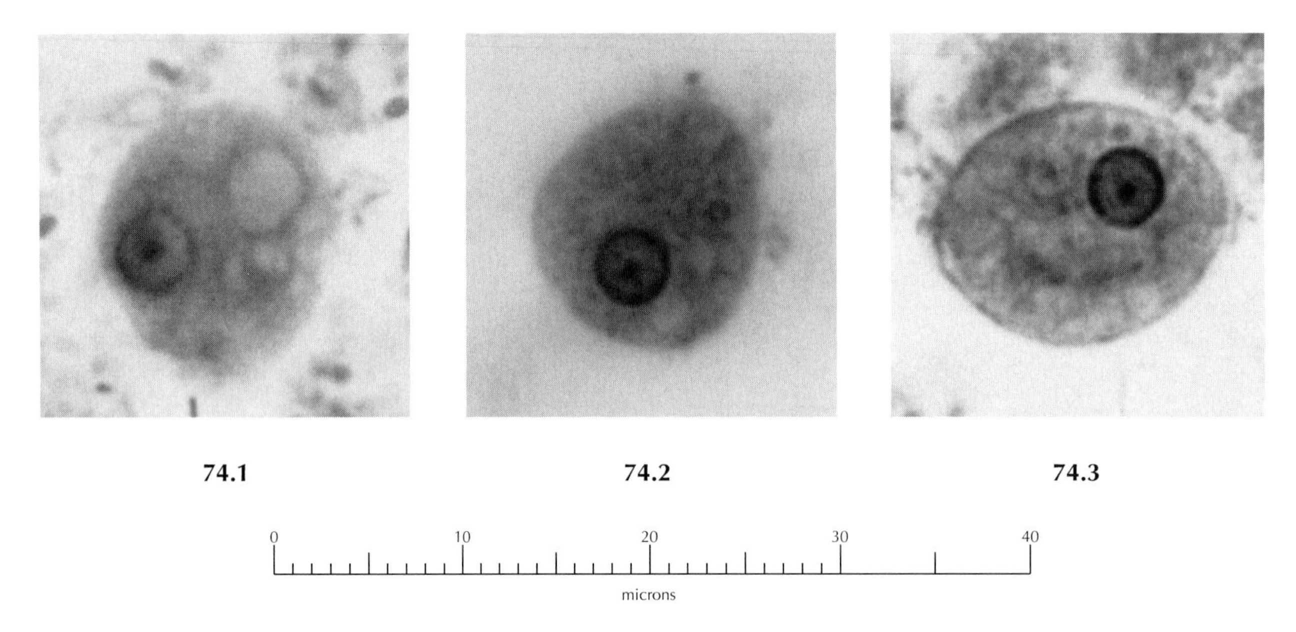

Plate 74. *Entamoeba histolytica*, on permanently stained fecal films. 1 and 2: T, PVA, T; 3: T, MIF, T; 4 and 5: T, MIF, P; 6: C, MIF, P; 7: C, PVA, T; 8: C, MIF, T; 9: C, MIF, P. The morphologic detail of the trophozoites of *E. histolytica* in photomicrographs (1–5) is excellent and identification can be made readily. Identification of the mononucleate cyst with the glycogen vacuole (6) is less certain but the cyst with chromatoid bars (7) can be readily identified. The two mature cysts (8, 9) also are easily identifiable. Trophozoites in some photomicrographs (1, 2) were fixed in PVA and stained with trichrome while others (3–5) were fixed in MIF. Note that the trophozoites fixed in PVA appear less crisp and do not stand out from the background as well as those fixed in MIF. Some trophozoites (1–3) were stained with trichrome and others (4, 5) were stained with Polychrome IV. The cytoplasm in the latter two photomicrographs is much more detailed and the nucleus remains distinct. The cyst (7) was fixed in PVA and the others were fixed in MIF. Again the PVA-fixed cyst is not as sharply distinct from the background as are the MIF-fixed cysts but the nucleus and chromatoid bars are sharp. The mononucleate cyst with the glycogen vacuole (6) was fixed in MIF and stained with Polychrome IV. The internal detail is less clear than in the trophozoites. One mature cyst (8) was stained with trichrome and one (9) was stained with Polychrome IV. They are obviously from different specimens and are stained about equally (see Plates 55, 56).

PHOTOMICROGRAPHS OF INTESTINAL PROTOZOA

Two series of photomicrographs of intestinal protozoa are presented. The first series is of protozoa as they appear in wet preparations (page 160), and the second series is of protozoa as seen in permanently stained fecal films (page 194).

Photomicrographs From Permanently Stained Fecal Films

In photomicrographs prepared from permanently stained fecal films, representatives of the amoebae, flagellates, and *Blastocystis hominis* are presented. The Plates are arranged by parasite species with both trophozoites and cysts, where they occur, on the same Plate. The fixing solution, PVA, MIF, or PIF, and the stain used for preparing the permanently stained fecal film is indicated in the legend for each photomicrograph. Three stains were used for preparing permanently stained films shown in the photomicrographs, the iron hematoxylin method of Tompkins and Miller (H), modified Wheatley's trichrome method (T), and the Polychrome IV method of Price (P). The stage (trophozoite T or cyst C), the fixative, and the stain are indicated for each photomicrograph. Several slides of stained fecal films were loaned by the Center for Disease Control (CDC) and photomicrographs taken from these films are indicated by a CDC at the end. Plates 74 through 84 are of protozoa on permanently stained fecal films. A standard scale was used for all photomicrographs, which appears in the legend.

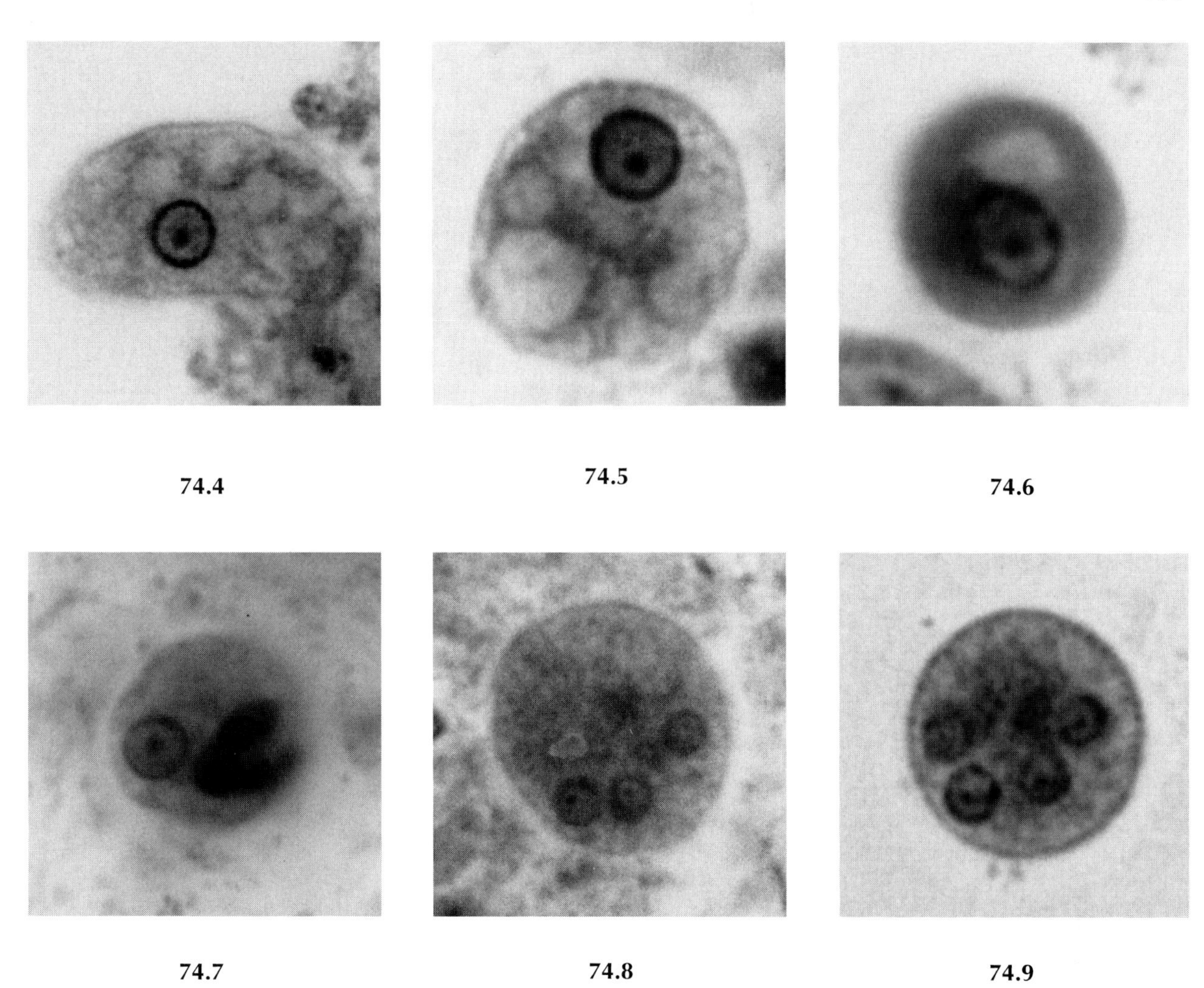

74.4 74.5 74.6

74.7 74.8 74.9

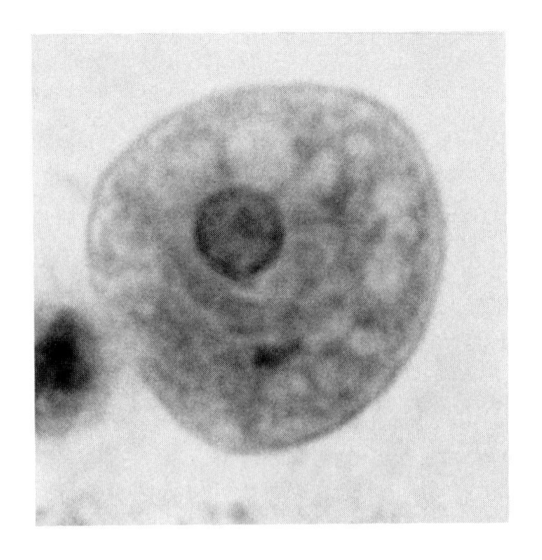

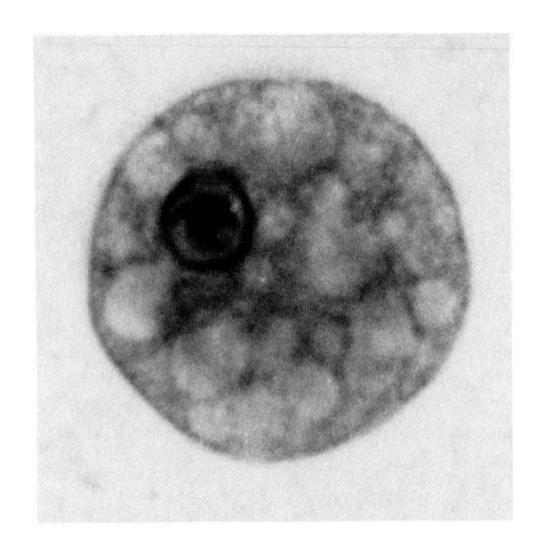

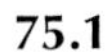

75.1

75.2

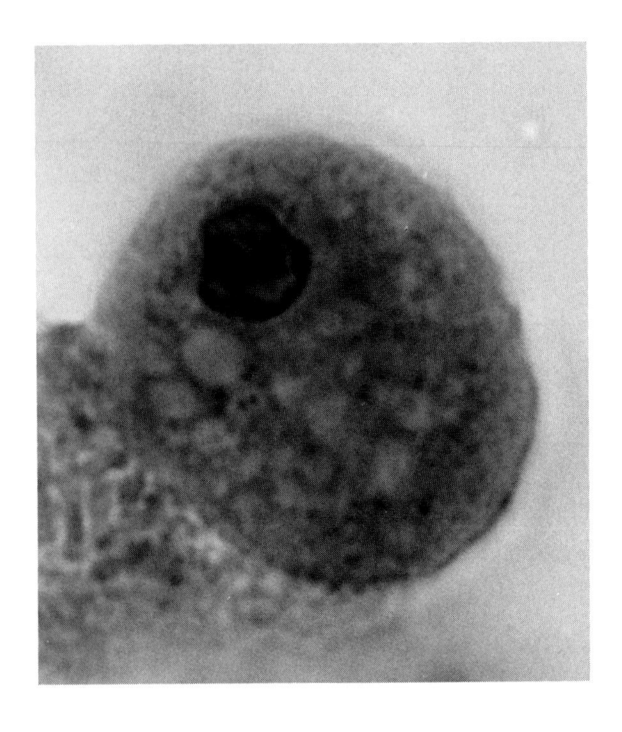

75.3

Plate 75. *Entamoeba coli* on permanently stained fecal films. 1: T, PVA, P; 2 and 3: T, MIF, P; 4: C, PVA, T; 5: C, MIF, T; 6: C, MIF, P. The characteristics of the trophozoites of *E. coli* are well shown in all three photomicrograph (1–3). All three were stained with Polychrome IV, which brings out the cytoplasmic detail. The vacuoles are more distinct in those fixed in MIF (2, 3) than in the trophozoite fixed in PVA (1). Since the karyosome is not centered, it is often difficult to show it in the nucleus when the entire organism is in focus. It can be seen in 1. The binucleate cyst with its large glycogen vacuole that presses the cytoplasm to the cyst wall (4) is typical of *E. coli*. The two mature cysts (5, 6) were both fixed in MIF but one (5) was stained with trichrome and the other (6) with Polychrome IV. The detail appears to be slightly better in the one stained with Polychrome IV.

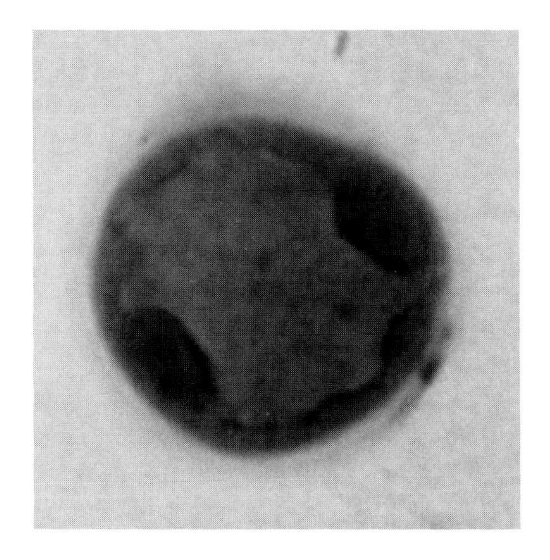

75.4

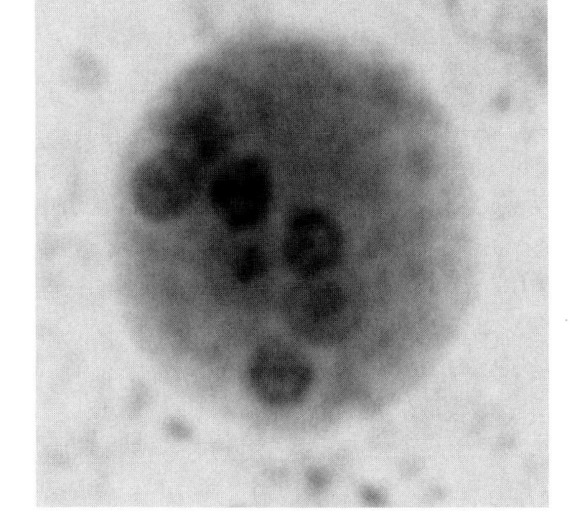

75.5

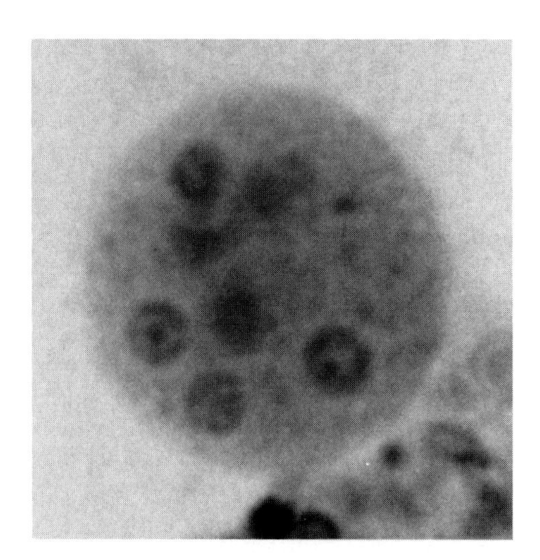

75.6

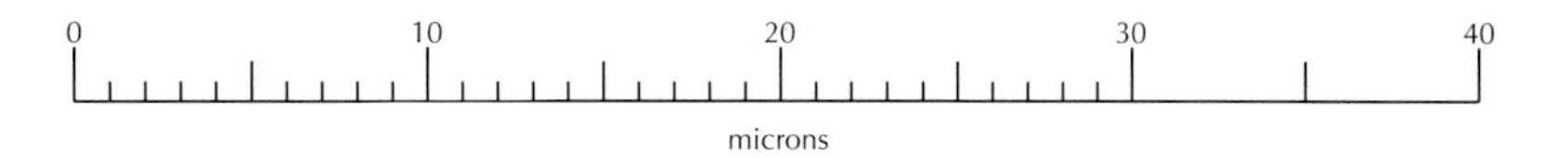

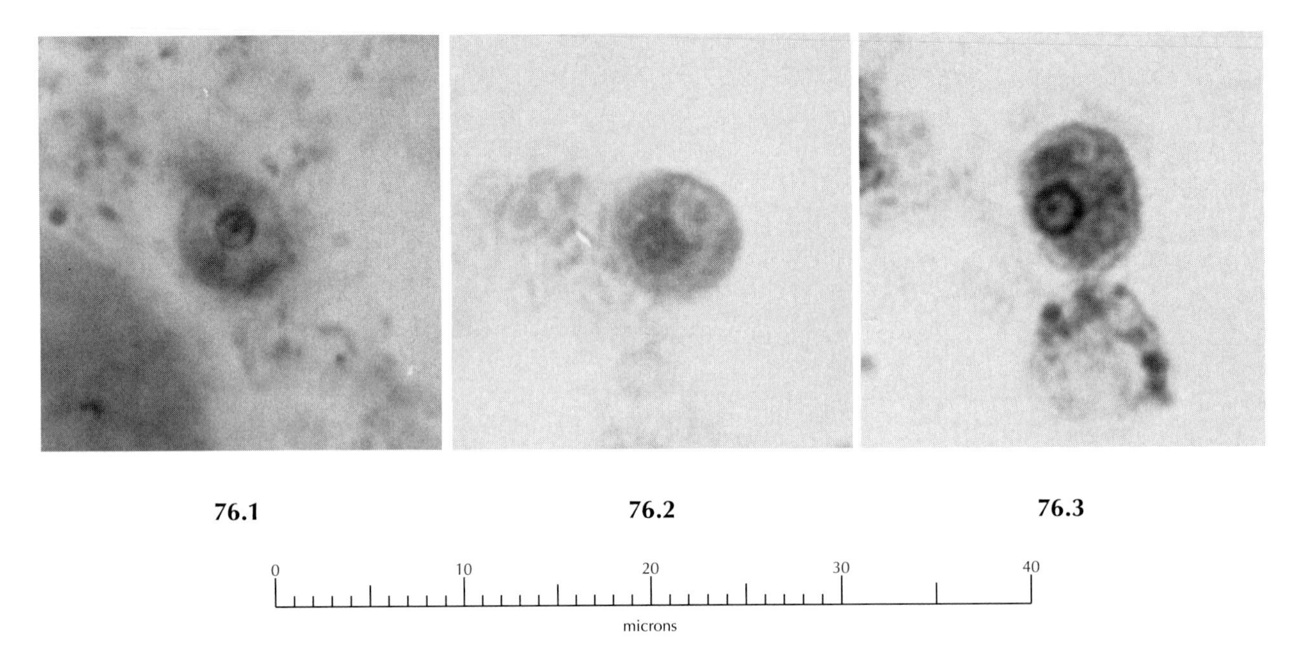

Plate 76. *Entamoeba hartmanni* on permanently stained fecal films. 1: T, PVA, T (CDC); 2 and 3: T, MIF, T; 4 and 5: T, MIF, P; 6: T, PVA, H (CDC); 7: C, PVA, T; 8: C, MIF, P; 9: C, MIF, T. The trophozoites of *E. hartmanni* (1–5) are usually very small and it is often difficult to clearly see the nucleus and identify the organism. The nucleus of each trophozoite shown here is distinct. The fixatives affect the staining results and those fixed in PVA (1) often tend to blend into the background more than those fixed in MIF (2–5). The cytoplasm in those fixed in MIF is often better defined, especially those stained in Polychrome IV (4, 5). Cysts appear to be more difficult to stain than trophozoites. Two cysts shown were from PVA-fixed specimens (6, 7) and two were from MIF-fixed specimens (8, 9) and each was stained differently. One cyst from a PVA-fixed specimen was stained with hematoxylin (6) and shows well-defined detail especially of the chromatoid bars. The two nuclei are not as well defined but the karyosomes stand out. The cyst stained with trichrome (7) does not appear to be as crisp. Only one of the three nuclei present is distinct at this focal plane. The chromatoid bar is less well defined. The cyst fixed in MIF and stained with Polychrome IV stands out even though surrounded by debris. Three nuclei are seen at this focal plane and in two the karyosomes can be seen. The cyst fixed in MIF and stained with trichrome has four nuclei rather poorly defined. It is especially important to note the size of the nuclei in the four nucleate cysts in relation to the diameter of the cyst. The four nuclei are very small and occupy only about one fourth of the body of the cyst.

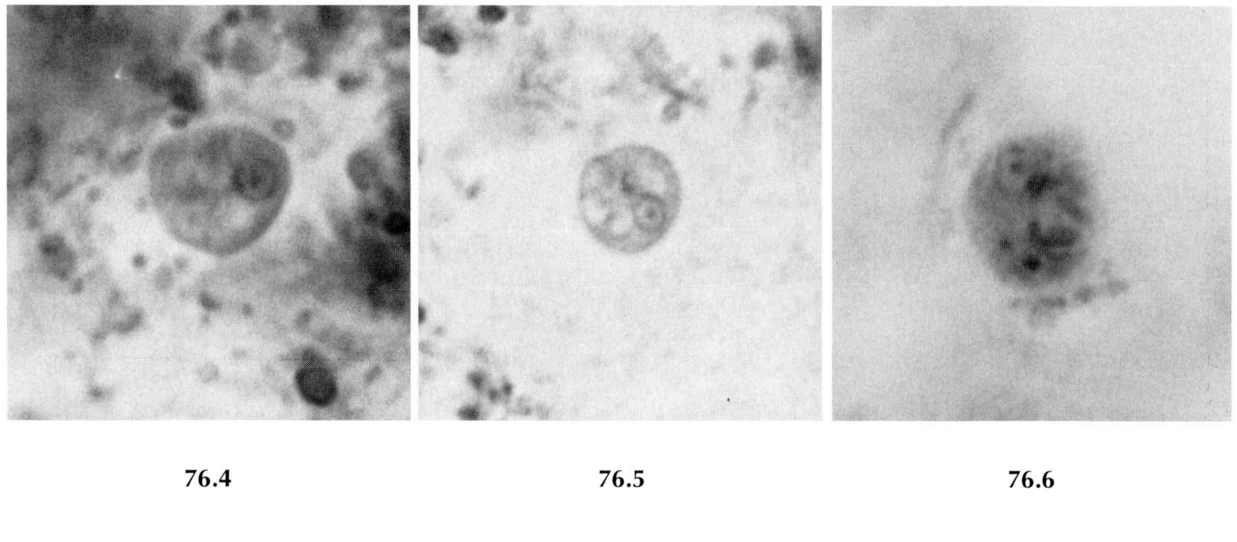

76.4 76.5 76.6

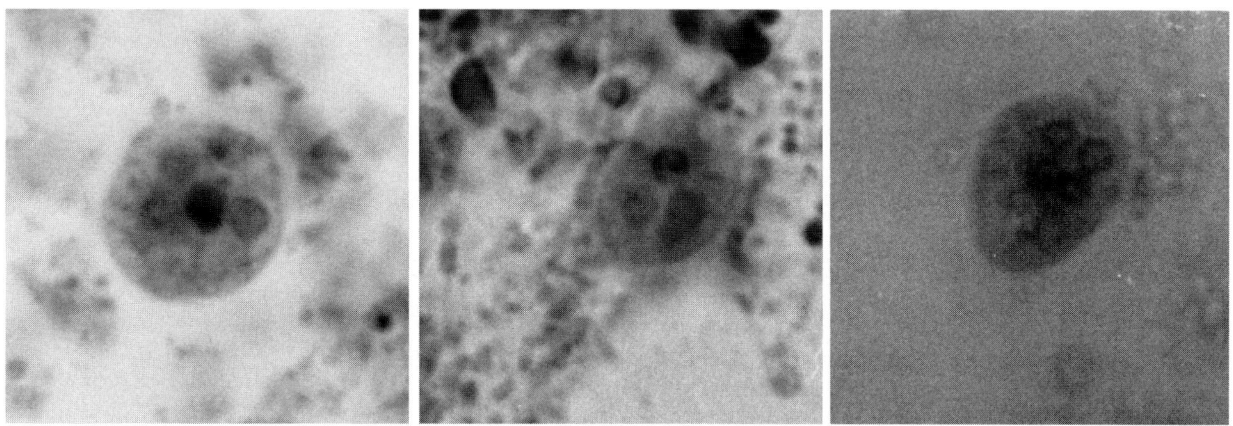

76.7 76.8 76.9

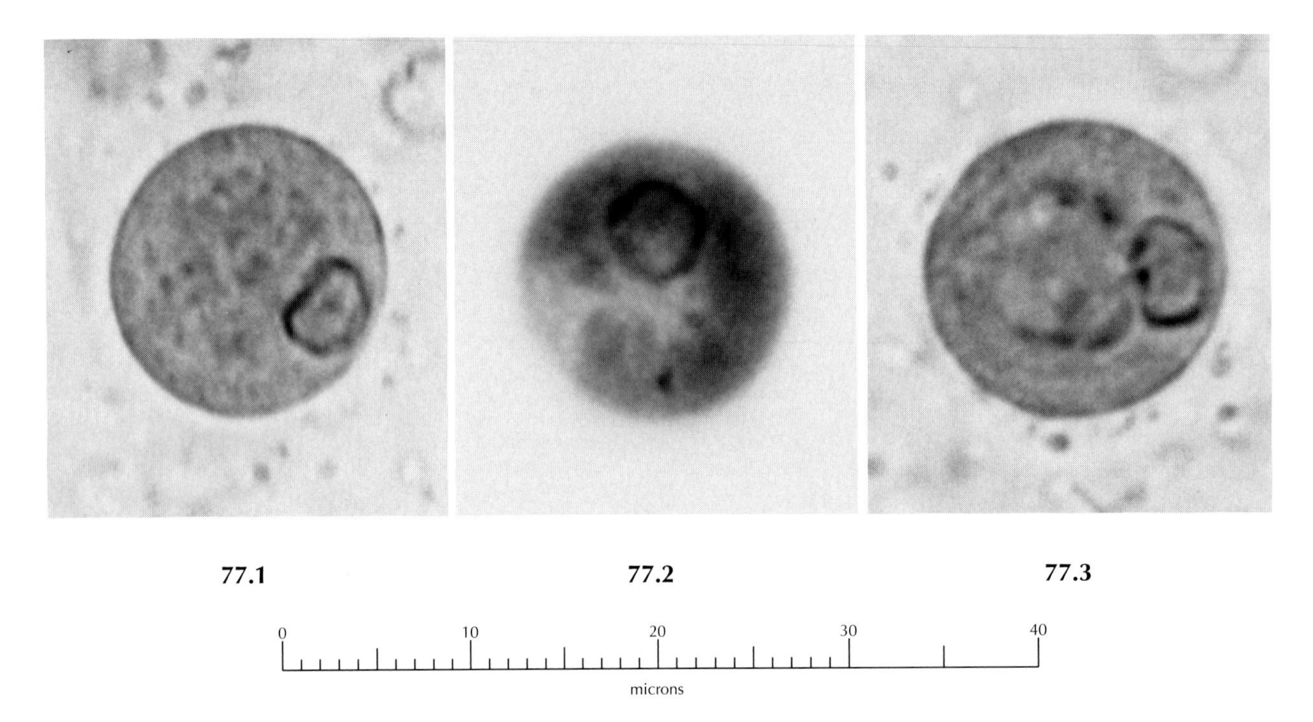

Plate 77. *Entamoeba polecki* on permanently stained fecal films. 1: T, MIF, P; 2–7: C, MIF, P; 8 and 9: C, PVA, T. The karyosome of *E. polecki* is small and the nucleoplasm is dense so that it is sometimes difficult to see (4–9). The nucleus is often misshaped, especially in cysts (1, 3, 4) but the peripheral chromatin is usually very evenly beaded and distributed uniformly around the nuclear membrane (7–9). The trophozoite usually has many vacuoles, much like *E. coli* but may have an evenly dense cytoplasm with some small vacuoles (1). The cytoplasm of the cysts varies greatly. It is usually irregular, may have large glycogen vacuoles (2–4) but more frequently it is seen with small vacuoles and dense, dark staining masses of various sizes and shapes in the cytoplasm (5–9). In fecal films stained with trichrome-type stains, these masses may be brightly stained (usually red). The frequently distorted shape of the nucleus and the irregular shape of the cyst are more commonly seen in *E. polecki* than in *E. histolytica* and the masses of condensed material in the cytoplasm is unique to *E. polecki.*

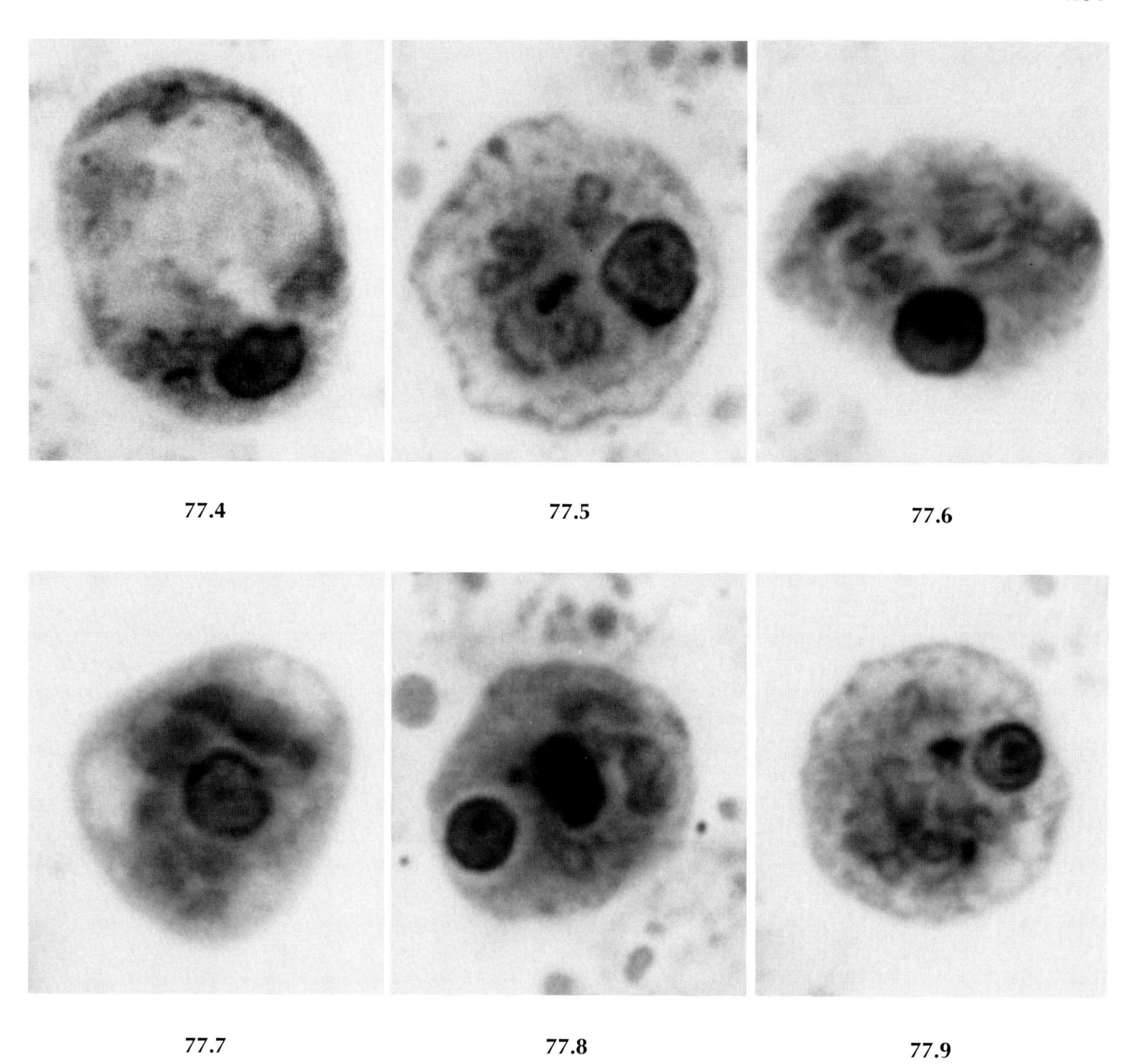

77.4 77.5 77.6

77.7 77.8 77.9

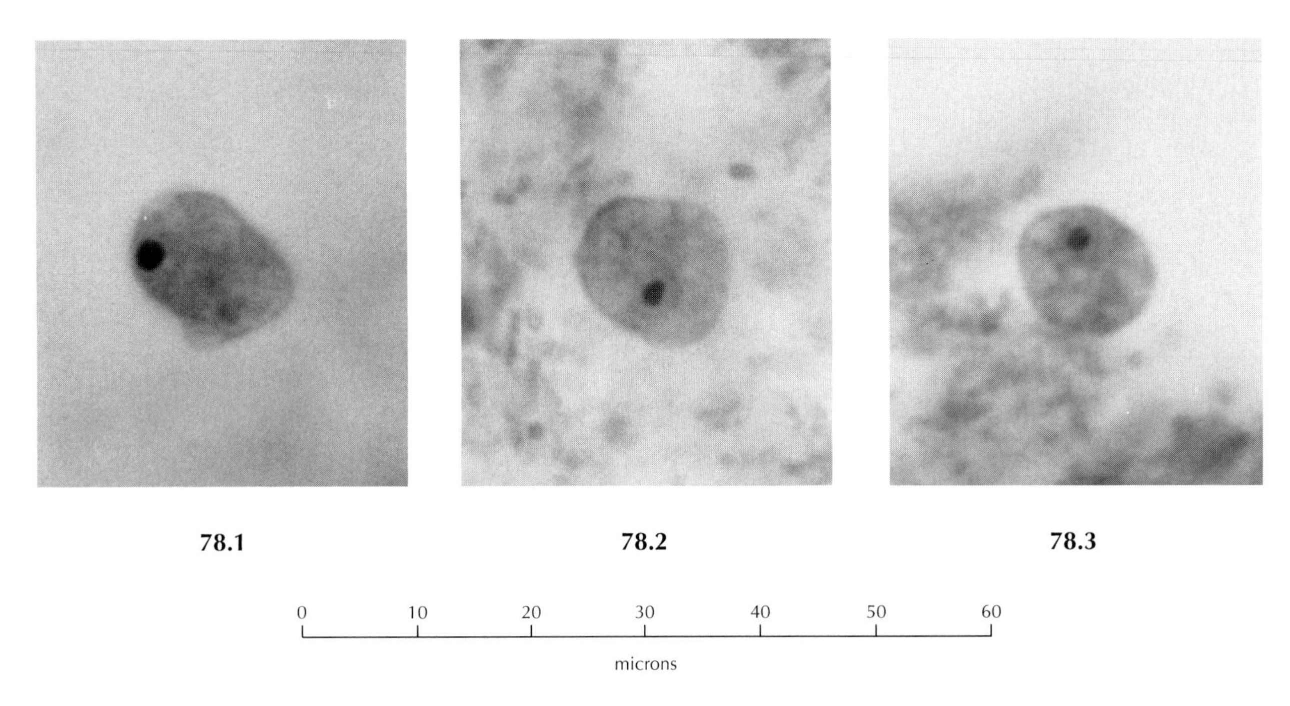

Plate 78. *Endolimax nana* on permanently stained fecal films. 1: T, PVA, T; 2: T, PVA, T (CDC); 3: T, MIF, T; 4: T, MIF, P; 5: T, MIF, T; 6: T, MIF, P; 7: C, PVA, T; 8 and 9: C, MIF, T. The karyosome of *E. nana* is large, somewhat flattened, and takes up about half of the inner space of the nucleus (1–3). It is often seen divided into two parts (5) or four parts (6). The nuclear membrane is often outlined by cytoplasmic granules that appear to adhere to it (2–5). The inner surface of the nuclear membrane is free of granules (2, 3). The cytoplasm is usually not as well defined in trophozoites fixed in PVA (1, 2) as it is in trophozoites fixed in MIF (3–6). It is better defined in trophozoites stained with Polychrome IV (4). In the cysts, the nucleus appears as a hole punched out of the cytoplasm with the darkly stained mass, the karyosome, inside (7–9). The cysts fixed in PVA and MIF stain equally well using trichrome stain.

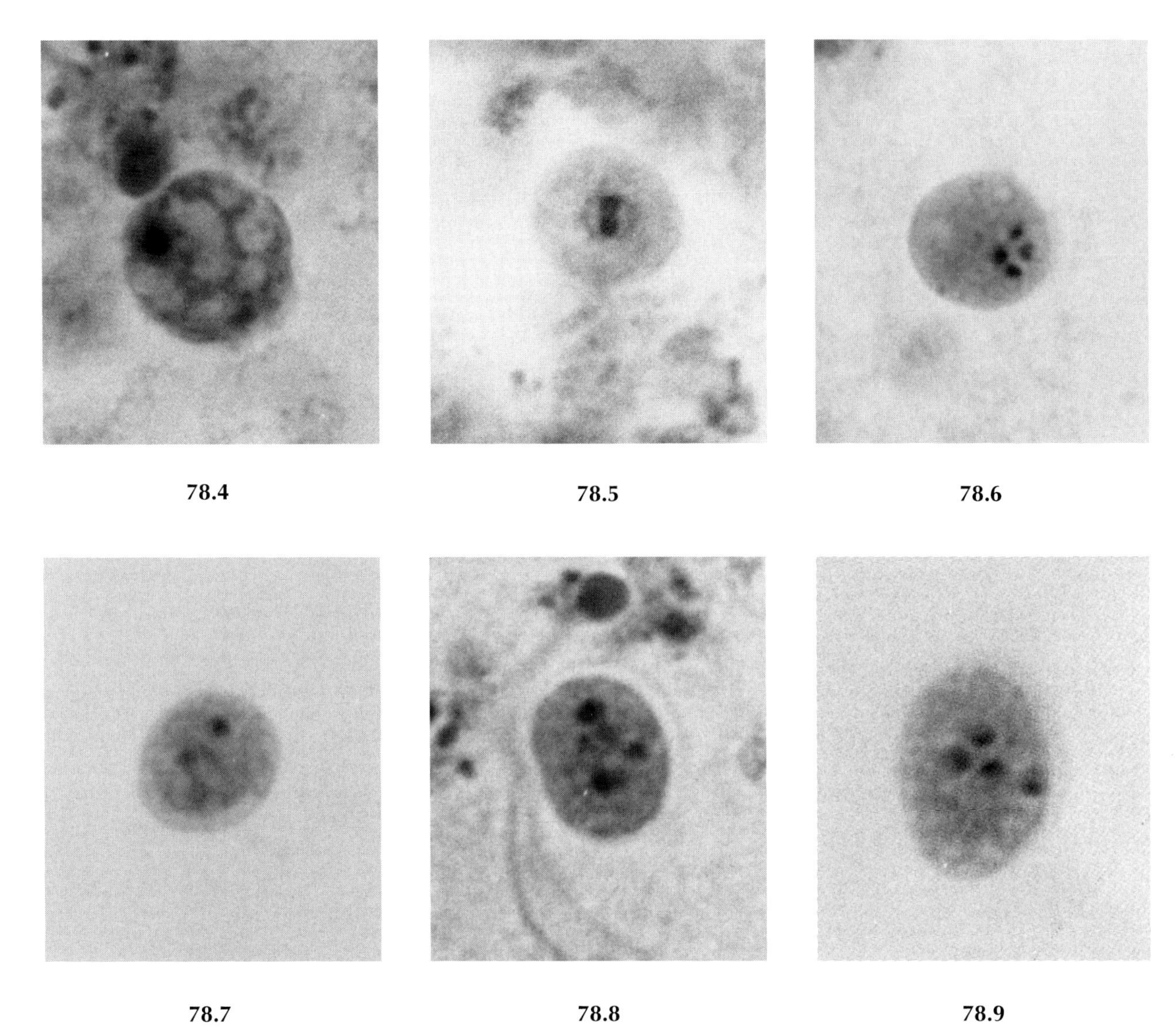

78.4 78.5 78.6

78.7 78.8 78.9

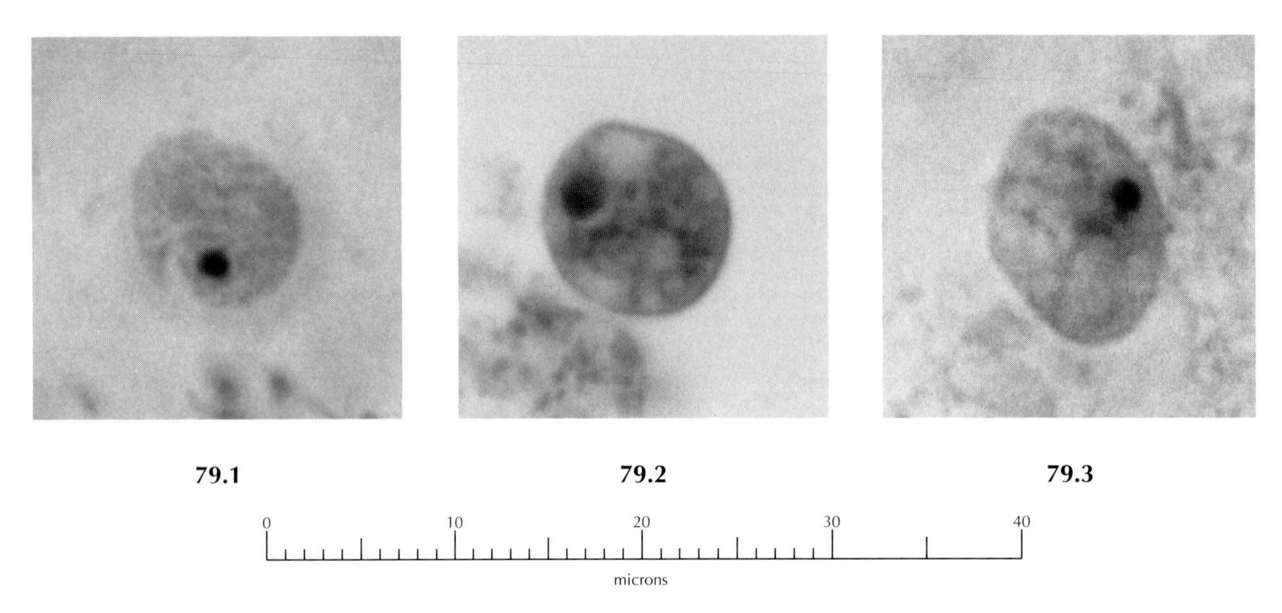

Plate 79. *Iodamoeba butschlii* on permanently stained fecal films. 1: T, PVA, H (CDC); 2: T, PVA, T (CDC); 3: T, MIF, T; 4: T, MIF, P; 5–7: C, PVA, H (CDC); 8–9: C, MIF, P. The trophozoites of *I. butschlii* vary greatly in size. The cytoplasm is dense with relatively coarse granules, usually with many vacuoles, and often with inclusions (1–4). The karyosome is large and is usually dense (1–4), sometimes with a less dense, vesicular portion (2). The nuclear membrane is often indistinct (2–4). Trophozoites are shown preserved in two different fixatives, two in PVA (1, 2) and two in MIF (3, 4). Those fixed in PVA were stained with hematoxylin (1) and trichrome (2). Those fixed in MIF were stained with trichrome (3) and Polychrome IV (4). The cytoplasmic detail is increased from 1–4. Cysts have a single nucleus with a large, dense karyosome and an indistinct nuclear membrane (5–9). The cysts have one large glycogen vacuole (5–7) or sometimes several smaller vacuoles (8, 9). There are usually large dense granules on the periphery of the vacuoles (8, 9) which are better seen in trichrome and Polychrome IV stained cysts.

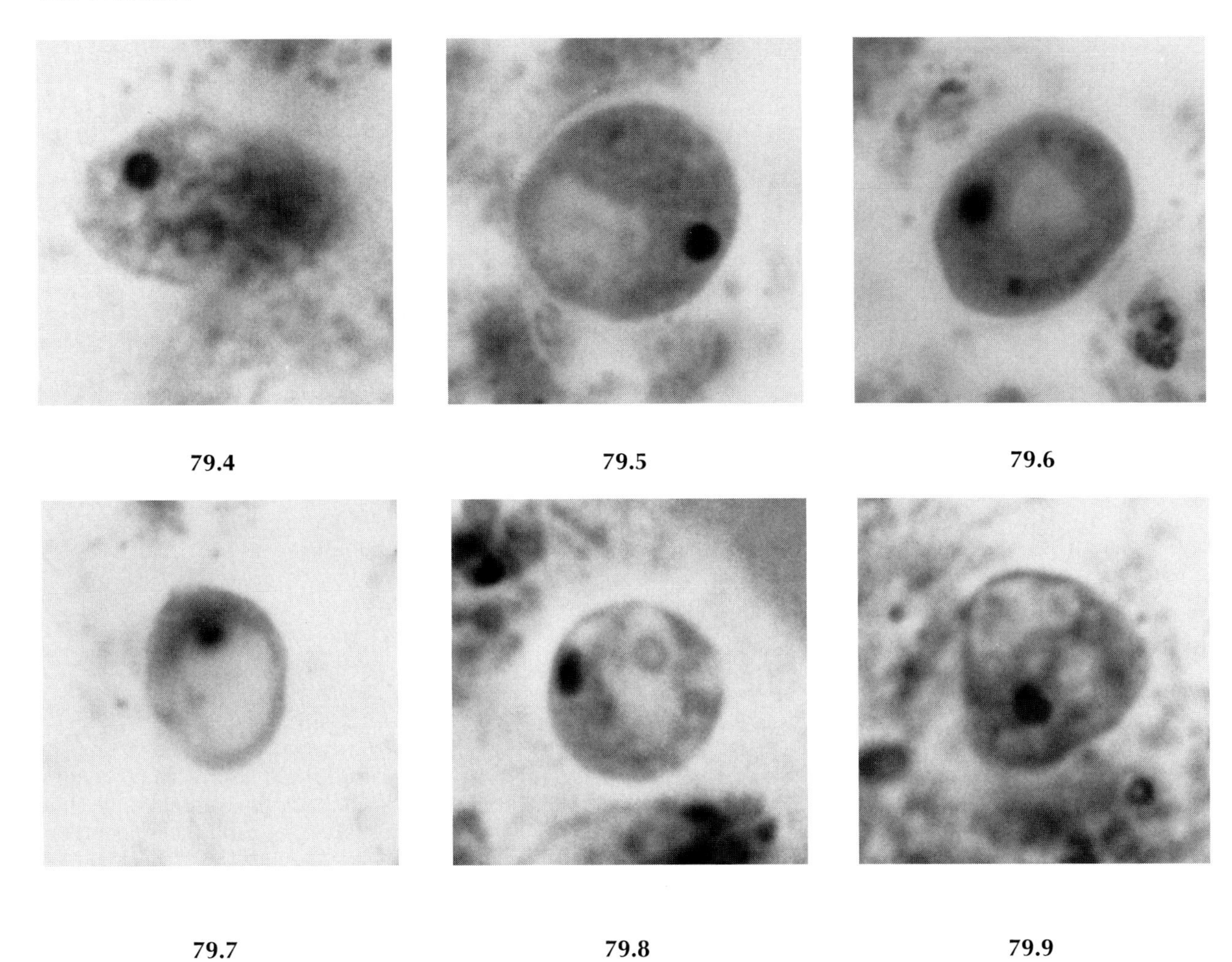

79.4 79.5 79.6

79.7 79.8 79.9

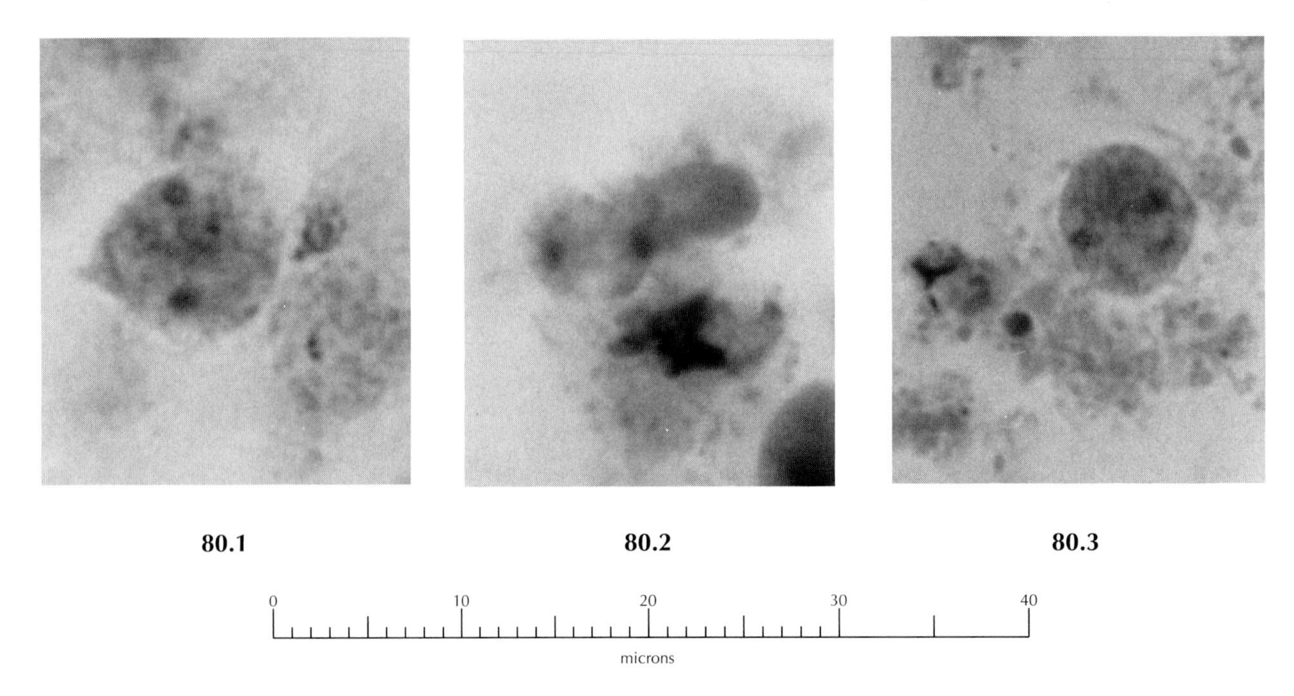

Plate 80. *Dientamoeba fragilis* on permanently stained fecal films. 1: T, PVA, T (CDC); 2–3: T, PVA, T; 4: T, PVA, P; 5: T, MIF, T; 6–9: T, MIF, P. *D. fragilis* is a delicate, active, amoeboid trophozoite that may vary greatly in size (compare 3 and 9 with 4 and 5). Trophozoites on permanently stained fecal films appear similar to those of *Endolimax nana* but the cytoplasm is always highly vacuolated. Binucleate forms are predominant (1–9) but some mononucleate forms are usually present in a positive fecal film. The nuclei stain well in hematoxylin (not shown), trichrome (1–4, 6), or Polychrome IV (5, 7–9) whether fixed in PVA (1–4), MIF (5–9), or PIF (not shown). Vacuoles with inclusions often obscure or mimic a nucleus and make it difficult to define (7, 8). The beaded structure of the karyosome is perhaps best seen in 8, center. The beading is rarely discernible because the nucleoplasm surrounding the karyosome is usually cloudy (1–9). In trichrome-type stains, the nuclear membrane is usually indistinct. The highly vacuolated, delicate appearing cytoplasm of the trophozoites is seen well in 4 and 9.

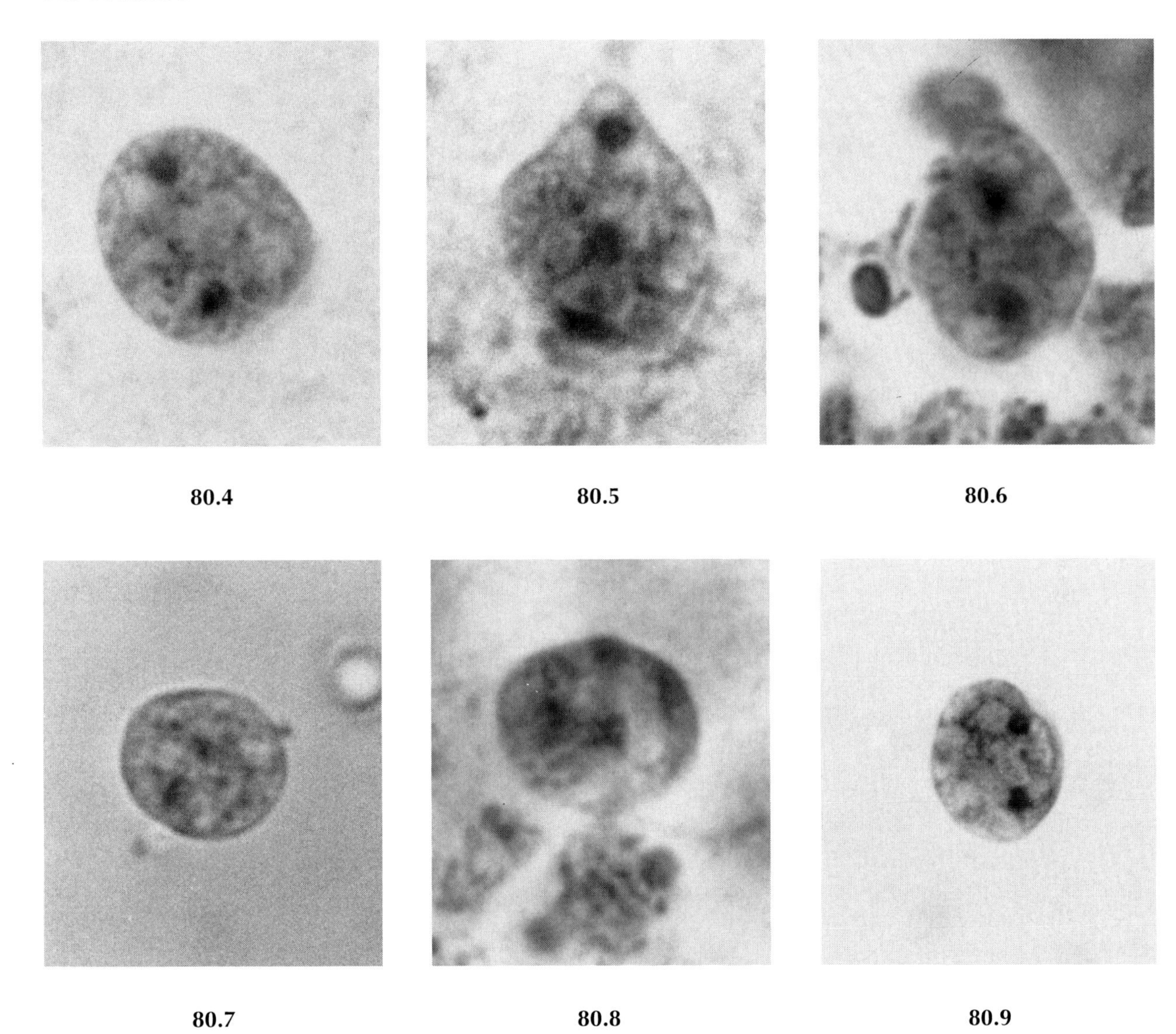

80.4 80.5 80.6

80.7 80.8 80.9

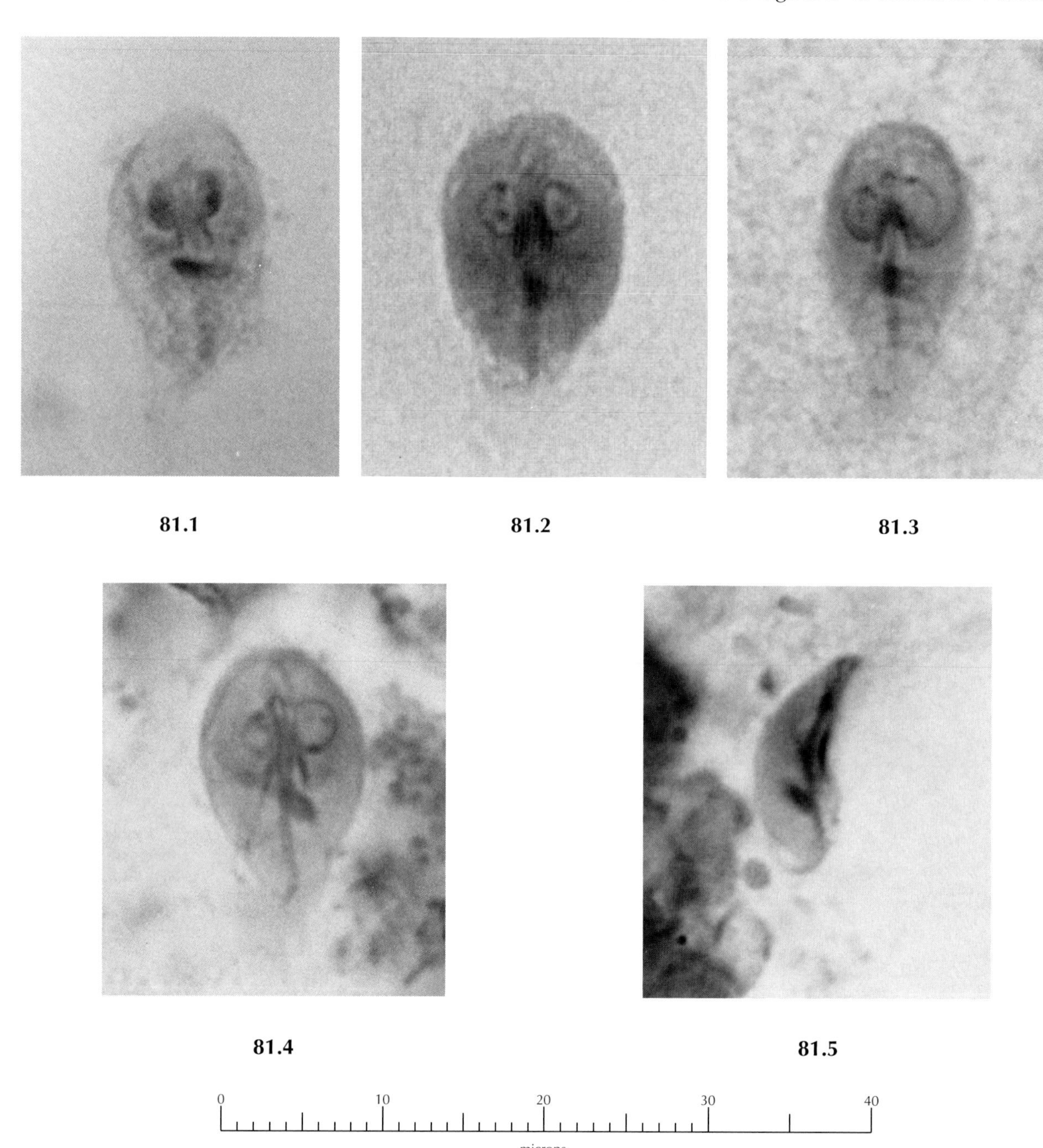

Plate 81. *Giardia lamblia* on permanently stained fecal films. 1: T, PVA, T; 2–3: T, MIF, T; 4: T, MIF, P; 5–6: T, MIF, T; 7: C, PVA, T; 8: C, PVA, P; 9: C, MIF, P. The trophozoites of *G. lamblia* (1–6) are bilaterally symmetrical, pear-shaped, with two nuclei (1–4), and a large ventral sucker disk used for attachment (3, 5). The axostyle is often discernible (3, 4). The parabasal bodies may appear as dark structures perpendicular to and at the center of the axostyle (1–4, 6). In a lateral view (5), the nuclei and flagella are not always visible and the trophozoite may be harder to recognize. Often, several trophozoites are clumped or attached together as are the three in 6. The cysts are usually distinctive and readily recognizable. The body of the organism is usually separated from the cyst wall but the cyst wall is not always discernible in permanently stained slides. The cyst wall may not be visible but is often detectable when surrounded by background material (7). The internal structures (organelles) appear to be more readily discernible when the fecal smear has been stained with Polychrome IV (8, 9). The nuclei, axostyle, parabasal body, and axonemes may be visible in a well-preserved and well-stained preparation (7–9).

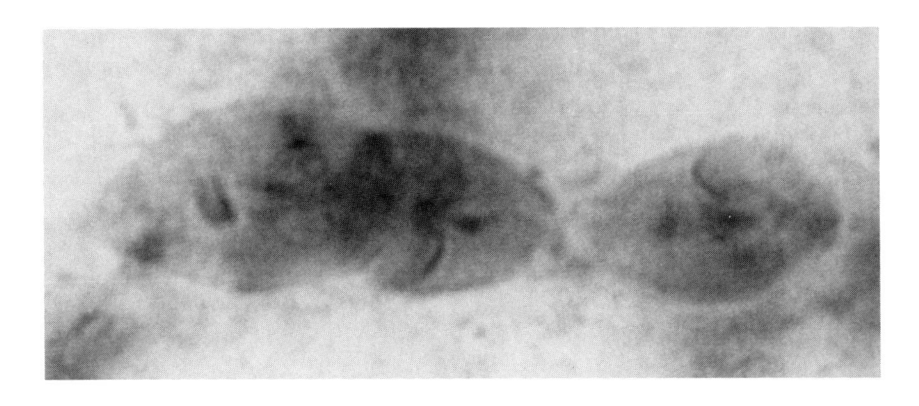

81.6

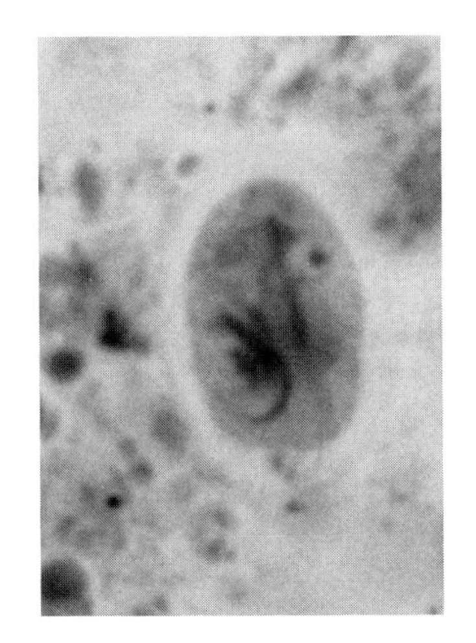

81.7

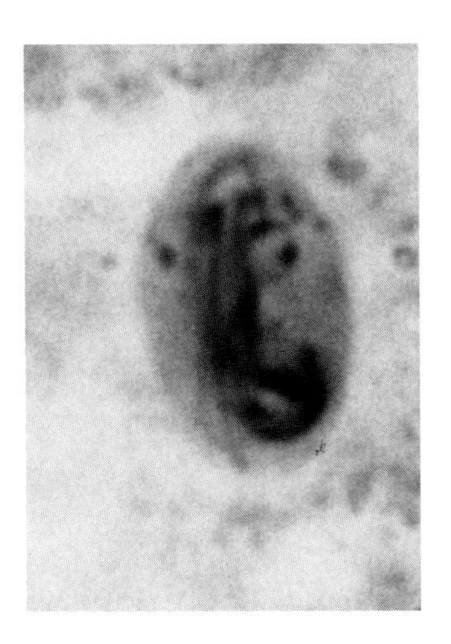

81.8

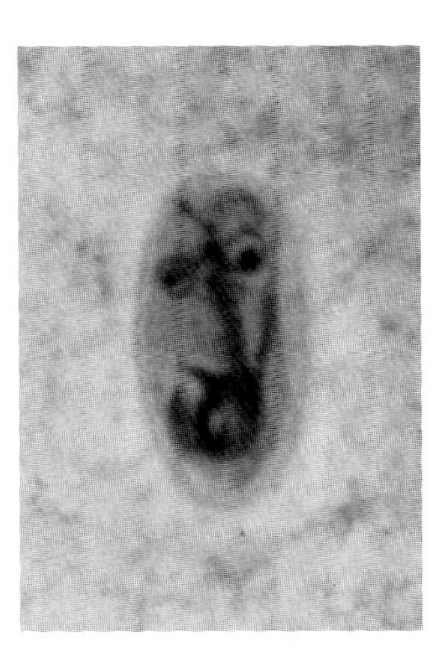

81.9

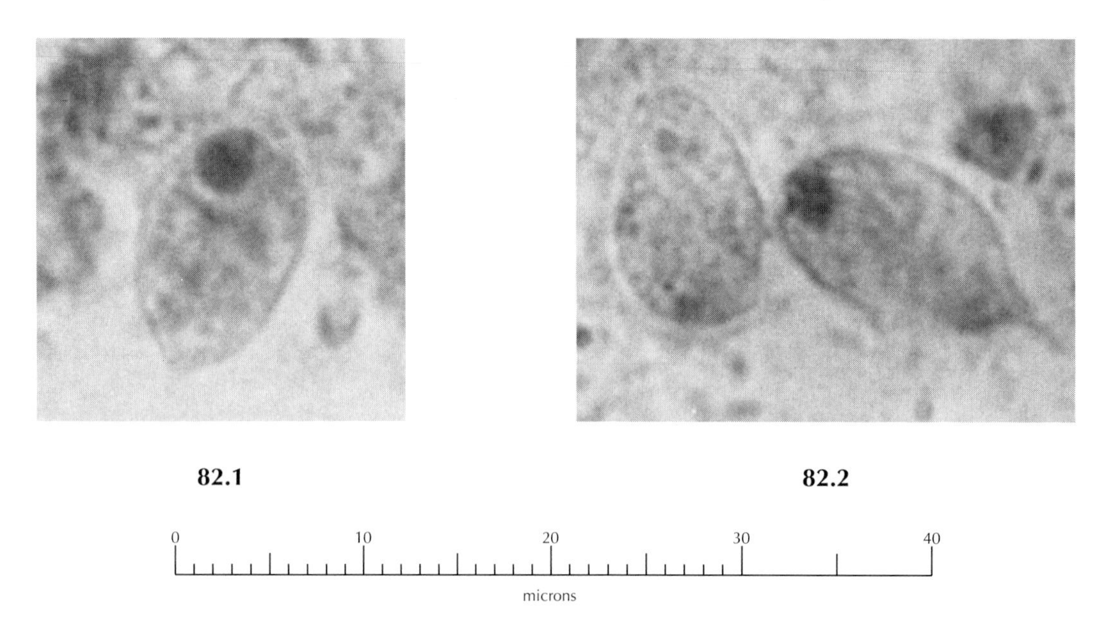

Plate 82. *Chilomastix mesnili* on permanently stained fecal films. 1–3: T, MIF, T; 4: C, PVA, T (CDC); 5: C, PVA, T; 6: C, MIF, T. The trophozoites of *C. mesnili* vary greatly in size, especially in life. The fixed trophozoite is usually from 10 to 15 µm in length depending on how much it is rounded (1–3). In the lateral view, it is pear-shaped, rounded at the anterior end, and usually pointed at the posterior end (1, 2). The morphology of the trophozoite is not well shown by the usual staining methods. The nucleus may be sharply defined but usually it stains darkly and the karyosome is not distinct (1–3). The cytostome is usually seen as an unstained area beginning at a level near the center of the nucleus and extending posteriorly about 5 µm (2, 3). The flagella may be visible but they are usually buried in the surrounding debris and are not distinct (1–3). The vacuolated cytoplasm is seen in most trophozoites. In 2, one of the two trophozoites is out of focus and not identifiable, which further emphasizes the importance of the depth of field when using the microscope. The cysts of *C. mesnili* (4–6) appear essentially as they do in wet preparations. The cyst is lemon-shaped with a raised area at one end when seen in the lateral view (4, 5) or oval (6). The cyst has a single nucleus similar to that of the trophozoite. The long, curved parabasal body that supports one side of the cytostome is often clearly seen (5) but the cytostome is not distinct.

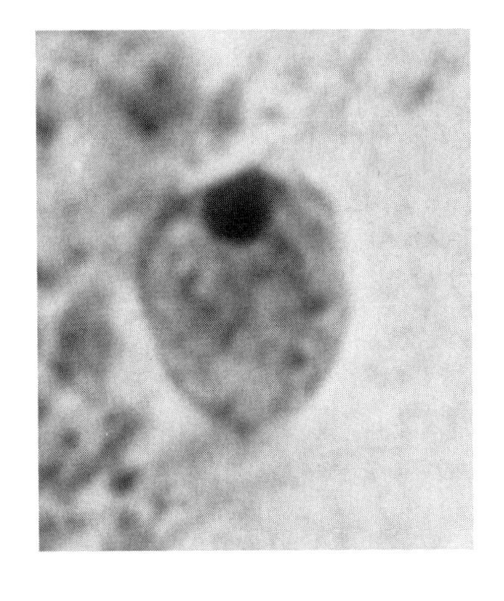

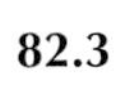

82.3

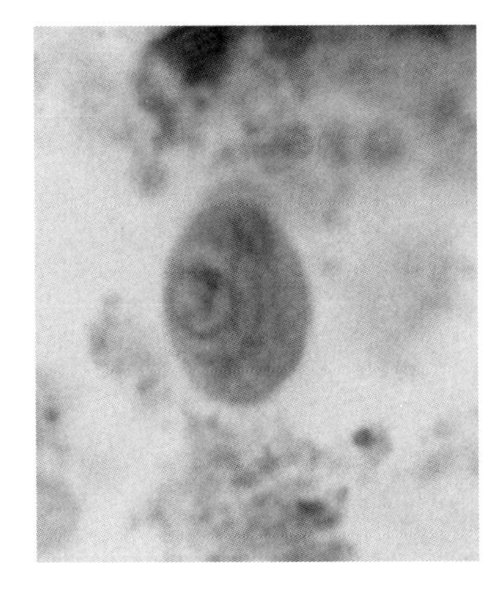

82.4

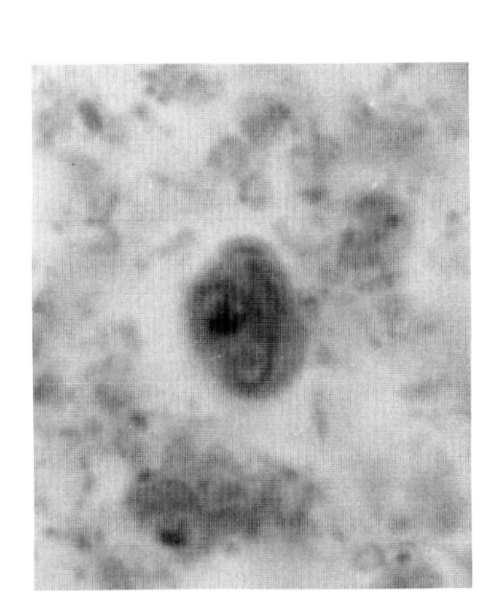

82.5

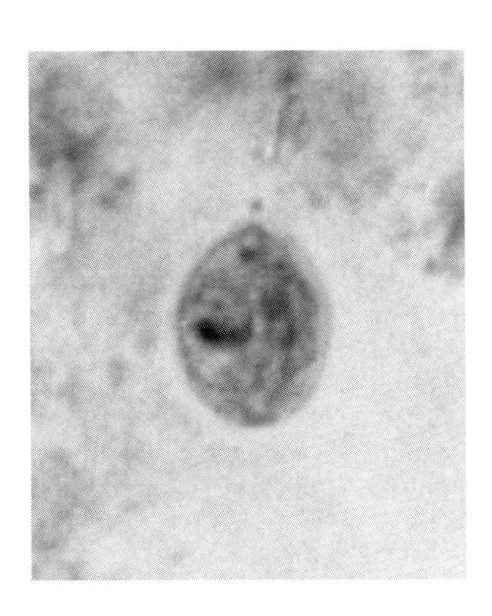

82.6

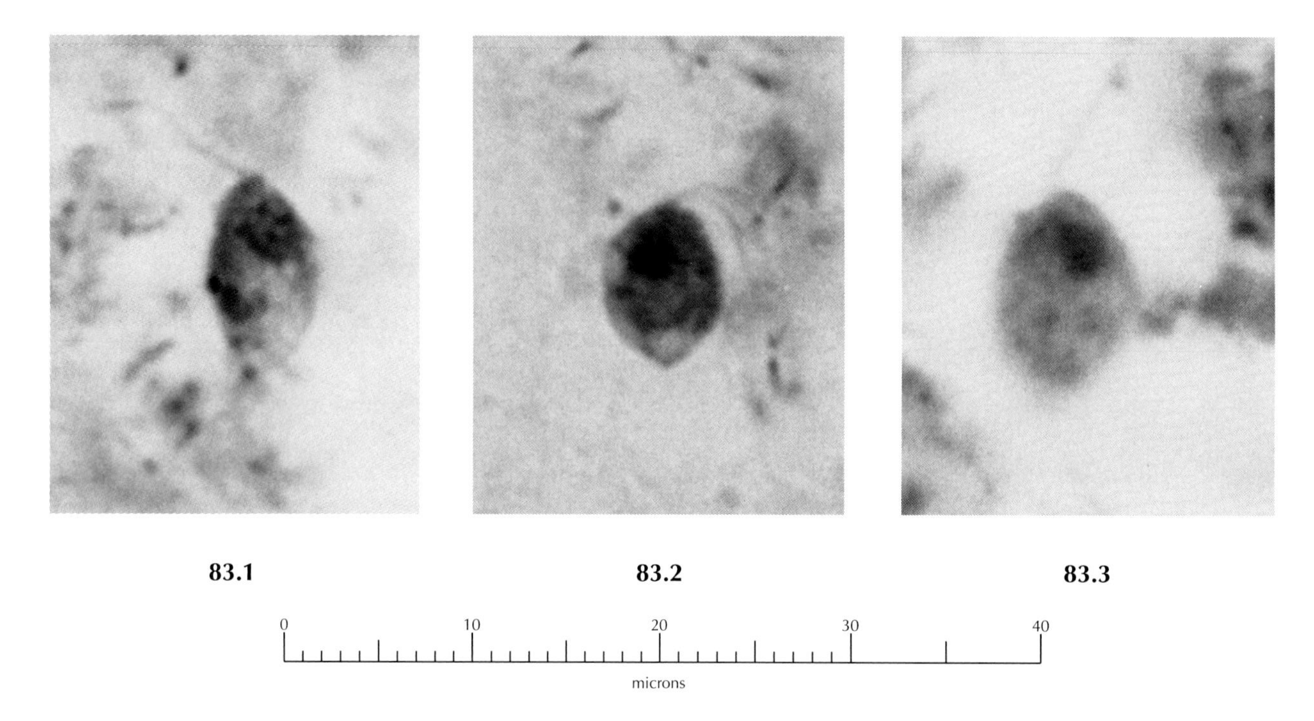

Plate 83. *Trichomonas hominis* on permanently stained fecal films. 1: T, MIF, P; 2: T, MIF, T; 3: T, MIF, P; 4: T, MIF, T; 5: T, MIF, P; 6: T, MIF, T; 7: T, MIF, P; 8: T, MIF, T; 9: T, MIF, P. The trophozoites of *T. hominis* may vary greatly in size (compare 2 and 7) sometimes even in the same fecal specimen. It may be elongated and lemon-shaped (2, 3, 7) or more rounded (5, 8). The nucleus may stain poorly and the central karyosome may or may not be visible depending on the plane of focus. The axostyle is clear and is usually not visible but the costa extending posteriorly from the level of the nucleus is sometimes visible (6). Parts of the undulating membrane can often be seen as a dark, wavy line near the nucleus where it apparently has its origin (3–5). Also the trailing flagellum, which is a continuation of the axoneme of the undulating membrane, can sometimes be seen (6, 7). The anterior flagella are frequently seen (1–3, 5, 7, 9) except when the anterior end is in debris (6). The cytoplasm is sometimes filled with small vacuoles (3, 5, 9). The cyst stage is not known and the mode of transmission is not clearly understood.

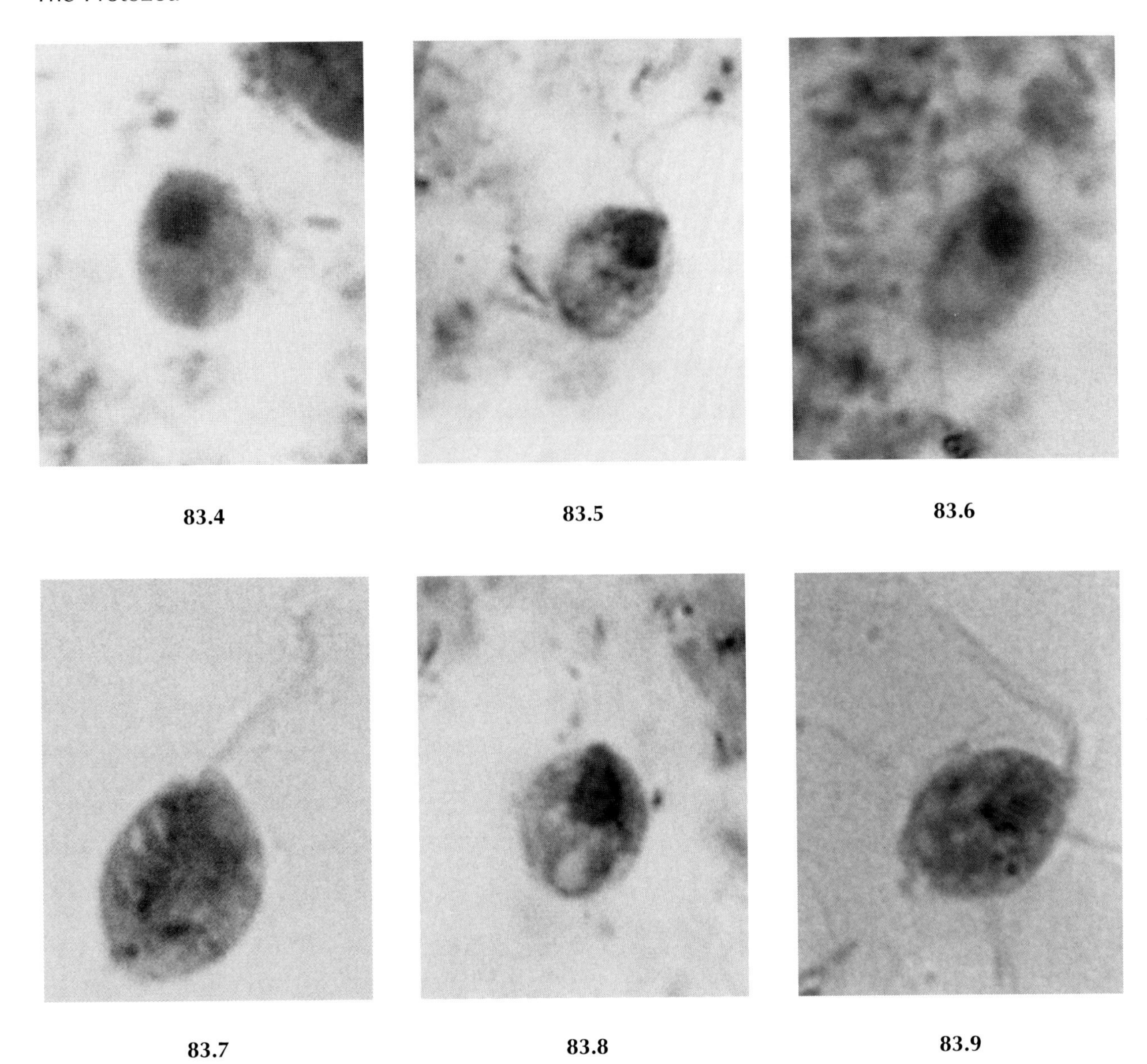

83.4 83.5 83.6

83.7 83.8 83.9

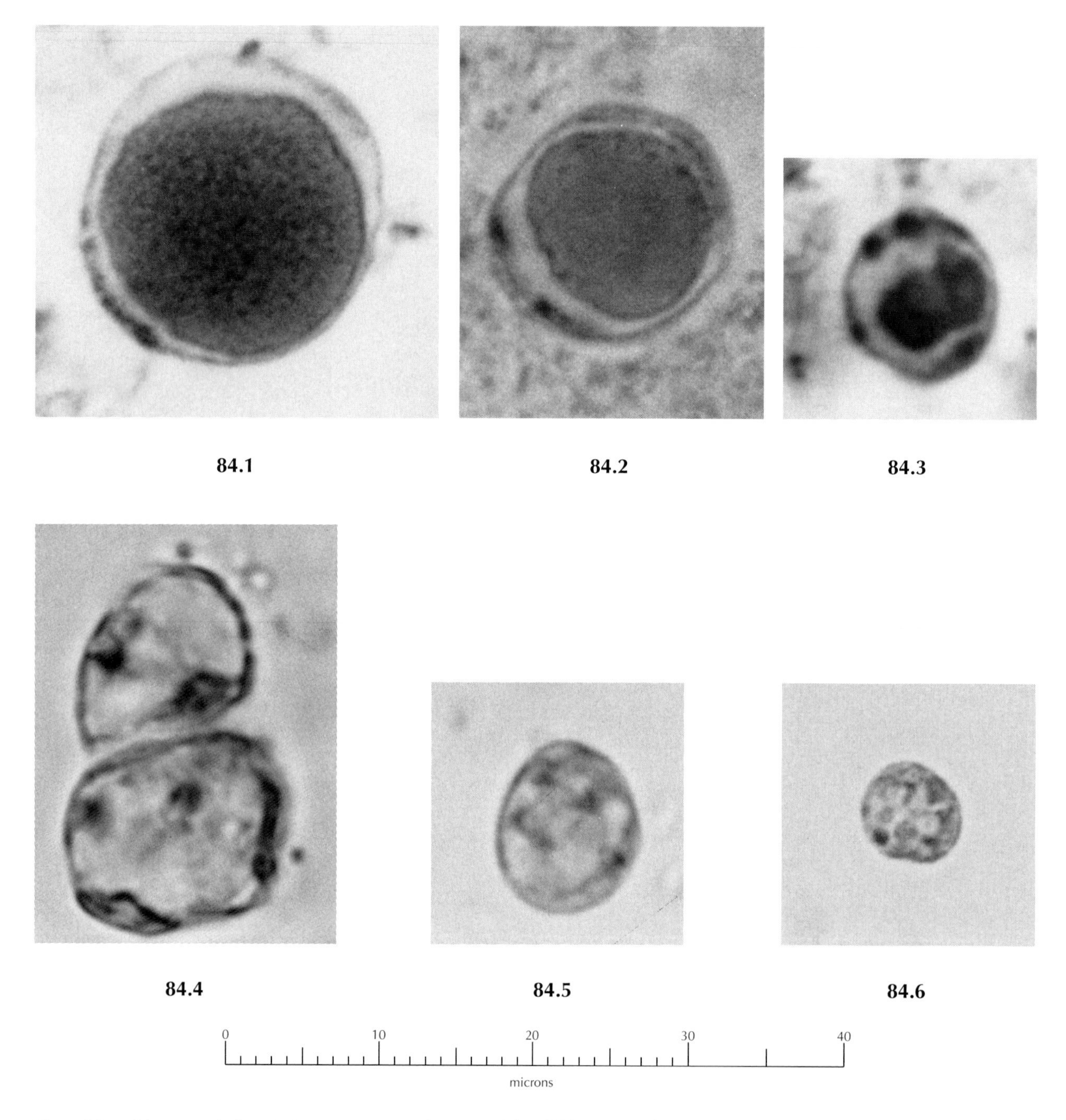

Plate 84. *Blastocystis hominis* on permanently stained fecal films. 1 and 2: MIF, T; 3–5: MIF, P; 6: PIF, P. *B. hominis* varies greatly in size and structure in fecal preparations (compare 2 and 5). The classical form is round, with a vacuole-like center, which is surrounded by a thin layer of cytoplasm (1, 2). Within the cytoplasm are one or several pale-staining nuclei and refractile, dark-staining globules called volutin (1–4). In some forms, dark areas form in the central area (3) which at times appear to be in several parts. The organism is more delicate and less refractile than cysts of amoebae. Some forms lose the vacuole-like central area and appear more like the amoebae (5, 6). They divide by binary fission (4).

Chapter 10

Nonhuman Parasites and Structures that Mimic Parasites

There are many things that may be in a fecal specimen that when first seen appear to be a protozoan parasite or an egg of a helminth related to a human infection. In addition, there may be cysts of protozoa, oocysts of coccidia, and eggs of some helminths that are not of human origin and are not related to human infections but their presence requires active investigation to make certain they do not pose a problem for the patient.

In the three Plates of photomicrographs that follow, several objects are shown that might raise questions regarding their identification and importance in the fecal specimen.

STRUCTURES RESEMBLING PROTOZOA

In Plate 85, some structures that resemble protozoa when found in a fecal preparation are presented. White blood cells, especially neutrophils, may resemble cysts of amoebae because of the way the nucleus stains (Plate 85:1). Usually, these can be identified easily on wet preparations because of their irregular shape but they may be more difficult to identify on permanently stained fecal films. The object in Plate 85:2 appeared to be an amoeba when first seen on a wet preparation. When the cover glass was gently tapped, the object was seen as flat with what appears to be a nucleus superimposed on top. Pollen grains are very common during certain periods of the year (Plate 85:3, 6) but can usually be identified (or at least not mistaken for parasites) because of their size and usual shape. Yeasts are very common in fecal specimens (Plate 85:5). Some yeast cells are 4 to 6 µm and may resemble oocysts of cryptosporidia. The budding yeast cell immediately differentiates it from cryptosporidia. Iodine will usually stain yeast cells but not cryptosporidia. Acid-fast stains may be used to identify cryptosporidia but there are some yeast cells that are also acid-fast. Care must be taken to use a stain and procedure that will differentiate between oocysts of cryptosporidia and acid-fast yeast cells. A staining artifact (4) that was present on a permanently stained fecal film resembled cysts of *Entamoeba coli* as sometimes seen on trichrome-stained films, but on closer examination, it showed no internal structure.

STRUCTURES RESEMBLING HELMINTH EGGS

Structures that resemble helminth eggs are probably most commonly seen. The structure of many plant cells appear similar to the structure of eggs of different helminths. All of the structures seen in Plate 86 are from plants. The object shown in Plate 86:1 appears to have an outer shell with embryo inside.

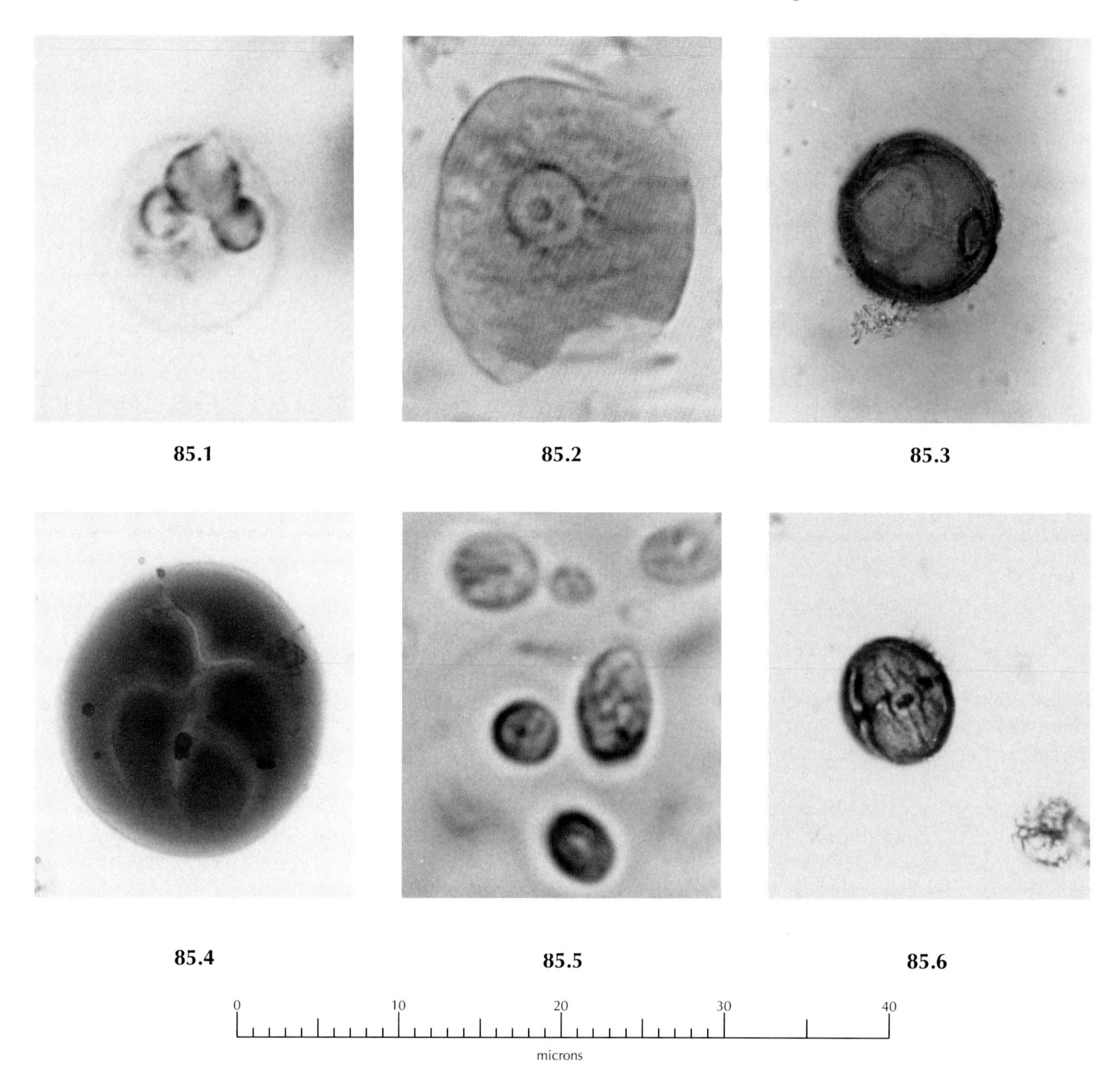

Plate 85. Structures that may resemble protozoan cells. 1: WBC, MIF, T; 2: artifact, MIF; 3: plant pollen, MIF; 4: staining artifact, MIF, T; 5: yeast cells, MIF; 6: plant pollen, MIF. White blood cells may stain in permanently stained fecal films so that the nuclei appear as they do in amoebic cysts (1). They are occasionally mistaken for a protozoan especially when very few are present. Some artifacts present in wet preparations may be mistaken for protozoa. An artifact appeared to be an amoeba with a nucleus with peripheral chromatin and a central karyosome (2). Upon more careful examination, the body was flat and the nuclear-like structure was superimposed on top of it. It was obviously an artifact. Pollen grains such as those in 3 and 6 have been mistaken for protozoan cysts more often in wet preparations than in stained fecal films. The fertilization pore can be seen on one of the other pollen grains in (3). The other pollen grain in (6) has a structure unlike any known human protozoan parasite. The structure in 4 was seen on a trichrome-stained fecal film and apppears similar to the cysts of *Entamoeba coli* in a poorly fixed and stained film. The internal structure is not that of an *E. coli* cyst. It is simply another artifact. Some yeast cells (5) are about 4 to 5 µm. They may become a questionable object when cryptosporidia are suspected in a patient but, if budding (as one is in the photomicrograph), it should be easily recognized as a yeast cell.

Careful examination quickly shows that what appears to be shell is simply the outer wall of a cell with pore openings between the inner and outer portion (arrow) which do not occur in helminth eggs. The object in Plate 86:2 resembles an egg of *Hymenolepis diminuta* at first glance, but there is no structure in the central circle, there is material between the circle and the wall, and there is a pore in the upper area of the wall. Again, the structure in Plate 86:3 appears similar to a helminth egg in structure. The outer wall appears to be layered unlike helminth eggs which are clear. The internal structure is not typical of a helminth egg.

The outer membrane in Plate 86:4 is thin and the structure is somewhat similar to that of some eggs of *Trichostrongylus* spp. There is no organization to the internal structure which is unlike eggs of human parasites. The plant cell (Plate 86:5) resembles an egg of *Ascaris* but the portion that would be the shell is not clear and there is no separation between it and the material inside. The small structure with its dark outer coating and the inner portion with something that resembles hooklets suggests an egg of *Taenia* sp. (Plate 86:6) but there are no striations in the outer covering as are found in the eggs of *Taenia*.

One must be cautious when identifying structures seen in a fecal specimen. An incorrect identification could lead to unnecessary and possibly unpleasant course of treatment for the patient for a parasite that is not present.

SPURIOUS INFECTIONS WITH EGGS AND CYSTS OF ANIMALS

There are many occasions when humans may ingest something contaminated with animal feces. Such contamination occurs on food, water, and utensils used in eating or preparing food. It is certainly not unusual for humans to handle something contaminated, then handle something they will eat without washing their hands. The eggs and the oocyst shown in Plate 87 are all from nonhuman sources and pose no threat to the individual who ingested them. It may be that at the time the individual ingested the egg or cyst he also ingested bacteria or viruses that did have an impact on his health and brought him to the physician for help.

In Plate 87, the eggs seen in 1 and 2 are from wild birds and were probably ingested when the individual cleaned or ate a game bird obtained for food. The source of the egg in 3 is not known but it is not from any known human helminth.

Arthropod eggs, such as those in 4 and 5, are often seen in human feces. They may be from any of a number of sources but are usually from cereal grains or nuts. The etchings on the egg shell in 4 usually suggest an egg of a mite, whereas the embryo in the egg in 5 appears to be developing a head and legs and is another kind of arthropod egg.

The oocyst seen in 6 has two sporocysts each with four sporozoites, which is probably at the infective stage. The structure of the oocyst places it in the same group as the human species but it is smaller than the oocysts of *Isospora belli* or *Sarcocystis hominis* and is probably from a mammal. The individual who passed the oocysts is a hunter and had killed and eaten squirrel a few days before providing the fecal specimen.

Eggs and cysts (or oocysts) of the parasites of animals can be ingested by man under certain circumstances and pass in the feces as spurious infections. These must be eliminated as being responsible for any illness in the patient. By knowing the general structure and special characteristics of parasites that infect or infest the intestinal tract and adjacent organs of man and by using reliable resources and methods for identification of the organisms found, one should be able it identify those organisms commonly infecting man and eliminate as unimportant those eggs and cysts present that are temporary visitors just passing through.

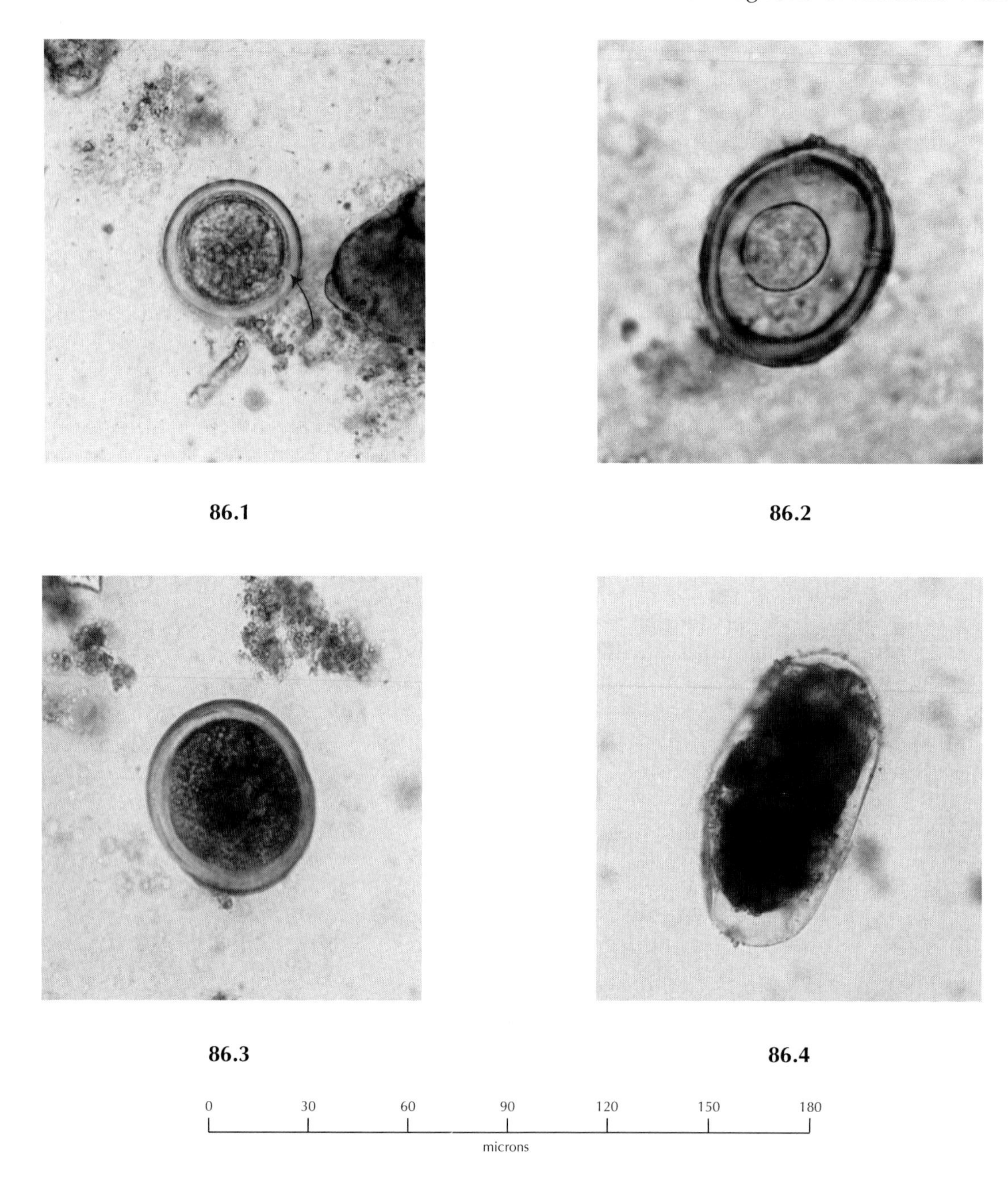

Plate 86. Structures that may resemble helminth eggs. 1–3: plant cell, MIF; 4: plant material, MIF; 5, 6: plant cell, MIF. There are many plant cells that resemble helminth eggs (1–6). Plant cells may look similar to decorticated eggs of *Ascaris lumbricoides* (1), resemble *Hymenolepis diminuta* (2), or late stages of hookworm (3). When a plant cell is lying free in the medium it may resemble an egg of *Trichostrongylus* spp. (4), or when the cell is surrounded by debris it may resemble an egg of *Ascaris lumbricoides* (5). Small plant cells or materials with a thick outer area and internal structures that appear similar to hooklets may be mistaken for eggs of *Taenia* spp. (6). None of the plant cells shown in the photomicrographs should be mistaken for helminth eggs when compared with the diagrams and photomicrographs presented in this manual.

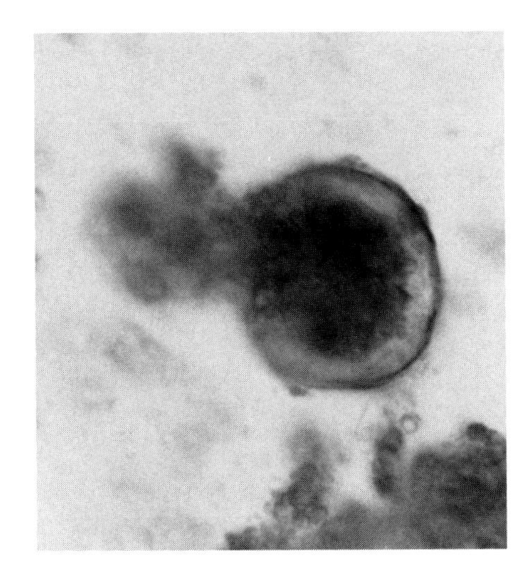

86.5

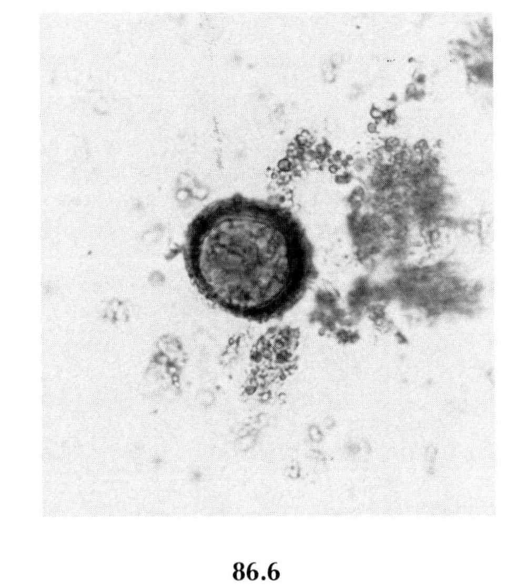

86.6

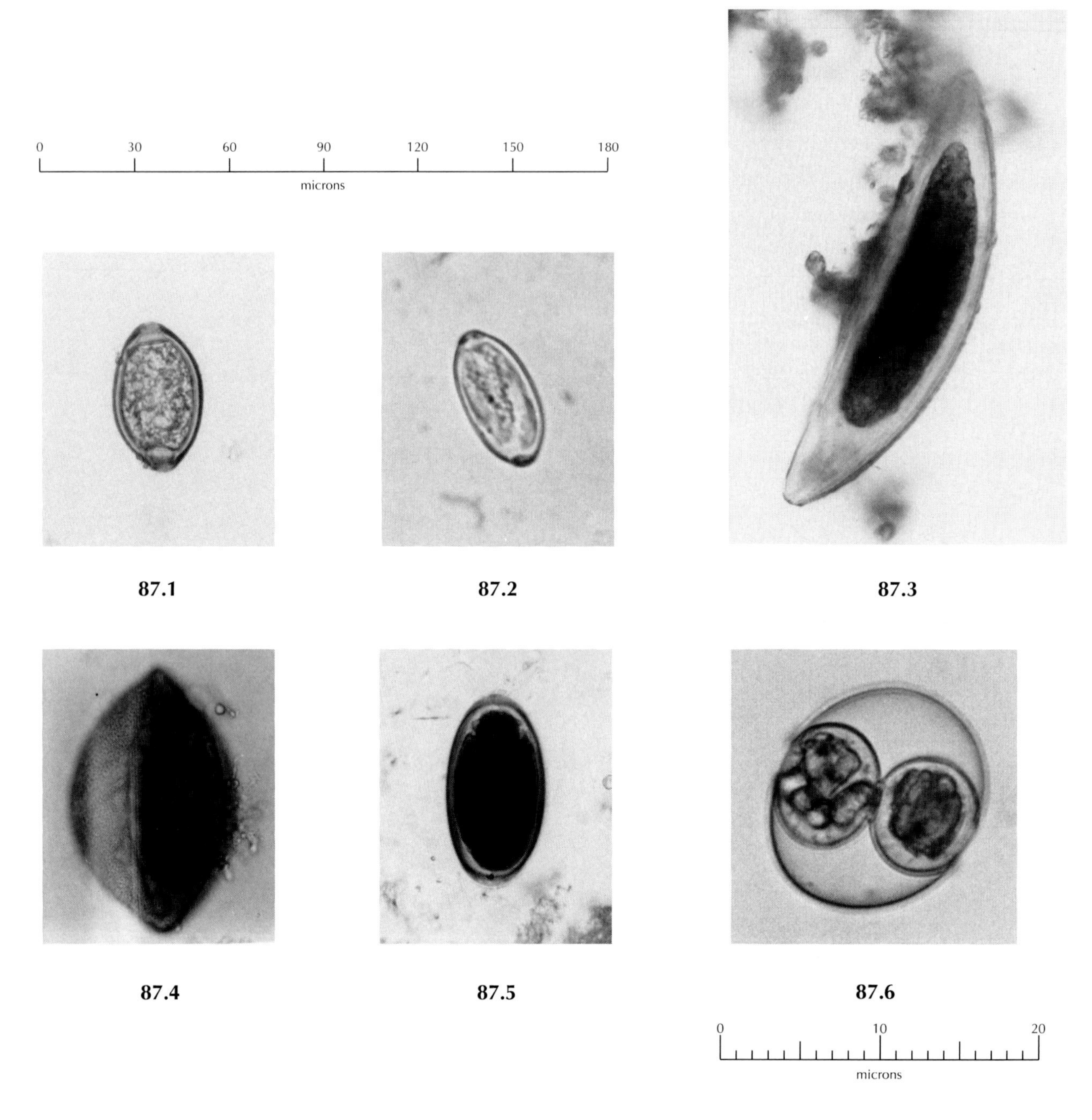

Plate 87. Spurious infections of man with parasites from animals. 1–2: helminth eggs from bird, MIF; 3: egg from animal or plant, MIF; 4–5: arthropod egg, MIF; 6: oocyst from animal, MIF. Many parasites infect plants and game animals and birds. When cleaning an animal obtained when hunting, care should be taken not to contaminate the meat with waste materials from the animal. The *Trichuris*-like eggs (1, 2) are from wild birds killed and eaten for food. The source of the egg in 3 is not known but it is not from any parasite known to infect man. Eggs of arthropods (4, 5) are usually ingested accidentally when eating grains, salads, etc. The etching on the outer part of the shell (4) usually identifies the egg as one from a mite. The internal structure of the egg in 5 is unlike that of any nematode. There is a developing head at the upper end and legs on either side. The oocyst shown in 6 is thought to be an *Isospora* from a squirrel, since the patient was a hunter and had recently killed and eaten squirrel. In each case, the egg or cyst could not be found in fecal specimens obtained subsequently. The cysts and eggs represent spurious infections. When an unrecognized egg, cyst, or oocyst is seen when examining a fecal specimen, measure it first, then compare it with those of similar structure and size found in human feces. If it cannot be identified, several additional fecal specimens should be requested to determine if it continues to be present in the feces. If none are found after several days, it was most likely a spurious infection. If unidentified eggs or cysts persist in the fecal specimens, the physician should request the help of a consultant.

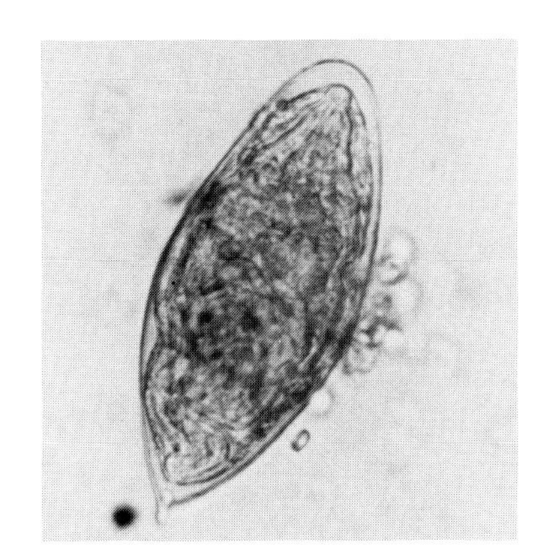

Part 5 — Appendices

Appendix 1

Technical Information

Most of the subject material that appears here was introduced in the major Parts of the manual. The information included in the Appendix is expanded or more detailed. Formulas for various chemical solutions, stains, and collecting/preserving solutions along with helpful information, are presented. The material may relate directly to the specimen, the microscope, or some procedure. In most cases, the subjects refer to the body of the text.

Following this Appendix is a Glossary of terms used in this manual. It provides ready definitions that apply directly to the terminology used and to the subject matter included in this manual, thus freeing the individual from going to a dictionary or other resource for definitions.

THE MICROSCOPE AND MICROSCOPY

The microscope is introduced in Chapter 2 and the major components used in its operation are pointed out and named (see Chapter 2, Figure 1). Some of the terms used are further defined and some topics related to procedures are included.

Bright-Field Illumination

In bright-field illumination the specimen appears dark against a bright field of light. In order to see detail, the specimen being examined must be translucent and usually fairly thin so as to allow light to pass through, much as it passes through frosted glass. If light cannot pass through the specimen, it will appear as a silhouette. For parasitology, bright-field illumination (in contrast to dark-field, phase contrast, or fluorescent illumination) is most commonly applied for the examination of fecal specimens.

The use of filters in bright-field microscopy and photomicrography is a topic in itself. Filters are sometimes called for in bright-field illumination. It is customary to use a blue filter with a tungsten filament lamp to compensate for the red rays that increase when the light intensity is reduced below the prescribed color temperature of the light source. With a halogen lamp, used in some microscopes, the light intensity should be set at the prescribed color temperature of the lamp and the intensity of the light for visual viewing controlled with neutral density filters. A blue filter is not appropriate for use with a halogen lamp.

For all photomicrography, the light source should be set at the color temperature for which the film is balanced (usually 3200°K for color film) and adjusted to the proper color temperature for the film with appropriate filters. For black-and-white photomicrography, a green filter is used to adjust color brightness and bring the light to the proper balance with the film. For color photomicrography, the use of filters is more complicated (for more detailed information, see Delly, 1980).

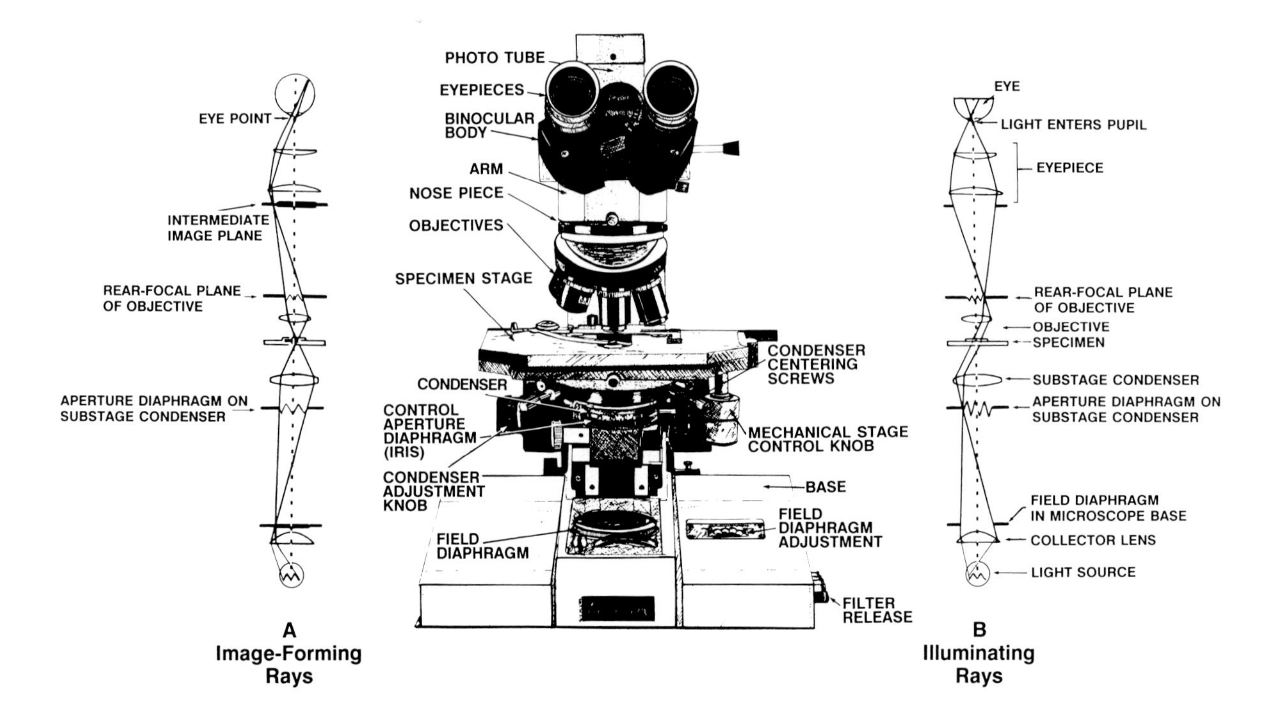

Figure 1. Köhler illumination: diagrams of the image-forming rays and the illuminating rays relative to the microscope. The most effective system for illuminating a microscope specimen for both visual work and photomicrography is Köhler illumination. In the system, the field condenser focuses an image of the lamp filament on the substage condenser. The condenser is focused to bring the image of the lamp filament into the plane of the specimen. The substage condenser becomes the source of illumination. The microscope is diagrammed in the center, the image-forming rays on the left (A) and the illuminating rays on the right (B). *The image-forming rays (A, diagram shows two light rays)*: light rays are delivered to the field diaphragm, are passed on to the substage condenser, and brought into sharp focus in the plane of the specimen. The focused light picks up an image of the specimen and projects it into the objective. As the rays travel through the microscope toward the eyepiece, an aerial image of the part of the specimen in focus is formed at the eyepoint. The eyepiece receives the image, magnifies it, and focuses the image of the specimen on the retina of the eye or film plane. *The illuminating rays (B, two light rays shown)*: a collector lens gathers an image of the lamp filament and projects this image through the field diaphragm and focuses it in the plane of the aperture diaphragm. The rays pass through the condenser and then a uniform bundle of light passes through the specimen, the objective, the tube of the microscope, and the eyepiece to the retina of the eye or film plane.

Critical Illumination

Critical illumination refers to a microscope setup in which light is focused at the level of the specimen. It is described in older texts but its use is no longer recommended for better quality microscopes that have built-in light sources.

Köhler Illumination

Köhler illumination refers to a microscope setup in which light is focused on the aperture diaphragm of the condenser, a system that satisfies all theoretical requirements for optimum light microscopy and provides homogeneous illumination from a non-homogeneous light source (see Figure 1 for a detailed presentation of Köhler illumination in relation to the microscope).

Index of Refraction (Refractive Index)

Index of refraction refers to changes in the speed of light passing from one substance to another of differing density. When light passes from air or water to a more dense substance such as a crystal, it is slowed down, and if the ray of light enters at an angle, it is refracted (bent) at the point of meeting.

The refractive index of air is 1.0 and water, glass, and immersion oil all have refractive indices of approximately 1.5. The ratio between the speed of light in these substances to its speed in another substance, such as an amoeba, is the index of refraction of the substance. When viewing a wet preparation microscopically, a substance or object with the higher index of refraction appears visually brighter than one with a lower index of refraction at the same plane of focus (see Becke Line); the greater the relative index of refraction, the brighter the object appears.

When a fecal specimen is suspended in a liquid medium, protozoan parasites and most helminth eggs have a relatively high index of refraction as compared to vegetable cells, bacteria, and many other materials. As the objective is moved up and down with the fine adjustment and objects come into view, those with a greater index of refraction appear momentarily brighter; therefore, the parasites in a wet mount of a fecal specimen appear brighter than most of the surrounding vegetable matter and other objects.

Becke Line

The Becke Line refers to the light plane at which refracted light that passed through an object leaves the object and is again refracted (Ehlers, 1987). If the index of refraction of the object is higher than that of the immersion liquid, refraction tends to concentrate the light rays above and toward the center of the object. One can observe this by microscopy with the aperture diaphragm of the condenser partially closed. The areas of light concentration move from the edges of the object inward toward its center as the microscope objective is raised. If the index of refraction of an object is less than that of the immersion liquid, then the areas of light concentration move away from the object as the microscope objective is raised (see Index of Refraction).

Resolving Power (Resolution)

The resolving power refers to the ability of a microscope to discern fine detail and usually is related to the ability to see two dots or lines very close together as separate rather than as one blurred image. The larger the numerical aperture (N.A.) of an objective, the higher the resolving power. The upper limit of magnification of a light microscope is considered to be 1000× the N.A. of the objective. Magnification above that level is considered to be "empty magnification" since resolution above that magnification is not improved. The resolving power (R.P.) of a lens can be calculated by applying the formula, R.P. = wavelength/(2 × N.A.) Visible light is made up of all wavelengths between approximately 380 and 760 nm, but for calculation purposes the wavelength of yellow-green light can be used, which is 550×10^{-9} m or 0.00055 mm. The N.A. of a 100× oil immersion objective is 1.25.

$$\text{R.P.} = 0.00055/(2 \times 1.25)$$

$$\text{R.P.} = 0.00055/2.5 = 0.00022 \text{ mm}$$

$$\text{R.P.} = 0.00022 \text{ mm or } 0.22\ \mu\text{m}$$

Under these conditions, two granules closer together than approximately 0.22 μm would appear as one slightly blurred granule.

Working Distance

Working distance refers to the distance between the front mount of the objective being used and the top of the cover glass. The greater the magnification of the objective, the smaller the working distance. All objectives are designed to be used with a cover glass, usually of a maximum thickness of 0.18 mm for the highest power objective. The working distance of the 40× dry objective is often limited and, depending on the thickness of the specimen, it may not be possible to use it when there is oil on the cover glass (see Chapter 2).

The Visual Depth of Field

Depth of field refers to the thickness of what is seen (or photographed) at a single plane of focus using the microscope. When examining a fecal film using a 40× objective and a 10× eyepiece (400× magnification), what can be seen in a single visual plane is 2.8 μm thick. When examining an amoeba 12 μm in diameter, the fine adjustment on the microscope must be used to change the level of the visual plane about four times so that structures at each depth can be resolved (seen clearly).

The depth of field changes when higher magnification is used. Using a 100× objective and a 10× eyepiece, the magnification is 1000× and the depth of field at the visual plane becomes 0.55 μm. About 22 visual levels are necessary to view an entire organism 12 μm in diameter (see Formulas, below).

Formulas for Calculating the Visual Depth of Field

T_v = depth of field for vision (in millimeters)

$T_v = T_g + T_w + T_a$

T_g = is the Geometric factor

T_w = is the Wave optic factor

T_a = is the Accommodation factor

(Accommodation of the eye changes with age. The average value is 250, which is used in the formula.)

$T_g = Z_v \times [1 + (1/\text{magnification})]/(\text{N.A.} \times \text{magnification})$

(Z_v is the Circle of Confusion, which is the limiting point of resolution, i.e., the limiting point at which two dots very close together can remain separate and do not come together as one slightly blurred dot. The value of Z_v for vision is 0.15 mm.)

$T_w = \lambda / 2 \times [(\text{N.A.})^2]$

(λ, the average wavelength for green light = 0.00055 mm)

$T_a = 250/[\text{magnification} \times (\text{magnification} + 1)]$

(Average distance of normal vision has a value of 250 mm, which is used to calculate the eye accommodation factor.)

The Depth of Field for Photomicrography

The depth of field for a photomicrograph is much less than that of a visual field and depends on the microscope and camera system used. The image scale (SCA) is the magnification of the image at the film plane, which is determined by multiplicative factors from the specimen to the film plane. The SCA is determined by multiplying magnifications of the objective, the photo-eyepiece, the microscope tube (tube factor), and the camera factor. For a Leitz laboratory microscope (Leica) and 35 mm camera, the camera factor is .32 (a factor of 1 is used when there is no magnification value). Other microscope and camera setups may have different values.

Formulas for Calculating the Photographic Depth of Field

T_p = The depth of field for photomicrography (in mm)

$T_p = T_g + T_w$

$T_g = Z_p \times [1 + (1/\text{SCA})]/(\text{N.A.} \times \text{SCA})$

(Z_p is the average value for the circle of confusion. For photomicrography, the value of $Z_p = 0.03$ mm.)

$T_w = \lambda /[2 \times (\text{N.A.})^2]$

(The value remains the same as in the visual depth of field.)

For example, if calculating the depth of field for photomicrography using a 40× objective, a 10× photo-eyepiece, a tube factor of 1, and a camera factor of 0.32, the values are:

SCA = 128

Z_p = .03

N.A. = .65

λ = 0.00055

$T_p = T_g = T_w$

T_p = 0.36 + 0.65 = 1.01 µm

The depth of field for photomicrography using the microscope and camera setup in the example is 1.01 µm. Using the 40× objective and a 10× eyepiece, the visual magnification is 400× and the depth of field is 2.8 µm.

If a 1000× objective is substituted for the 400× objective and all else remains the same, the SCA becomes 320 and the N.A. is 1.25. The depth of field becomes .26 µm. Using a 100× objective and a 10× eyepiece, the visual magnification is 1000× and the depth of field for a photomicrograph is .26 µm.

At 400× the depth of field for a photomicrograph is about 36% of what is seen in the visual field, and at 1000× the depth of field of a photomicrograph is about 47% of what is seen in the visual field. Getting exactly what is wanted in the photomicrograph is sometimes very difficult because one sees much more than will appear on the photographic film. The depth of field should be taken into consideration when viewing the photomicrographs.

SPECIFIC GRAVITY

Specific gravity is defined as the ratio of the weight or mass of a given volume of a substance to that of an equal volume of a standard, water for liquids and solids and air for gases. Although the specific gravity of individual species of parasites and eggs may vary, the range of variation is very close. When a sample of a fecal specimen is mixed in water, the trophozoites, cysts, eggs, and juvenile worms (larvae) sink to the bottom along with much of the materials in the fecal mass. In the bottom layer there is a gradient of various substances with the lightest (that with the lowest specific gravity) at the top and the heaviest at the bottom. The parasites and eggs tend to group at a particular level.

Cysts and eggs of certain species of parasites will float to the surface of a solution if the specific gravity of the solution is raised to above that of the parasites. For example, in the zinc sulfate flotation procedure the specific gravity of the solution is raised to 1.8, which allows cysts and eggs of certain species to float. The eggs of species that have a higher specific gravity than the solution will sink to the bottom of the tube. In the case of sedimentation procedures, the specific gravity of the solution must be less than that of the parasites and eggs.

GRAVITY SEDIMENTATION

Gravity sedimentation refers to a method of concentrating solid or semisolid substances or objects at specific levels in a medium (usually a solution) relative to their respective specific gravities (see Chapter 2). In parasitology, it applies to the settling of trophozoites, cysts, eggs, and/or juvenile worms at specific levels within the base material after a specimen has been mixed in a liquid medium and allowed to settle. The eggs and parasites must have a higher specific gravity than the liquid medium with which they are mixed in order to become a part of the base layer. Ideally, the container in which gravity sedimentation takes place should have straight sides to the bottom such as a conical sedimentation flask or a straight sided, flat bottom vial. The time it takes for materials to settle to the bottom depends on differences between the specific gravities of the solution and the materials being concentrated. The closer the specific gravities are to one another, the longer the settling time.

THE CENTRIFUGE

The centrifuge is used routinely in the clinical laboratory for the concentration of eggs and parasites. If the information provided with the centrifuge in use does not give the Relative Centrifugal Force (RCF) at various revolutions per minute (rpm), that value should be calculated. RCF is expressed as xg where g equals the force of gravity.

First, there must be a way of determining the rpm at various settings. If the rpm is known at a particular setting, the RCF can be calculated. The nomogram provided is used for this calculation (see Figure 2).

CHEMICALS, STAINS, AND STAINING SOLUTIONS

Precautions in Handling Chemicals

There are many chemicals that are used in staining and in other types of procedures related to parasitology that are potentially harmful. Material Safety Data Sheets (MSDSs) should be available for all chemicals that are used and the proper precautions should be taken to minimize the danger and protect personnel from exposure to toxic, corrosive, and otherwise dangerous substances that are used in the laboratory including formaldehyde, phenol, acids, some dyes, etc. These are usually obtained from the manufacturer or distributor (see also Safety in the Laboratory).

Alcohols Used in Staining Procedures

All staining procedures require the use of alcohol and pure ethyl alcohol is considered to be the proper alcohol of choice. Because of regulations established by the Federal Government on the use and storage of ethyl alcohol and the impact of these regulations on the smaller laboratories, other alcohols have been substituted. In some staining procedures isopropyl alcohol appears the be satisfactory. Alcohol sold as "Reagent Alcohol, Absolute" is specifically denatured ethyl alcohol, Formula 3A, and contains 5 parts per volume of both methyl and isopropyl alcohol. There appears to be little difference between the results obtained using solutions of reagent alcohol and those obtained using pure ethyl alcohol. For laboratories not having a Federal Treasury Permit for ethyl alcohol, Reagent Alcohol Absolute purchased from a major supplier is recommended for use where the procedure calls for ethyl alcohol or does not specify which alcohol.

Hematoxylin Stains

Hematoxylin stains are especially good for staining protozoa in tissues and in fecal smears. Probably the best staining method for protozoa in fecal smears is Heidenhain's-Iron-Hematoxylin staining method. To perform the procedure properly takes considerable time since in the long method the staining step alone requires 12 hours and level of destaining, if done properly, must be checked by microscopy. The time

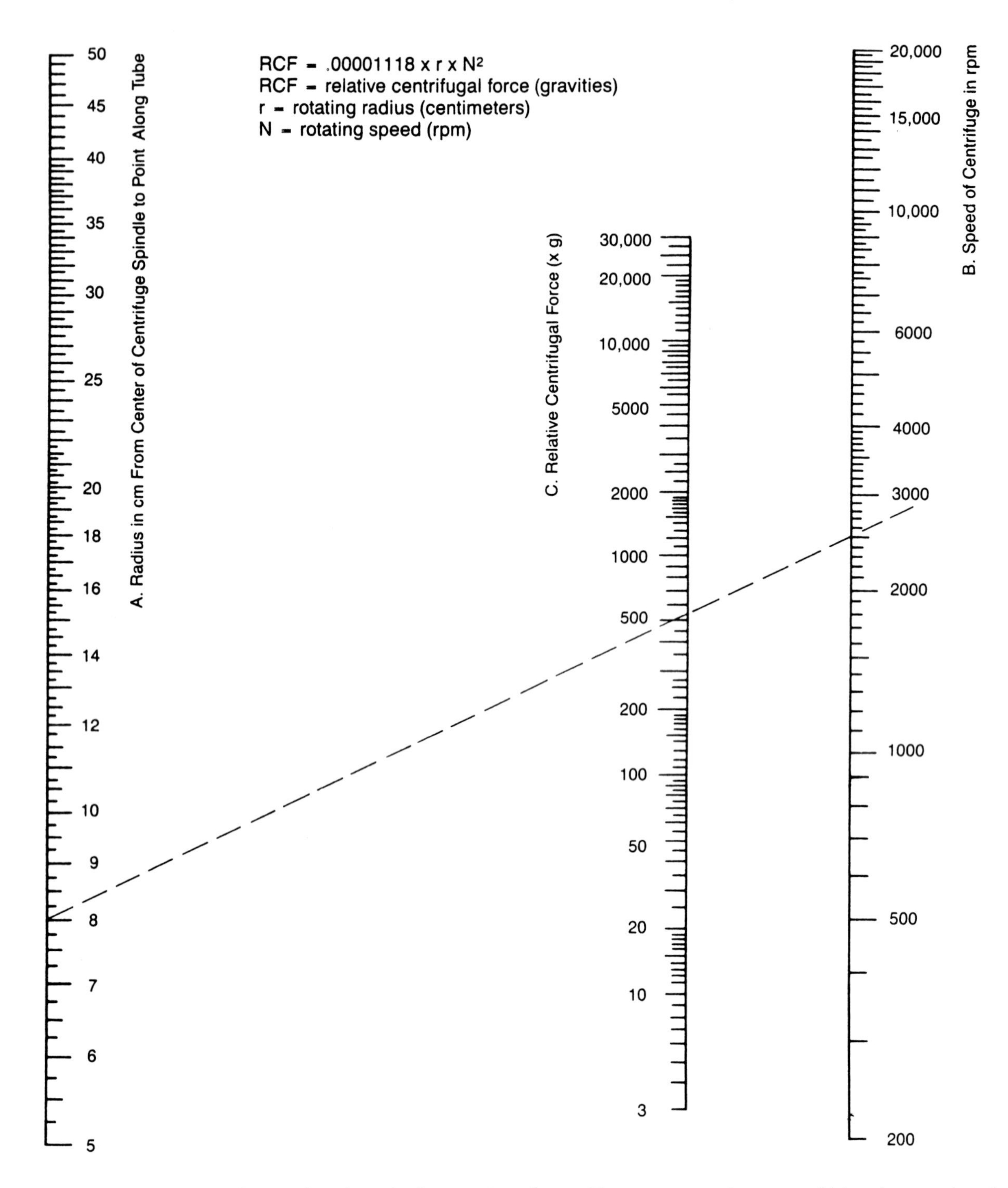

Figure 2. Nomogram and procedure for calculating RCF (g force). The nomogram (nomograph) has three scales: the left is the rotating radius (r), the right is the rpm, and the center is the RCF (relative centrifugal force or g force) scale. To determine the rotating radius (r), measure the distance from the center of the drive shaft to the center of the column of the material (suspension) being centrifuged using a centimeter scale. The tube containing the suspension must be fully extended horizontally when the measurement is taken. To determine the g force, place the left end of a straightedge at the measurement determined for r on the left scale, and the right end at the rpm to be used on the right scale; where the straightedge crosses the center scale, read the RCF, g force. If r = 8 cm and rpm = 2500, then RCF = 540 xg (or 540 times the force of gravity). (Nomogram supplied by Scientific Products Division, Baxter Diagnostics, Inc., Ocala, FL, 32674.)

can be shortened by heating the fixative, mordant, and stain. A method for PVA-fixed specimens has been described which takes overnight unless certain solutions are heated (Melvin and Brooke, 1982). Because of the considerable time and detailed manipulation of the slide during destaining, the method is not considered practical for the clinical diagnostic laboratory. Several more rapid methods have been developed and perhaps the most commonly used is that developed by Tompkins and Miller (1947) which follows.

Phosphotungstic Acid Iron-Hematoxylin Stain

Hematoxylin Stock Solution

Materials (Stock Solution)

Hematoxylin dye (crystals or powder)	10 g
Ethyl alcohol (100 or 95%)	100 mL

Preparation

1. Dissolve the hematoxylin in the alcohol.
2. Place in a bottle with a loose stopper and allow the solution to stand for 2 months or more to ripen. The stock solution improves with age.

Iron-Hematoxylin Stain (Working Solution)

Materals

Stock solution of hematoxylin	5.0 mL
Distilled water	95.0 mL

Preparation

1. Add hematoxylin stock solution to distilled water and mix.

The working solution should be prepared on the day it will be used and discarded at the end of the work day.

Mordant Solution (Ferric ammonium sulfate)

Materials

Ferric ammonium sulfate (violet crystals)	4.0 g
Distilled water	100.0 mL

Preparation

1. Dissolve crystals in water and store in the refrigerator.

Used as a mordant in the phosphotungstic acid hematoxylin method.

Phosphotungstic Acid Solution

Materials

Phosphotungstic acid crystals	2.0 g
Distilled water	100.0 mL

Preparation

1. Dissolve phosphotungstic crystals in water and store in a refrigerator.

Used for destaining in the phosphotungstic acid hematoxylin method.

Lithium Carbonate (Stock Solution)

Materials

Lithium carbonate crystals	1.5 g
Distilled water	100.0 mL

Preparation

1. Dissolve lithium carbonate in distilled water. A saturated solution is approximately 1.25% in distilled water so a small amount of the crystals should remain in the bottom of the container. A few drops are added to the 70% alcohol following the destaining step in the phosphotungstic acid hematoxylin method.

Carbol-Xylene

Materials

Phenol (liquefied crystals)	25.0 mL
Xylene 7	5.0 mL

Preparation

1. Place about 30 g of phenol crystals in a jar with a loosened cap and heat in a water bath until liquefied.
2. Pour 25 mL into a 100-mL glass graduate and add 75 mL of xylene. Store in a glass-stoppered bottle.

The solution is stable for up to a year if no moisture is absorbed. Used in some trichrome and hematoxylin staining procedures as a first clearing solution. Use particular caution when handling phenol since it is very corrosive and is absorbed by the skin.

Trichrome-Type Stains

These stains are water based and are therefore very stable. The stains are used full strength and may be reused over and over again as long as they remain debris-free, and a control slide is included in each run to check staining results. A 250-mL volume of these stains used repeatedly should be satisfactory for staining up to about 1000 slides; or, after being used for the first time, the lot should remain reusable for 2 months or longer if not used daily. Stains used once a week will remain satisfactory for much longer.

Following each use, the stain may be filtered through coarse filter paper to remove any accumulated foreign materials; this may be repeated many times during the period of time the stain is used. If the stain is allowed to remain on the bench top between periods of usage, even in a covered container, some evaporation may occur. If only a small amount of the stain has evaporated between periods of usage and the exact volume of the stored stain is known, a small amount of new stain from the stock bottle may be added to bring the stain back to its original volume. This should not be done more than a few times. It is better practice to discard the used stain and begin fresh with unused stain.

The quality of these stains usually improves with age as long as they remain in the original bottle and nothing is introduced into it. Any portion of stain from the original bottle that is used should not be poured back and mixed with the stain that remains in the bottle. If not left on the bench top in a staining container, it should be poured into a separate, labeled, clean bottle for temporary storage between times of usage. When the stain is returned to the storage bottle after use, it should be passed through coarse filter paper so as not to introduce any foreign material into the storage bottle. Stains of different lots should not be mixed together.

When preparing a stain, close attention should be paid to the method used because variations in preparation can alter the results obtained during staining. Dyes put into solution separately, with the dye solutions then added together may give different results than when the same dye powders are all added to a flask and put into solution together.

Wheatley's Trichrome Stain

Materials

Chromotrope 2R C.I. No. 16570	6.0 g
Light Green SF C.I. No. 42095	1.5 g
Fast Green FCF C.I. No. 42053	1.5 g
Phosphotungstic acid (reagent grade crystals)	7.0 g
Acetic acid (glacial, reagent grade)	10.0 mL
Distilled water	1000.00 mL

Some workers recommend using only 3.0 g of Light Green and omitting Fast Green, which is essentially the same as Gomori's One Step Trichrome Stain formula.

Preparation

Method A

1. Add dry materials to a small flask and mix together.
2. Add the acetic acid to dry materials, mix, and allow the mixture to stand about 30 minutes.
3. Add the distilled water and thoroughly mix.

Method B

1. Add acetic acid to the distilled water.
2. Dissolve the phosphotungstic acid crystals in acetic acid solution and divide the solution into three parts.
3. Dissolve each dye separately in one part. After 30 minutes mix the three dye solutions together and allow to stand overnight before using.

Polychrome IV Stain*

The formula is not available at this time.

---

* The stain is available from Alpha-Tec Systems, Inc., P.O. Box 17196, Irvine, CA, 92713.

Gomori's One Step Trichrome Stain

Materials

Chromotrope 2R C.I. No. 16570	0.6 g
Light green SF C.I. No. 42095	0.3 g
Phosphotungstic acid	0.8 g
Glacial acetic acid	1.0 mL
Distilled water	100.0 mL

Preparation

Follow Method B for preparing Wheatley's Trichrome above.

The staining procedure used for Wheatley's Trichrome stain can be used with Gomori's Stain (see Chapter 4, Tables 1, 2, and 4), however, the procedure used in the histopathology laboratory often involves post-fixing the tissues (or fecal smears) in Bouin's fluid and these steps are included by some laboratories for fecal smears. The steps given below can be inserted between the 70% alcohol after the iodine alcohol step and the trichrome staining step in the Wheatley's procedure.

1. Place fecal smear in Bouin's fluid in an oven at 56°C for 1 hour
2. Wash in running tap water for 3 minutes
3. Place in trichrome and continue as in Wheatley's trichrome staining procedure.

Bouin's Fluid

Materials

Picric acid, saturated aqueous solution	75.0 mL
Formaldehyde (commercial 37 to 40%)	25.0 mL
Glacial acetic acid	5.0 mL

Preparation

Use particular caution in preparing Bouin's fluid, since the chemicals can be particularly hazardous.

1. Prepare a saturated aqueous solution of picric acid.
2. Mix picric acid solution and formaldehyde.
3. Add glacial acetic acid to the solution.
4. Store in a refrigerator.

Note: For staining procedures for fecal films see Chapter 4, Tables 1–6.

Iodine Solution for Removing Mercuric Chloride Before Staining

When staining fecal smears fixed in Schaudinn's fixative or prepared from PVA-fixed specimens, the mercuric chloride is removed from the smears by immersing them in a working solution of iodine/alcohol (iodine in 70% alcohol). Otherwise, crystals will be present in the smears obscuring parasites that may be present.

The working solution can be prepared from a stock iodine/alcohol or alternately substituting either Lugol's or D'Antoni's iodine solution for the stock solution (see "Iodine Solutions").

Freshly prepared working solutions of iodine are preferable and whenever possible they should be made up shortly before use and discarded at the end of the day they are used.

ACID-FAST STAINING

A brief treatment of the acid-fast stains that have been available and are used for staining intestinal parasites is presented in Part 3, Chapter 6. The formulas for several stains are given in Chapter 6, Table 1, and a recommended staining procedure is given in Table 2.

Price's Acid-Fast Stain*

The stain was developed over several years especially for staining oocysts of cryptosporidia in fecal smears. It had limited availability since first introduced following approval by the Food and Drug Administration in March 1987. The stain was supplied to several hospital laboratories for general trial where it was found to be preferable to the stains that

* The stain is available from Alpha-Tec Systems, Inc., P.O. Box 17196, Irvine, CA, 92713.

had been used regularly in that the oocysts of cryptosporidia were better defined and more easily recognized.

Acids Used for Destaining

Acids

Either sulfuric or hydrochloric acid may be used for destaining. The concentration of acid used may vary from 1 to 10%. Some procedures call for an aqueous solution and others for an alcoholic solution and the latter from 70 to 95% alcohol. The author recommends an acid alcohol destaining solution.

Acid Alcohol

Strength of acids used for destaining vary greatly depending on the procedure and the organisms being stained. For cryptosporidia in fecal smears, a 5% acid content is usually sufficient to properly decolorize fecal smears for coccidia. For tissue sections, 1% acid in 70% alcohol is generally used. The author recommends 5% acid in 95% alcohol for coccidia in fecal smears.

Materials

Sulfuric (or hydrochloric) acid, concentrated	5.0 mL
Alcohol, 95% (ethyl or reagent)	95.0 mL

Preparation

1. Slowly add 5 mL of concentrated sulfuric (or hydrochloric) acid to 95 mL of 95% reagent alcohol to prepare a working solution. If a stronger solution is preferred, add 10 mL of concentrated acid to 90 mL of 95% reagent alcohol.

Counterstains For Acid-Fast Staining

Methylene Blue Counterstain

Materials

Methylene blue chloride	0.3 g
Glacial acetic acid	1.0 mL
Distilled water	99.0 mL

Preparation

Concentrated Stock Solution

1. Add 1.0 mL of glacial acetic acid to 99 mL of sterile distilled water.
2. Dissolve 0.3 g of methylene blue chloride in the 1% acetic acid solution.
3. Store the solution in a refrigerator until needed for preparing a working solution.

Working Solution

1. Dilute 7 mL of the concentrated stain with 93 mL of sterile distilled water.
2. Store the working solution in a sealed container in the refrigerator between uses.

Counterstain may be reused until it becomes contaminated by fungi or accumulated debris. It is good practice to filter the counterstain through coarse filter paper between uses.

Loeffler's Alkaline Methylene Blue

Materials

Methylene blue chloride (90% dye content)	0.30 g
Ethyl alcohol (95%)	30.00 mL
Potassium hydroxide (KOH) 0.01% solution in distilled water	100 mL

Preparation

1. Place dye powder in a dry flask, add alcohol, and stir until dye powder is completely dissolved.
2. Add 100 mL of 0.01% KOH solution and mix.
3. Store in a glass bottle in a refrigerator.

Use as an alternate counterstain for acid-fast staining.

Malachite Green Counterstain

Materials

Malachite green	3.0 g
Distilled water	100.0 mL

Preparation

1. Dissolve malachite green in sterile distilled water using a magnetic stirrer.
2. Store in a refrigerator in a well-sealed bottle between uses. Use as an alternate counterstain for acid-fast staining.

QUALITY CONTROL FOR STAINING

Quality control slides should always be used with any staining procedure to check the results of staining and to monitor the solutions being used. When results of staining the quality control slides are not comparable to the previously stained control slides, the problem may be the stain or one or more of the solutions used in the staining process. When the control slide stains well but the test slide does not, it is possible that the protozoa in the specimen were not properly fixed and may not stain properly.

When a problem is encountered, older or questionable, previously used solutions should be replaced with fresh ones and a test control slide stained using the same staining solution. If, after a test run, results do not compare with the control slide standard, the stain could be the problem and should be replaced with a new unused stain or a stain of known quality.

Quality Control and Test Slides

At least two quality control slides are needed for stain control: (1) a slide (fecal smear) stained previously using the same stain and method (control slide standard) and (2) a test slide, one prepared similarly to the first slide from the same specimen (test control slide). A test slide should always be run with each batch of slides being stained, even if only one or two slides are included in the batch.

Fecal films prepared from PVA-fixed specimens should not be used as control slides for those prepared from specimens fixed in SAF, MIF, or PIF, and vice versa. A positive control specimen should be used for preparing the control slides for the latter fixatives. If prepared slides are coated with a cytology fixative as used in preparing Pap smears, several test slides can be prepared from the same specimen the first of each week for daily test slides for that week.

It is good procedure to use control slides with several species of protozoa. Whenever possible, the author uses a specimen with cysts of *Entamoeba coli* (difficult to stain well), *Endolimax nana*, (intermediate), and *Giardia lamblia* (relatively easy to stain) as the control slide organisms.

FIXATIVES AND SOLUTIONS

Fixatives include the collecting/preserving solutions used for obtaining fecal specimens for laboratory examination. There are other solutions that have specific functions in collecting, processing, and examining fecal specimens and some of these are included also.

Schaudinn's Fixative

There are a number of formulas for Schaudinn's fixative. The original is given first.

Part a. A saturated solution of mercuric chloride

Materials

Mercuric chloride crystals	75 to 80 g
Sodium chloride solution (0.85 to 0.9%)	1000 mL

Preparation

1. Prepare a 1000 mL solution of sodium chloride.
2. Dissolve the mercuric chloride crystals by heating at 56°C in the sodium chloride solution. (Excess crystals will precipitate out on cooling.)
3. Store the saturated solution in a glass-stoppered bottle.

Part b. Schaudinn's fixative (working solution)

Materials

Saturated mercuric chloride (from Part a)	2 parts
Ethyl alcohol, 95%	1 part
Glacial, acetic acid 1%	

Preparation

1. Combine mercuric chloride and alcohol and thoroughly mix.

2. Add acetic acid to mixture.

(*Note*: Mercuric chloride/alcohol solution can be stored until needed. Add acetic acid immediately before use.)

Modified Schaudinn's Fixative

In the modified fixative, acetic acid was increased to 5% and 1.5% glycerin was added. The formula is otherwise the same.

Modified PVA Fixative (Collecting/Preserving Solution)

Burrows (1967) modified the solution again and his modification is currently used to prepare PVA collecting/preserving solution (fixative). The recommended method is in three parts:

Part a. Schaudinn's Fixative

Materials

Mercuric chloride crystals	4.5 g
Ethyl alcohol, 95%	31.0 mL
Acetic acid, glacial	5.0 mL

Preparation

1. Dissolve mercuric chloride in alcohol in a glass-stoppered flask on a magnetic stirrer.
2. Add acetic acid and mix using a magnetic stirrer.
3. Store in a glass-stoppered bottle.

Part b. PVA Solution

The PVA mixture is prepared separately, then incorporated with the Schaudinn's fixative.

Materials

Glycerin	1.5 mL
PVA powder	5.0 g
Distilled water	62.5 mL

Preparation

1. Add glycerin to the PVA powder in a small beaker and mix to a paste with all particles coated with glycerin.
2. Scrape the PVA paste into a 125-mL flask.
3. Add distilled water and allow to stand for several hours (usually overnight) with occasional swirling.

Part c. PVA Fixative (Collecting/Preserving Solution)

Materials

Schaudinn's fixative (from Part a.)	36.0 mL
PVA solution (from Part b.)	64.0 mL

Preparation

1. Place a loosely stoppered flask containing the PVA solution on a hotplate/magnetic stirrer adjusted to 70 to 75°C with the stirrer set at a moderately low speed.
2. When the PVA powder appears to be melted, add the Schaudinn's fixative, recap, and continue to mix at the same speed.
3. Continue stirring and heat until the mixture is free of bubbles and the solution is clear.
4. Remove the flask from the heat and when cool transfer the PVA fixative to a glass-stoppered bottle until ready to dispense into vials.

Personnel in a number of laboratories have cited inconsistencies found in the preparation of this collecting/preserving solution in-house. If prepared in the laboratory, the modification recommended by Burrows (1967a) gives the best results; however, to obtain consistency from lot to lot, the author recommends purchasing the product from a reputable commercial company where their quality control insures a more consistently good product.

Formalin Fixatives (Collecting/Preserving Solutions)

Buffered formalin solutions appear to work better than unbuffered formalin solutions and cold solutions give more satisfactory results. Formaldehyde is a gas (HCOH) which is soluble to 40% in water. Commercial products may vary from 35 to 40% formaldehyde in water. Percentages given for formalin are usually based on the percent of the commercial solution of formaldehyde (not the actual amount of formaldehyde present). Some workers prefer formol-saline to aqueous formalin. To change a 10% formalin solution to formol-saline, add 1.0 g of sodium chloride to each 90 mL of distilled water.

Phosphate Buffered Formalin, 10%

Materials

Commercial formaldehyde (35 to 40%)	100 mL
Distilled water	900 mL
Sodium phosphate monobasic (NaH_2PO_4)	0.15 g
Sodium phosphate dibasic (Na_2HPO_4)	6.10 g

Preparation

1. Combine phosphate buffer powders and dissolve in distilled water.
2. When completely dissolved, add formaldehyde.

Calcium Carbonate Buffered Formalin, 10%

Many workers feel that formalin solutions can be adequately buffered by adding marble chips to the prepared solution.

Materials

Commercial formaldehyde (USP)	100 mL
Distilled water	900 mL
Calcium carbonate chips	100 g

Preparation

1. Mix formaldehyde and water.
2. Add dry calcium carbonate.
3. Store in the refrigerator.

The prepared solutions can be stored at room temperature if refrigerator space is not available. Any of these formalin solutions can be used in the formalin/ethyl acetate concentration method or for collecting and preserving fecal specimens.

SAF Fixative (Collecting/Preserving Solution)

SAF is a collecting/preserving solution for collecting fecal specimens.

Materials

Sodium acetate	1.5 g
Acetic acid, glacial	2.0 mL
Formaldehyde (commercial, 37% to 40%)	4.0 mL
Distilled water	92.5 mL

Preparation

1. Add acetic acid to distilled water.
2. Dissolve sodium acetate in the acid solution.
3. Add formaldehyde to the acid solution.
4 Dispense into vials, 15 mL per vial for patient use.

For collection of fecal specimens, add 1 to 2 g of feces to the vial and mix thoroughly. Concentrate the preserved specimen using a formalin/ethyl acetate method or the CONSED system. Use the method for preparing fecal films for permanent staining described for specimens fixed in MIF and PIF in Part 3, Chapter 4, and stain slides using either Tompkins an Miller's phosphotungstic acid iron-hematoxylin method (preferred) or with a trichrome-type stain.

PAF Fixative (Collecting/Preserving Solution)

Fecal material is collected in the collecting/preserving solution then subsequently stained with thionin or Azure A stains. The procedure is given in Part 3, Chapter 3 (see Preparation of PAF below).

Materials

Phenol crystals (liquefied by heating)	23.0 mL
Saline solution (0.85%)	825.0 mL
Ethyl alcohol (95%)	125.0 mL
Formaldehyde (commercial, 40%)	50.0 mL

Preparation

1. Mix liquefied phenol in the saline solution.
2. Add ethyl alcohol and mix.
3. Add formaldehyde and mix.
4. Store at room temperature.

Use particular caution when handling phenol since it is very corrosive and is absorbed by the skin.

Usually 2 to 3 g of feces are added to 15 mL of the collecting/preserving solution and thoroughly mixed. Fecal specimens should be mixed with the solution immediately after passage to provide good trophozoite morphology.

Thionin Stain

Materials

Thionin (dye powder)	10.0 mg
Distilled water (sterile)	100.0 mL

Preparation

1. Completely dissolve dye powder in water.
2. Store dye solution in a refrigerator in a tightly stoppered glass bottle.

Azure A Stain (or Methylene Azure A)

Materials

Azure A (dye powder)	10.0 mg
Distilled Water (sterile)	80.0 mL

Preparation

1. Completely dissolve dye powder in water.
2. Store dye solution in a refrigerator in a tightly stoppered glass bottle.

MIF Fixative (Collecting/Preserving Solution)

The MIF Collecting/Preserving method that is used for routine collection of fecal specimens has two components, a vial of the MF Stock Solution and a unit of Lugol's iodine.

MF Stock Solution

A stable MF Stock Solution can be prepared as follows:

Materials

Distilled water	250 mL
Formaldehyde (USP)	25 mL
Tincture of merthiolate (1:1000)	200 mL
Glycerol (reagent grade)	5 mL

Preparation

1. Add glycerol to distilled water and mix.
2. Add tincture of merthiolate and mix well.
3. Finally add the formaldehyde and mix.
4. Store in a brown, glass bottle at room temperature, keep away from sunlight.

For collecting and preserving fecal specimens, 13 mL of the MF stock solution is placed in an appropriate vial and 1 mL of Lugol's iodine is added immediately before adding a proper portion of the fecal specimen to the vial.

PIF Fixative (Collecting/Preserving Solution)

PIF functions as does MIF in all circumstances where a collecting/preserving solution is needed (see Introduction). The product is available from Alpha-Tec Systems Inc., Irvine, CA. The formula is not available at this time.

The Importance of Iodine in MIF-Type Methods

As a part of the MIF-type methods, Lugol's iodine is added to the stock solution immediately before adding the fecal sample. The reason for this procedure is that a slow but immediate chemical reaction takes place when the iodine and the stock solution come together and optimal results are obtained only when parasites or eggs contained in the feces are present during the reaction.

When the feces is added to the stock solution and there is a delay in adding the iodine until after the parasites and eggs are fixed, the iodine does not take part in the fixation of the organisms but simply functions as a stain as it does when added to formalin-fixed specimens. Preservation of parasites in a well-mixed specimen does occur with only the stock solution present. The later addition of iodine does color the organisms to some extent. In specimens fixed without the presence of iodine, the specific gravity of the organisms fixed is different, the gravity sedimentation for direct examination is altered, the staining of protozoa is not as intense, and the overall quality of the resulting specimen is not as good.

For cost or convenience, some users of the MIF-type procedures have chosen to collect the specimen in the stock solution and add the iodine after the specimen vial reaches the laboratory. A delay in adding the iodine can reduce the effectiveness of the procedure, so whenever possible this practice should be avoided. If Lugol's iodine is not present when the fecal specimen is added or if the sample is not adequately mixed with the preserving solution, cysts of *Entamoeba coli* may not be adequately fixed and the nuclei of *Dientamoeba fragilis* and *Iodamoeba butschlii* may not stain adequately.

Iodine Solutions

Lugol's Iodine:

Materials

Iodine crystals	5 g
Potassium iodide	10 g
Distilled water	100 mL

Preparation

1. Dissolve potassium iodide in 30 to 40 mL of distilled water.
2. Add iodine crystals to this solution and shake or stir the mixture to dissolve the crystals.
3. When all iodine crystals are dissolved, add the remaining distilled water.
4. Place the expiration date on the label (2 weeks from preparation date) and store in a tightly capped, brown bottle in the refrigerator.

For the MIF and PIF procedures, add 1 mL of Lugol's iodine to the 13 mL of stock solution immediately before adding feces to the vial. Lugol's iodine is usually supplied in ampules when MIF or PIF kits or bulk packs are obtained from manufacturers or suppliers.

Lugol's iodine can be used to prepare iodine/alcohol for removing mercuric chloride crystals in a procedure for staining fecal smears fixed in Schaudinn's fixative or prepared from PVA-fixed specimens for staining in either hematoxylin or trichrome type stains. Sufficient Lugol's iodine solution is added to freshly prepared 70% alcohol to give the appropriate, orange, "strong-tea" color.

Dobell and O'Connor's Iodine Solution

Materials

Iodine crystals (or powdered)	1 g
Potassium iodide	2 g
Distilled water	100 mL

Preparation

1. Dissolve potassium iodide in about 30 to 40 mL of distilled water.
2. Dissolve iodine crystals in potassium iodide solution.
3. When all iodine crystals are dissolved, add additional water. Follow the exact procedure as for Lugol's iodine.

D'Antoni's Iodine Solution

Materials

Potassium iodide	1.0 g
Iodine (crystals or powder)	1.5 g
Distilled water	100.0 mL

Preparation

Same as Lugol's and Dobell and O'Connor's iodine.

All iodine solutions are unstable and fresh solutions must be prepared every 14 to 21 days.

The life of iodine solutions can be greatly increased by sealing them in a glass ampoule in an inert environment (usually under nitrogen gas) and protecting them from light. Such sealed ampules are stable indefinitely and are usually supplied by the manufacturer when iodine droppers and/or PIF or MIF kits are purchased.

Iodine/Alcohol Solution

Materials

Iodine crystals or powder	3 to 5 g
Alcohol (70%)	100 mL

Preparation

Stock Solution

1. Dissolve iodine crystals in the alcohol. A very dark solution should result.
2. Store the stock solution in an amber glass bottle (preferably glass-stoppered) in a refrigerator until needed to prepare a working solution. The stock solution appears to be stable for up to 2 months after which it may need to be replaced.

Working Solution

1. Add sufficient stock solution to 70% alcohol to make a dark orange solution, the color of

strong tea (see Chapter 4, Stains and Staining Methods).

Iodine/alcohol solution is used to remove crystals that form in fecal films prepared from fixatives containing mercuric chloride. The crystals often obscure organisms that are present on the fecal film.

MAINTAINING A POSITIVE SPECIMEN COLLECTION

Specimens in which parasites, eggs, etc. are found should be retained for subsequent examination and use in the laboratory for review, quality control, and teaching. A specimen to be added to the "Positive Specimen Collection" should be reprocessed as described below and brought to a standard ratio of solids to fluid.

After the specimen has been vigorously shaken and allowed to stand undisturbed overnight, most of the fluid standing over the sediment should be poured off, the specimen remaining in the vial mixed thoroughly to break up any remaining fecal lumps, and fresh stock solution added to the vial, to yield a feces/liquid ratio of 1:4 to 1:5. The vial should be tightly capped, shaken vigorously, adequately labeled, and placed in the collection.

Replacing the initial fluid will help remove chemicals or other materials that might be damaging to the organisms in the specimen. If MIF-type fixatives were used initially and iodine was added when the specimen was preserved, it is usually not necessary or advisable to add more iodine. If iodine was not added initially, the specimen should be reexamined after about a week to make certain that the nuclei of protozoa have retained their stain and dilute iodine added as necessary.

The same process should be followed after 6 months and again after 1 year for specimens that are held in the collection for longer periods. Some eggs and parasites will deteriorate in certain of the collecting/preserving solutions regardless of the precautions taken. By following a regular maintenance schedule and checking to make certain that the proper fluid level is maintained in the vial, good specimens may be maintained for several years. Some parasites and eggs in positive specimens in the author's collection, fixed initially in MIF or in 10% buffered formalin, have retained their original characteristics without distortion since collected in 1959 and 1960, more than 30 years ago at this writing.

When planning to examine specimens that have remained in the collection for a month or more, the following process should be followed because fecal masses stored in vials often tend to form clumps. A glass rod with a flat base can be used to crush clumps and separate small particles that tend to adhere together. These clumps need to be broken up before the specimen is examined. If the specimen is less than 5 months old, the supernatant fluid can be poured into a clean container and poured back into the vial after particles are broken up. After mixing the specimen, it should stand for an hour or more to allow the solids to adequately settle before being examined. Preferably, a sample of the base layer to be examined is drawn into the pipette by capillary action.

In specimens fixed in the MIF-type fixatives, certain organisms, e.g., *Balantidium coli* trophozoites and cysts, and *Ascaris* and hookworm eggs tend to become overstained after standing in a vial for a long time. To see the structures more clearly, transfer the portion to be examined to a small tube, add 1 or 2 drops of "409 Household Cleaner" to the tube to partially destain the organisms, and allow it to stand for about an hour. The sample can then be transferred to a slide for examination.

If the prepared slide is to be held for reexamination or is for permanent staining, the 409 should be removed before preparing the slide. To accomplish this, place the sample to be examined in a centrifuge tube, add the 409, and allow the tube to stand for about an hour. Then, mix the sample in the tube with 10 mL of the collecting/preserving solution used, centrifuge the tube at about 500 xg for 2 minutes, and pour off the supernatant fluid as in a sedimentation procedure. Add a few drops of the original stock solution to the plug and transfer the sample to a slide for examination or for subsequent staining.

FECAL COLLECTION KITS

Fecal collection kits containing any of the mentioned solutions for collecting and preserving feces for laboratory examination can be purchased from a number of sources or can be prepared in the laboratory. In preparing such kits, an appropriate vial should be selected to hold the stock solution. For ease in mixing the fecal specimen with the solution, a vial with a flat bottom, straight sides, and without an indented neck can be used most efficiently and effectively. Such vials are especially good for kits with PIF and MIF collecting/preserving solutions because of their suitability for concentrating eggs and parasites at specific levels in the solids layer formed by gravity sedimentation.

Vials made of glass, polypropylene (preferably clarified), or PET (polyethylene tetraphthalate) are usually considered suitable. Some other plastics may leach into the solution and adversely affect its quality. Clarified polypropylene is usually the best choice for collecting/preserving solutions used for collecting fecal specimens since it is the most inert and fecal material does not readily adhere to it as it does to some other materials.

Special consideration should be given to the vial cap for these solutions. Chemicals in the solution may affect the cap liner or the cap itself. Solutions with alcohol tend to creep, i.e., move into any narrow space, and vials without tightly sealing caps frequently leak, especially when they are not stored upright.

Most users of the kits prefer caps with an attached spatula for collecting the sample of feces to be placed in the vial rather than a separate spatula that can be misplaced.

A fecal specimen collection kit to be given to a patient should include:

- A vial of collecting/preserving solution having a cap with spatula attached for picking up the sample of feces
- 3 wooden applicator sticks for mixing
- A label for the patient's identification
- A set of directions for collecting, mixing, and submitting the specimen

For PIF and MIF kits, an ampoule of Lugol's iodine is also included. A specimen collection container may be included but, if instructed properly, the patient should have materials at home that can be used.

Vials of both PVA and formalin are included in a two vial Kit since both are needed to recover protozoa and helminth eggs. Some requests from physicians and/or laboratories include an empty vial for an unpreserved specimen.

SOME HELPFUL HINTS AND MATERIALS

Removing Immersion Oil From Slides

There are many grades of lens paper but only the thinner, more porous grades, usually referred to as lens gauze, are suitable for floating immersion oil off slides. The very porous lens gauze is cut into strips about the width of a glass slide. To remove oil from a slide, a strip of lens gauze is laid on the slide over the oil, then a single drop of xylene is added to one side of the oil to float the oil, and the lens paper is slowly pulled over and off the slide. Only the smallest amount of xylene is used, just enough to wet the filter paper to both edges of the strip. The oil will float on the xylene and remain in the filter paper as it is pulled across the slide leaving behind a clean, oil-free surface. By removing the oil in this manner, the slide is not damaged and it can be stored without the mess that can be created by old oil on slides.

If carefully done, oil can be removed in this manner from a slide while it is on the stage of the microscope which may be desirable when changing from an oil immersion lens to a 40× dry lens during the examination of a slide.

The lens paper that is most useful is not that usually available from the large suppliers but is sold by companies that produce fine tissue. The special lens paper (lens gauze) has other uses and should be available in every laboratory where fine lenses and mirrors are used. Lens gauze is available from Starr Corporation, 438 W. 37 Street, New York, NY 10018.

Useful Cleaning Solutions

The solutions mentioned are in addition to the regular solutions used in the laboratory and are not intended to replace them. They have worked well in specific situations.

409 Household Cleaner

A spray bottle of 409 can come in handy in the laboratory. It will quickly remove the red/orange stain caused by the eosin dyes in PIF, MIF, and CONSED. A light spray on a surface, glassware, or even on the hands will remove most of the stain immediately.

- It is useful in cleaning materials stained with other water soluble dyes but should be used as soon as possible after the stain occurs.
- It functions for destaining parasites and eggs that have been overstained with eosin dyes or iodine.

ERADO-SOL Biological and Chemical Stain Remover

This chemical removes stains such as Wright's blood stain and most medical and histological stains from hands, glassware, lab coats, and equipment. It should be applied as soon after the stain occurs as possible for best results. It is considered safe for most materials, and can be purchased from Cambridge Diagnostic Products, Inc., 6880 N.W. 17 Avenue, Ft. Lauderdale, FL 33309.

Lysol Household Liquid Cleaner

A solution of the cleaner is very useful for cleaning work benches and surfaces and for soaking glassware contaminated with materials and solutions in the laboratory. For glassware, use 1 oz in 1 qt of distilled water and allow glass pipettes, bottles, etc. to soak. Then, when convenient, go back and wash them. Lysol is not good for soaking plastic since it will etch some plastic surfaces after a short time, but it can be used for plastic that will be washed the same day.

SAFETY IN THE LABORATORY

In some laboratories, safety and first aid are considered only when preparing for an inspection. At other times, only very limited information is available to the worker. In most laboratories, there are eye wash stations and showers suitable for many emergency conditions. There are times when other information and items are appropriate that are not covered by OSHA regulations.

Material Safety Data Sheets (MSDSs) on all chemicals used in the laboratory should be obtained from the manufacturer or supplier. Where important for safety, pertinent information should be posted in appropriate places in the laboratory so that personnel have ready access to immediate steps to be taken and to emergency telephone numbers.

In a microbiology laboratory, there is always danger of contamination with infectious agents. Viruses, bacteria, and fungi that may be present in unpreserved feces could pose a problem. Viruses such as hepatitis, polio, and possibly the HIV viruses of AIDS are certainly potential hazards. Infection control is one of the usual duties of the microbiology laboratory and personnel should be aware of appropriate precautions. OSHA has established regulations on bloodborne pathogens (Paragraph 1706), which should be consulted.

There are only a few species of parasites that can be transmitted directly from fecal specimens. When performing some procedures, it is appropriate to wear protective gloves and aprons, and even masks may be advisable under certain conditions.

A "Safety and First Aid Handbook" can be extremely valuable and can be prepared by laboratory personnel and placed where it is readily accessible when needed. Each product and each chemical or substance that is used in a particular section of the laboratory and may be involved in an emergency situation should be listed alphabetically. If a chemical or substance has more than one name, it may be listed under each name. Much of the information needed to prepare the handbook can be obtained from Material Safety Data Sheets. Some information may appear on the product container or in the information handouts supplied with the product. A relatively small effort to collect the information and prepare a handbook for each laboratory section before it is needed can make a major difference in the outcome when an emergency actually occurs.

Appendix 2

Glossary

A GLOSSARY OF TERMS APPLIED TO INTESTINAL PARASITES

Abopercular end — In eggs having an operculum, the opposite end is referred to as the "abopercular end".

Acetabulum — A muscular organ of attachment, commonly called a "sucker"; usually referring to those on the scolex of tapeworms and the ventral and dorsal suckers of trematodes.

Adult stage — The stage that is sexually mature and in which procreation occurs.

AIDS — Acquired immune deficiency syndrome; a condition resulting from infection with one of a particular group of viruses (see HIV) in which the immune system does not function to prevent or reduce the effects of various disease entities.

Arthropod — Any of a large group of bilaterally symmetrical, invertebrate animals having segmented (jointed) bodies and hollow, jointed appendages, such as legs, wings, and antennae in opposing pairs; e.g., mites, ticks, spiders, insects.

Autoinfection — Reinfection by a parasite derived from within the host. The source of the infection is the same host and the parasite is not exposed to the outside environment e.g., in infections with *Strongyloides stercoralis* reinfection by infective juveniles that are not evacuated with the feces.

Axoneme (rhizoplast) — In flagellates, an internal fibril arising from a **Blepharoplast** and passing through the cytoplasm. An axoneme may leave the body of the flagellate with a small sheath of cytoplasm to become a flagellum or run along the surface of the body lifting the periplast (cell membrane) to form an undulating membrane.

Axostyle — A rod-like structure that gives rigidity to the bodies of some flagellates, e.g., *Trichomonas* spp.

Basal granule — In ciliates, the granule-like body from which each cilium arises. Comparable to the blepharoplast from which a flagellum arises.

Binary fission — The division or splitting of a cell, especially protozoa, into two equal parts, each having the structures and characteristics of the parent cell.

Binuclear — A cell having two nuclei.

Blepharoplast — A small granule-like body, usually appearing in the cytoplasm, from which an axoneme arises. **Axonemes** may form rod-like structures in the cytoplasm, cilia, or flagella.

Bothrium (-a) — A sucker (organ of attachment) in the form of a groove on the scolex of some tapeworms, e.g., *Diphyllobothrium latum.*

Buccal cavity — In nematodes, the mouth chamber that joins the mouth opening with the esophagus. A structure used in differentiating juvenile worms of *Strongyloides* and hookworms.

Caecum (cecum) — A sac-like extension of the intestine that is open only at one end. Similar to a diverticulum. Seen in organisms with a true intestine, e.g., nematodes and arthropods.

Cell membrane — The superficial, resistant, outer membrane of the cell formed by the ectoplasm, the **periplast**.

Cephalic (cytolytic) gland — In trematodes, the gland in a miracidium that produces a fluid which enables the miracidium to penetrate the tissues of its snail host. A penetration gland.

Cephalic (cytolytic) gland duct — the duct that carries penetration fluid from the gland to the tissue of the host.

Cercaria (-ae) — The free-swimming larva of a trematode (usually possessing a tail) which escapes from a sporocyst or redia generation within the intermediate, molluscan host and constitutes the transfer stage to the next host.

Cercomer — In a tapeworm embryo, the caudal vestige of the onchosphere, containing six hooklets.

Charcot–Leyden crystals — A slender crystal, pointed at both ends, formed from the breakdown products of eosinophils usually following some type of immune response.

Chitinous shell — The hard shell of nematode eggs lined with the vitelline membrane which encases the embryo.

Chromatin — The darkly-staining portion of the nucleus forming a network of nuclear material within the achromatic nucleoplasm of the nucleus, sometimes adhering to the inner surface of the nuclear membrane (see **Peripheral chromatin**). The portion of the nucleus containing the DNA.

Chromatoid (chromatoidal) bar or body — A bar, rod, or splinter-shaped body in the cytoplasm of an amoeba that stains darkly and resembles chromatin.

Chromatoid basal rod — The rod-like structure that forms the base of the undulating membrane of flagellates.

Cilia — Minute, axial fibrils that arise from granules in the ectoplasm of Mastigophora (ciliates) and function in locomotion.

Cirrus — The retractile muscular organ at the outer end of the male reproductive system of species of Platyhelminthes.

Coccidia — A taxonomic grouping of certain members of the subphylum Apicomplexa (formally Sporozoa) in which both merozoites and gametocytes are produced by asexual reproduction (**schizogony**) in the same host. Four parasites of man, *Cryptosporidium parvum, Isospora belli, Sarcocystis hominis*, and *S. suihominis* are included in this group. Also the individual parasites, a coccidium or coccidia.

Coenurus — A larval cystic stage of a tapeworm containing an inner germinal layer producing multiple scolices within a single cavity (*Multiceps multiceps*).

Commensal — An organism that lives on or in another usually larger organism (the **host**) and derives its nourishment from its host organism without causing damage to the latter.

Concentration (method) — Increasing the strength of or numbers in a medium, e.g., 10% acid is more concentrated than 5% acid. A method or procedure that increases the intensity or numbers within a medium usually by reducing the volume of the medium or some component of it. A procedure carried out on a fecal specimen that increases the numbers of organisms found in a given unit over examination of a similar size unit before concentration.

Contractile vacuole — In *Balantidium coli* and many free-living protozoa, especially ciliates, a vacuole that is associated with removing liquid wastes from the body of the organism. In living ciliates, the contractile vacuole near the surface of the protozoa can be seen to fill with clear fluid, then by sudden contraction, discharge the fluid through the surface of the body and disappear, only to reappear in the same place as a small vacuole that begins to fill again.

Coracidium — In pseudophyllidian tapeworms, the onchosphere enclosed in its ciliated embryophore after hatching from the egg shell. It is free swimming, and is the precursor of the first stage larva (**procercoid**) of pseudophyllidian tapeworms.

Cortex — In helminths, the outer, mammillated coating of an egg of *Ascaris lumbricoides*.

Costa — In flagellates, a rib-like body; the chromatoid basal rod supporting the base of an undulating membrane (a **parabasal**).

Crustacea — A class of arthropods (phylum Arthropoda) composed of crustaceans.

Crustacean — Any of a group of primarily aquatic arthropods having a hard exoskeleton, jointed bodies and appendages, and gills e.g., lobsters, shrimps, crabs, etc.

Cuticle — In helminths, the outer covering layer secreted from the hypodermis or subcuticular layer.

Cyclophyllidian — Refers to an order of tapeworms having sucker discs encircling the scolex as in *Taenia solium*.

Cyst — An organism together with the enveloping membrane or wall secreted by that organism; the stage of a protozoan in which the organism is encased in a "cyst wall"; an encysted organism; a protected or more resistant stage that may be involved in transmission to a new host. Also, any thin-walled, sac-like body or capsule with a liquid or soft center, e.g., a tapeworm cyst.

Cyst wall — The outermost part of the coating protecting an encysted protozoan.

Cysticercoid — A larva of tapeworms in which the scolex is invaginated into a greatly reduced cystic cavity almost devoid of fluid (e.g.: *Dipylidium caninum*).

Cysticercus (-i) — A larva of tapeworms in which the scolex is invaginated into a bladder filled with fluid (e.g.: *Taenia* spp.).

Cytopharynx — The chamber behind the mouth (cytostome) in protozoa into which food may be taken then passed into the cytoplasm via food vacuoles formed at its base. **Cytostome** and cytopharynx are often considered synonymous.

Cytoplasm — The protoplasm of a cell surrounded by the cell membrane exclusive of the nucleus.

Cytoplasmic granules — Granular-appearing materials within the cytoplasm, especially in the amoebae.

Cytopyge — In *Balantidium* and some other ciliates, the permanent opening in the ectoplasm through which the residue of digestion of food materials is discharged (anal opening). In other organisms, the wastes may be discharged through a number of points in the body surface.

Cytostome — In some protozoa, especially flagellates and ciliates, the cavity that opens by way of the peristome (the lips) to allow solid food particles to enter (the mouth). In some species, a food vacuole may form directly behind the cytostome or it may open into a **cytopharynx** (esophagus-like structure) and the food vacuole is formed at the base of the cytopharynx.

Cytostomal flagellum — In *Chilomastix*, a flagellum that lies in a groove behind the cytostome (usually within the cytopharynx) and directs solid food particles to the posterior end where food vacuoles are formed.

Density gradient — Refers to the varying sequence of densities of materials suspended in a liquid column with the most dense material at the bottom and the least dense material at the top. A phenomenon that results from gravity sedimentation.

Diecious (dioecious) — Female and male reproductive organs in different individuals.

Digenetic — Three or more generations (literally "two", adult and larval) required for completion of one life cycle (or generation), as in digenetic trematodes.

Diverticulum — A sac-like structure extending out from a tubular organ such as a blood vessel or intestine.

Ectoplasm — The more hyaline, outer peripheral cytoplasm of a protozoan. The portion of the cytoplasm from which cilia usually originate. The surface of the ectoplasm apparently gives rise to the cell membrane, the **periplast**.

Egg — The female reproductive cell or ovum before fertilization, or the complex sex product following fertilization (if this occurs) with the addition of yoke and other nutritive materials and the addition of the embryonic membrane and other shell layers (see **Ovum**; the terms egg and ovum are not synonymous).

Embryo — The stage in development following cleavage of the egg up to, but not including, the first larval or first juvenile stage.

Embryophore — In tapeworms, the envelope immediately surrounding the onchosphere and derived from it.

Endoplasm — That portion on the cytoplasm in which vacuoles form and in which internal organelles and inclusions are suspended.

Feces — The excrement (waste matter) evacuated from the intestinal tract which includes unabsorbed food materials, sloughed cells, mucus, bacteria, etc.

Filariform (juvenile) — A post-feeding stage of a nematode characterized by its delicate, elongate structure and its slim, capillary esophagus. The infective stage of hookworm, filarial worms, and some other nematodes.

Flagellum (-a) — A filament, arising from a granule-like body (the **blepharoplast**) and covered by a thin sheath of cytoplasm that usually projects from the body of an organism and functions as an organelle of locomotion. When lying in a groove in the cytostome (**cytostomal flagellum**), it causes movement of the fluid medium in a certain direction. In sessile flagellates, the flagella create currents in the medium to bring food particles to the mouth of the organism and to move wastes away from it.

Flame cell — A structure in a primitive excretory system in Platyhelminthes. Cilia in flame cells apparently move fluid wastes from the body into the protonephritic tubules to be excreted (see **Protonephridia**). Sometimes flame cells can be seen in living schistosome eggs.

Fluke — A common name for trematodes.

Free-living — Living in a free and unrestrained manner in the environment; living free of a host.

Gamete — A mature reproductive cell capable of uniting with another reproductive cell to form a fertilized cell, the zygote, that can develop into a new individual (animal or plant).

Gametogony — Asexual reproduction (a form of schizogony), in which macrogametocytes and microgametocytes are formed (see **schizogony**).

Genital atrium — In Platyhelminthes, the antechamber to the genital tubules.

Genital primordium — A group of cells in juvenile nematodes that are the precursors of a reproductive system. Seen in rhabditoid juveniles of *Strongyloides* but usually not detectable in those of hookworms.

Germ cell — Cells in an egg from which the embryo grows.

Ghost cyst — A cyst of a protozoan in which the body has shriveled to become unrecognizable. *Endolimax nana* and *Giardia lamblia* cysts may become ghost cysts when fixation is delayed.

Glycogen vacuole — A vacuole, usually in a cyst, in which glycogen (carbohydrate storage material) is accumulated until needed in the development process. Such vacuoles usually form early in the development of an amoebic cyst and are usually present in cysts of *Iodamoeba butschlii*.

Gravid — Filled with eggs, as a gravid pinworm or gravid proglottid of a tapeworm.

Helminth — From the Greek, *helmins, -inthos,* meaning worm. Originally referred to intestinal worms. Any of the worms, Platyhelminthes, Nematodes, et al.

Hemocoele — A body cavity in mollusks and arthropods through which the blood (hemocoele fluid) circulates carrying nutrients etc. to the organs.

Hermaphroditic — Containing both male and female reproductive organs in one individual (see **Monecious**).

Heterogonic — Development in which both females and males are present in the colony.

Hexacanth embryo — "Six-hooked" embryo, the mature embryo within the egg of many tapeworms, including all species that parasitize man in the adult stage.

HIV (human immunodeficiency virus) — A group of viruses that impact on the immune system and severely reduce its effectiveness in fighting disease. Infection may lead to **AIDS**. Infection occurs through transfer of blood, tissue fluids, and body fluids from an infected to an uninfected individual.

Hologonic — Development in which only one sex (usually the female) is present in a colony, as in *Strongyloides stercoralis* infections.

Hooklet — In tapeworms, the small hook-like organ of attachment present on the **rostellum** of the tapeworm scolex. A small hook.

Host — An organism that harbors and nourishes another.

Aberrant host — One in which the parasite cannot complete its development (life cycle or appropriate phase of its development).

Accidental host — One in which the parasite is not commonly found but is suitable for continuation of its life cycle.

Alternate host — One that alternates with another host in the life cycle of a parasite. (Snails and man are alternate hosts of schistosomes.)

Definitive host — One in which the terminal (usually sexually mature) stage occurs. (Man is the definitive host of *Taenia saginata*.)

Helping host — An intermediate host (usually mechanical) on which or in which a stage of the parasite comes to rest or attaches but does not develop further. An intermediate host such as the water chestnut (*Eliocharis tuberosa*) upon which the metacercaria of *Fasciolopsis buski* encysts.

Intermediate host — One that alternates with the definitive host and harbors the larval stage (s) of the parasite. (Man is the Accidental, Intermediate host of *Taenia solium* and pigs are the typical intermediate host.)

Reservoir host — One in which the parasite lives and is available for transmission to another host. In parasitology, the term usually refers to a host which harbors a stage of the parasite that is found in the typical host (e.g., the dog is the reservoir host of *Trichuris vulpis* for man).

Typical host — One in which the parasite is commonly found and in which it can continue its development (or the appropriate phase of its development) necessary for subsequent completion of its life cycle.

Hydatid cyst — A cystic larval stage of *Echinococcus* spp. containing an inner germinal layer that produces many scolices, which, when set free into the cystic cavity, can develop into daughter cysts, in which further production of scolices can take place.

Hyperinfection — Infection superimposed upon an existing infection by the same parasite in which the parasite reaches high numbers. The term usually refers to internal autoinfection e.g., strongyloidiasis, oxyuriasis (pinworm), or hymenolepiasis nana (see **Autoinfection**).

Infected source — An animal or plant (an intermediate host) in which a parasite has established an infection and can act as the source of infection for another host. Cattle infected with the cysticerci of *Taenia saginata* function as the source of infection for the adult parasite for man. Infection occurs when the infective stage is ingested along with the intermediate host (or part of it).

Infection — Containing the property of producing infection. Invasion of a host (tissues or cells) resulting in injury and reaction to that injury, i.e., disease.

Infective stage — The stage of a parasite capable of initiating infection.

Infestation — The term is usually applied to ectoparasites and describes a host/parasite relationship in which the parasite lives on the surface of the host. In some instances, e.g., scabies, the parasite may invade and inhabit the superficial tissues. Some workers apply the term to commensal organisms, such as *Entamoeba coli*, that do not invade tissues or cause disease but live in a part of the body (the large intestine) with direct access to the outside environment.

Infested source — A source of the infective stage of a parasite when that stage is essentially free living, e.g., cercariae of schistosomes that infest water (swim about freely outside of the host) are capable of initiating the infection of a suitable host. Cercariae infest water, the source from which infection is initiated.

Inner shell — Eggs of some helminths have an inner and an outer shell, e.g., *Hymenolepis nana*. *Ascaris* eggs have an outer mammillated coat (the **cortex**) that covers the thick hyaline shell.

Juvenile stage — Any stage in the development of a helminth parasite (usually in reference to nematodes) between the egg and the mature adult stage that appears similar in shape and structure to the adult. The term implies a form less developed but similar in structure to the adult. Trematodes pass through larval stages; nematodes pass through juvenile stages. The term "**larva**" for juvenile nematodes has become so entrenched that the correct term is rarely used.

Karyosome — A structure (body) within the nucleus having a relatively constant size and location in each species and made up of two components, one achromatic (not staining) and another that stains similarly to chromatin and appears as a granule or bundle of granules. The size and location of the karyosome, especially the staining portion, is frequently used as an aid in differentiating species of amoebae that are similar in structure.

Key — A "biological Key" is a guide to the identification of individual organisms of a group of plants or animals having specific determining characteristics arranged in a systematic way; a series of questions arranged in a format designed to lead the users to an accurate identification of species within a biological group of organisms.

Larva (-ae) — The postembryonic stage in which internal organs are developing or are partially developed and are at least partially functioning. Any preadult stage (after the embryonic stage) in the life cycle of a parasite that is morphologically distinct from the adult stage (see **Juvenile**). The term larva should probably not be used for immature nematodes since they are not morphologically distinct from the adults but simply smaller and less developed.

Linin fibrils — Very delicate fiber-like strands that may appear in the nuclei between the karyosome and the peripheral chromatin of some protozoa after staining, especially in some trophozoites of *Entamoeba histolytica*.

Longitudinal cords (lines) — In nematodes, four cords (lines) that extend from the anterior to the posterior end (one dorsal, one ventral, and two lateral) which enclose the longitudinal nerves and, in the lateral cords, the longitudinal excretory tubules.

Lumen — The space within a tubular organ such as the intestine.

Macronucleus — In some ciliates, the large kidney-shaped nucleus. Its function is not clearly understood but it is usually in association with a "**micronucleus**".

Macrophage — One of a group of mononucleate, migrating, phagocytic cells that play a role in cellular immunity. Sometimes mistaken for an amoeba in fecal specimens.

Merogony (schizogony) — Asexual reproduction in which the nucleus of the parasite divides and daughter nuclei divide repeatedly within the cytoplasm. When nuclear division is complete, each nucleus moves to the surface where it becomes surrounded with a small amount of cytoplasm to form a bud-like **merozoite** which pinches off leaving behind a residual body.

Merozoite — The resulting product of asexual reproduction (**merogony**) that initiates infection of a new cell or tissue (see **Schizogony**).

Mesentery — A sheet of tissue or membrane that enfolds and supports an internal organ by attaching it to the body wall or another organ. Adults of schistosomes live and female worms pass their eggs in the mesenteric venules associated with the intestine and urinary bladder.

Metacercaria (-ae) — The encysted stage of a monecious trematode succeeding the cercaria. The cercaria invades or attaches to an animal or plant (the second intermediate host) where it encysts to await transfer (usually by ingestion) to the definitive host. In blood flukes, the cercaria does not encyst but directly invades the definitive host (see **Schistosomule**) at which time the tail of the cercarial stage is left behind.

Micrometer — An instrument for measuring units equal to a micron, 0.001 mm (1 μm).

Micronucleus — In ciliates, a small nucleus closely associated with the **macronucleus**. The macronucleus and micronucleus perform differently than do the nuclei in most other protozoa during division. (See Wenyon, 1965.)

Micropyle — In coccidia, a pore in the cyst wall closed by a plug of material which is more easily dissolved than the cyst wall and through which the structures formed in the cyst emerge.

Miracidium — In trematodes, the larva, usually free swimming, that emerges from the egg.

Mollusca — A phylum composed of a large group of animals having no backbone, soft unsegmented bodies, usually covered with a hard shell. The shell is secreted by a covering mantle and is formed on snails, oysters, clams, and whelks but not on slugs, octopuses, or squids.

Molluscan Host — A member of the phylum Mollusca that acts as an intermediate host for the class Trematoda.

Monecious (monoecious) — Male and female sex organs in the same individual. **Hermaphroditic**.

Monogenetic — A single generation constituting a complete life cycle.

Morula stage — The cleaving stage of an egg in which it forms a mulberry-like, solid mass of cells.

Multilocular cyst — A type of cestode cyst with many cavities in which scolices develop, e.g., in *Echinococcus multilocularis*.

Nucleoplasm — The protoplasm of the nucleus not including the **karyosome**, **plastin**, and **chromatin**.

Nucleus — A spheroid body within the protoplasm of a cell, distinguished from the rest of the cell by its dense structure and presence of chromatin. The nucleus controls growth, cell division, and other activities of the cell; and contains DNA, the basic substance controlling the genetic characteristics of the cell or organism.

Occult blood — Blood in such small quantities that it is not readily detectable except by chemical means.

Onchosphere — The stage that escapes from the egg shell and later from the embryophore of tapeworms. In human tapeworms it is a six-hooked (hexacanth) embryo.

Oocyst — The stage of a coccidian protozoa that is produced by sexual reproduction; the product of fertilization, which is evacuated with the feces. Sporocysts develop within the oocyst and, as the oocyst matures, sporozoites develop within sporocysts (except cryptosporidia where sporozoites develop free in the oocyst). Mature oocysts containing sporozoites, or rarely sporocysts freed from oocysts, become the infective stage.

Operculum — Eggs of some trematode and cestode helminths have a cap-like structure at one end through which the embryo (larva) emerges. Protrusions surrounding the base of the operculum, called shoulders, are present in eggs of some species.

Oviparous — Egg-laying organisms (see **Ovoviviparous**).

Ovoviviparous — Refers to species in which eggs hatch *in utero* and the freed young are released from the female worm, e.g., in *Trichinella spiralis.*

Ovum — The mature, naked, female reproductive cell preceding combination with the male gamete and the addition of an embryonic membrane and other shell layers. An unfertilized egg (see **Egg**).

Parabasal (parabasal body) — In flagellates, a heavy fiber (rhizoplast) present in some flagellates. It supports the cytostome in *Chilomastix* and is the basal fibril (**chromatoid basal rod**) of the undulating membrane in *Trichomonas.* The function of parabasals in *Giardia* is unknown.

Parasite — An organism that lives on or within and at the expense of another organism.

Facultative — One that may employ either a free-living or parasitic mode of life during the course of its life cycle.

Obligatory (obligate) — One that must live a parasitic existence.

Opportunistic — An organism that is not typically a parasite but may become parasitic under specific conditions. Amoebae of the genus *Naeglaria* are usually free living but may become opportunistic parasites.

Parenchyma — The soft, undifferentiated tissue composing the general substance of the body of some invertebrates e.g., members of the phylum Platyhelminthes. Also, the essential or functional components of organs as differentiated from the connective or support components, the stroma.

Parasitism — A symbiotic relationship in which one partner, the parasite, lives on or within its host, and inasmuch as it derives nourishment from its host, it is potentially harmful since it may either deprive or damage the host.

Parthenogenesis — Reproduction without fertilization with the male element. In infections by *Strongyloides stercoralis,* no male is present in the intestine and eggs are produced by the virgin female.

Pathogen — A parasite that injures or deprives its host, i.e., is capable of producing disease.

Pathogenic — Giving origin to disease or symptoms of disease.

Pellicle — In some protozoa, the thin, tough membrane covering the **ectoplasm** which allows the organism to maintain its shape.

Peripheral chromatin — That portion of the nuclear chromatin adhering to the inner surface of the nuclear membrane as in *Entamoeba histolytica* and *E. coli.*

Periplast — The limiting, outer membrane of protozoan (cell membrane) formed from the ectoplasm.

Peristalsis — Rhythmical, wave-like constrictions of the wall of the intestine that move the contents.

Peristome — Any parts or set of parts around the mouth or oral opening of invertebrates; comparable to lips in higher animals.

Phasmid — One of a pair of caudal chemoreceptors in certain nematodes (i.e., Phasmidia).

Plastin — Achromatic substance within the nucleus other than the nucleoplasm. A portion of the nucleus that stains only with special stains and is probably the material from which chromosomes are formed.

Plerocercoid (larva) — A tapeworm larva in which the scolex is embedded in a greatly enlarged tail (see **Sparganum**) as in *Diphyllobothrium latum.*

Proboscis — In tapeworms, an anterior protrusile organ, typically studded with hooks as in the dog tapeworm *Dipylidium caninum.*

Polar filaments — Filaments arising from the opposite poles of the onchosphere membrane of *Hymenolepis nana.*

Polar plugs — Mucoid plugs that are located at both ends of eggs of *Trichuris* and other members of the family Trichuridae.

Prepatent period — The span of time from infection until the infection is detectable. The biological incubation period, usually between infection and initial clinical manifestations.

Primitive gut — The structure that gives rise to the gut (intestine).

Procercoid (larva) — The first larval stage of pseudophyllidian tapeworms which develops from the onchosphere; it contains a body proper and a caudal vestige of the onchosphere, the **Cercomer**.

Proglottid — One complete unit of a tapeworm, containing male and female organs, located below the **Scolex**, and commonly called a "segment".

Protonephridia — A primitive excretory system consisting of flame cells and tubules for the elimination of liquid wastes, in Platyhelminthes.

Pseudophyllidian — Refers to an order of tapeworms in which the scolex has a single terminal or two, opposite, lateral organs of attachment, the **bothria**, e.g., in *Diphyllobotrium latum.*

Pseudopodium (-ia) — A clear projection of the ectoplasm of protozoa, especially amoebae, that is usually associated with movement and/or food gathering.

Retracted flagellum — A flagellum that at some stage in the life cycle of an organism extends beyond the body membrane but rests within the confines of the body of the organism in the cyst stage or in a resting stage.

Rhizoplast (axoneme) — In flagellates, a fibril arising from a **blepharoplast** and running through the cytoplasm.

Rhabditoid (rhabditiform) juvenile (larva) — The first, feeding stage of a juvenile nematode that emerges from the egg in which the esophagus is functional, is usually muscular, and has an enlarged, bulbous posterior end.

Rostellum — The somewhat protuberant apical portion of the **scolex** of certain tapeworms, frequently bearing hooklets (a circlet of hooklets in *Taenia solium*, seven transverse rows in *Dipylidium caninum*).

Schistosomule — The immature stage of schistosomes (blood flukes) from the time of entry into the definitive host until the fluke reaches sexual maturity (see **Metacercaria**).

Schizogony — An asexual reproduction process seen in menbers of the subfamily Apicomplexa in which the nucleus of a parasite within or attached to its host cell divides repeatedly without immediate division of the cytoplasm to form a multinucleate body, the **schizont**. Each nucleus with a small amount of cytoplasm separates to form a daughter cell, the **merozoite**, which becomes free of the parent cell and the host cell to initiate infection of another host cell. The process occurs in the coccidia of man (see also **Gametogony, Merogony,** and **Sporogony**).

Schizont — A trophozoite of a member of the coccidia in which the nucleus has gone through a number of divisions without division of the cytoplasm. In the asexual process, **Schizogony**.

Scolex (-ices) — The attachment end (head with organs of attachment) of a tapeworm from which the neck arises and, in turn, gives rise to the proglottids.

Sparganum — The second larval stage (see **Plerocercoid**) of pseudophyllidian tapeworms that is characterized by its elongated shape and lack of a cystic cavity.

Sporocyst — In protozoa, a cyst that develops within an oocyst of a coccidian protozoa in which sporozoites develop. In some species of coccidia, the oocyst may rupture within the intestine of the host and free sporocysts are evacuated with the feces. In such cases, the small sporocysts must be found on fecal examination, e.g., *Sarcocystis* spp.

Sporocyst — In trematodes, the first larval stage in the developmental cycle in the snail, intermediate host. A sac-like structure with a germinal lining that produces secondary larval stages that develop in the snail host.

Sporogony — An asexual reproductive process (**schizogony**) in which **sporozoites** are the end product.

Sporont — A cell or zygote that through subsequent development and division will form sporozoites. It may form sporocysts, as in *Isospora* or may not, as in *Cryptosporidum*.

Sporozoite — The form that develops within an oocyst that, when freed from the oocyst after ingestion by a suitable host, can enter an intestinal cell to initiate an infection. The infective stage for man in some members of the subfamily Apicomplexa.

Spurious infection — False infection. An organism or egg found on fecal examination that is from a source outside of the individual and is not related to an infection of the individual being examined; often derived from eating part of a host infected with a non-human parasite. Just passing through! Eggs of *Dicrocoelium dendriticum* may represent a spurious infection but an actual infection must be ruled out since man may become an accidental host of the parasite.

Strobila — A complete tapeworm consisting of scolex, neck, and immature, mature, and usually gravid proglottids.

Strobilization — Asexual production from the neck of a tapeworm of a series of sexual reproductive units, the **Proglottids**.

Superinfection — A new infection of a host superimposed on an existing one by the same species of parasite (see **Hyperinfection**).

Trophozoite — The active, vegetative stage of a protozoan.

Unilocular — In tapeworms, an intermediate larval stage (cyst) having only a single cavity.

Uninuclear — Having only one nucleus.

Vacuole — A space or cavity in the cytoplasm of a protozoan usually functioning in collecting and digesting food taken into the organism and eliminating wastes (see **Contractile vacuole**).

Villus (-i) — The small intestine is lined with small thread-like folds of the mucosa that project into the intestinal lumen (the villi), greatly increasing intestinal surface. The cells that cover the villi are enterocytes.

Vitellaria (vitelline glands) — The glands in Platyhelminthes that produce yolk material and (probably) the shell of the egg.

Vitelline membrane — The innermost layer in the shell of fertilized eggs of helminths.

Volutin granules — Bodies in the peripheral cytoplasm of *Blastocystis* that stain darkly.

Zoonosis — Infection in man with a parasite that typically infects animals.

Zygote — An individual organism resulting from the fusion of male and female gametes. The stage in the life cycle produced by fertilization.

Appendix 3

References

Alicna, A. D. and Fadell, E. J. Advantage of purgation in recovery of intestinal parasites or their eggs. *Am. J. Clin. Pathol.* 31:139–142, 1969.

Alves, W. The eggs of *Schistosoma bovis, S. mattheei* and *S. haematobium*. *J. Helminth* 23:127–134, 1949.

Ash, L. R. and Orihel, T. C. *Parasites: A Guide to Laboratory Procedures and Identification*. Chicago: ASCP Press, 1987.

Bartlett, M. D., Harper, K., Smith, N., Verbanac, P., and Smith, J. W. Comparative evaluation of a modified zinc sulfate flotation technique. *J. Clin. Microbiol.* 7:524–528, 1978.

Beaver, P. C., Gadgil, R. K., and Morera, P. *Sarcocystis* in man: a review and report of five cases. *Am. J. Trop. Med. Hyg.* 28:819–844, 1979.

Beck, J. W. and Barrett-Connor, E. *Medical Parasitology*. St Louis: C.V. Mosby Co., 1971.

Blagg, W., Schloegel, E. L., Mansour, N. S., Khalaf, G. I. A new concentration technic for the demonstration of protozoa and helminth eggs in feces. *Am. J. Trop. Med. Hyg.* 4:23–28, 1955.

Brooke, M. M. and Goldman, M. Polyvinyl alcohol-fixative as a preservative and adhesive for protozoa in dysenteric stools and liquid materials. *J. Lab. Clin. Med.* 34:1554–1560, 1949.

Bullock-Iacullo, S. L. *Laboratory diagnosis of intestinal cryptosporidiosis: Course No. 8008-C*. Atlanta: U.S. Dept Health and Human Services, Public Health Service, CDC, 1988.

Bundy, D. A. P., Foreman, J. D. M., and Golden, M. H. N. Sodium azide preservation of faecal specimens for Kato analysis. *Parasitology* 90:463–469, 1985.

Burrows, R. B. Morphological differentiation of *Entamoeba hartmanni* and *E. polecki* from *E. histolytica*. *Am. J. Trop. Med. Hyg.* 8:583–589, 1959.

Burrows, R. B. Improved preparation of polyvinyl alcohol-$HgCl_2$ fixative used for fecal smears. *Stain Technol.* 42:93–95, 1967a.

Burrows, R. B. A new fixative and technics for the diagnosis of intestinal parasites. *Am. J. Clin. Pathol.* 48:342–346, 1967b.

Burrows, R. B. Other surface active agents for use with the PAF sedimentation technic for intestinal parasites. *Am. J. Clin. Pathol.* 51:155–156, 1968.

Camp, R. R., Mattern, C. F. T., and Honigberg, B. M. Study of *Dientamoeba fragilis* Jepps & Dobell. I. Electron microscope observations of the binucleate stages. II. Taxonomic position and revision of the genus. *Protozoology* 21:69–82, 1974.

Carroll, M. J., Cook, J., and Turner, J. A. Comparison of polyvinyl alcohol- and formalin-preserved fecal specimens in the formalin-ether sedimentation technique for parasitological examination. *J. Clin. Microbiol.* 18:1070–1072, 1983.

Chitwood, B. G. and Chitwood, M. B. *Nematology*. Baltimore: Monumental Pub, 1950.

Danciger, M. and Lopez, M. Numbers of *Giardia* in the feces of infected children. *Am. J. Trop. Med. Hyg.* 24:237–242, 1975.

Delly, J. D., ed. *Photography through the Microscope*, 7th ed. Rochester: Eastman Kodak Co., 1980.

Dunn, F. L. The TIF direct smear as an epidemiological tool with special reference to counting helminth eggs. *Bull. WHO Geneva* 39:439–440, 1968.

Ehlers, E. G. *Optical Mineralogy: Theory and Technique*. Palo Alto: Blackwell Scientific Publications, 1987.

Faust, E. C. *Human Helminthology, A Manual for Physicians, Sanitarians, and Medical Zoologists*, 3rd ed. Philadelphia: Lea & Febiger, 1949.

Garcia, L. S. and Ash, L. A. *Diagnostic Parasitology Clinical Laboratory Manual*, 2nd Ed., St. Louis, C. V. Mosby Company, 1979.

Garcia, L. S., Shimizu, R. Y., Brewer, T. C., and Bruckner, D. A. Evaluation of intestinal parasite morphology in polyvinyl alcohol preservative: comparison of copper sulfate and mercuric chloride bases for use in Schaudinn fixative. *J. Clin. Microbiol.* 17:1092–1095, 1983.

Garcia, L. S., Shimizu, R. Y., Shum, A., and Bruckner, D. A. Evaluation of intestinal protozoan morphology in polyvinyl alcohol preservative: Comparision of zinc sulfate- and mercuric chloride-based compounds for use in Schaudinn's fixative. *J. Clin. Microbiol.* 31:307–310, 1993.

Garcia, L. S. and Bruckner DA. *Diagnostic Medical Parasitology*. New York: Elsevier Science Publishing Co., 1988.

Georgi, J. R., Sprinkle, C. L. A case of human strongyloidosis apparently contracted from asymptomatic colony dogs. *Am. J. Trop. Med. Hyg.* 23:899–901, 1974.

Goldman, M. Identification and diagnosis of *Entamoeba histolytica*. *Am. J. Gastroenterol.* 41:362–365, 1964.

Hunter, G. W., III, Frye, W. W., and Swartzwelder, J. C. *A Manual of Tropical Medicine*, 3rd ed. Philadelphia: W. B. Saunders, 1960.

Isenberg, H. D., ed. *Clinical Microbiology Procedures Handbook, Vol. 2, Sec. 7, Parasitology*. Washington, D.C.: American Society for Microbiology, 1992.

Jenrikhen, S. T. and Pohlenz, J. F. L. Staining of cryptosporidia by a modified Ziehl-Neelsen technique. *Acta. Vet. Scand.* 22:594–596, 1981.

Juniper, K., Jr. Acute amebic colitis. *Am. J. Med.* 33:377–386, 1962.

Junod, C. Technique coprologique nouvelle essentiellement destinée a la concentration des trophozoites d'amibes. *Bull. Soc. Pathol. Exot. Filiales* 65:390–398, 1962.

Kinyoun, J. J. A note on Uhlenhuth's method for sputum examination for tubercle bacilli. *Am. J. Pub. Health* 5:867–870, 1915.

Kuntz, R. E. Biology of schistosome complexes. *Am. J. Trop. Med. Hyg.* 4:383–413, 1955.

Levine, J. A. and Estevez, E. G. Method for concentration of parasites from small amounts of feces. *J. Clin. Microbiol* 18:786–788, 1983.

Levine, N. D. *Protozoan Parasites of Domestic Animals and of Man.*, 2nd ed. Minneapolis: Burgess Publishing Co., 1973.

Levine, N. D. The taxonomy of *Sarcocystis* (Protozoa, Apicomplexa) species. *J. Parasitol.* 72:372–382, 1986.

Lillie, R. D. *Histopathologic Technique and Practical Histochemistry*, 3rd ed. New York: McGraw-Hill, 1965.

Markel, E. K., Quinn, P. M. Comparison of immediate polyvinyl alcohol (PVA) fixation with delayed Schaudinn's fixation for the demonstration of protozoa in stool specimens. *Am. J. Trop. Med. Hyg.* 26:1139–1142, 1977.

Markel, E. K., Voge, M., and John, D. T. *Medical Parastology*, 6th ed. Philadelphia: W. B. Saunders Co., 1986.

Markel, E. K. and Udkow, M. P. *Blastocystis hominis*: Pathogen or fellow traveler. *Am. J. Trop. Med. Hyg.* 35:1023–1026, 1986.

Marsden, A. T. H. Detection of cysts of *E. histolytica* in feces by microscopic examination. *Med. J. Aust.* 1:915–916, 1946.

Martin, L. K. and Beaver, P. C. Evaluation of Kato thick-smear technique for quantitative diagnosis of helminth infections. *Am. J. Trop. Med. Hyg.* 17:382–391, 1968.

Melvin, D. M. and Brooke, M. M. *Laboratory Procedures for the Diagnosis of Intestinal Parasites*, 3rd Ed., Atlanta: U.S. Department of Health and Human Services, Public Health Service, CDC, 1982.

Montessori, G. A. and Bischoff, L. Searching for parasites in stool: Once is usually enough. *Can. Med. Assoc. J.* 137:702, 1987.

Peters, C. S., Hernandez, L., Sheffield, N., Chittom-Swiatlo, A. L., and Kocka, F. E. Cost containment of formalin-preserved stool specimens for ova and parasites from outpatients. *J. Clin. Microbiol.* 26:1584–1585, 1988.

Peters, C. S., Sable, R., Janda, W. M., Chittom, A. L., and Kocka, F. E. Prevalence of enteric parasites in homosexual patients attending an outpatient clinic. *J. Clin. Microbiol.* 24:684–685, 1986.

Pitchford, R. J. Observations on a possible hybrid between the two schistosomes *S. haematobium* and *S. mattheei*. *Trans. Roy. Soc. Trop. Med. Hyg.* 55:44–51, 1961.

Pitchford, R. J. Differences in the egg morphology and certain biological characteristics of some African and Middle Eastern schistosomes, genus *Schistosoma*, with terminal-spined eggs. *Bull. WHO Geneva* 32:105–120, 1965.

Price, D. L. Hepatic, intestinal, and pulmonary trematodes, in *CRC Handbook Series on Clinical Laboratory Science*. Section E: Clinical Microbiology. Vol. II, von Graevenitz, A., Sect. Ed. Cleveland: CRC Press, 1977.

Price, D. L. Culturette Brand, MIF Procedure Kit. Kansas City, MO: Marion Scientific Corporation, 1978.

Price, D. L. *Reference Manual of Intestinal Protozoa in MIF*. Kansas City, MO: Marion Scientific Corporation, 1979.

Price, D. L. Comparison of three collection-preservation methods for detection of intestinal parasites. *J. Clin. Microbiol.* 14:656–660, 1981.

Reese, N. C., Current, W. L., Ernst, J. V., and Bailey, W. S. Cryptosporidiosis of man and calf: A case report and results of experimental infections in mice and rats. *Am. J. Trop. Med. Hyg.* 31:226–229, 1982.

Rendtorff, R. C. The experimental transmission of human intestinal protozoan parasites. *Am. J. Hyg.* 50:209–220, 1954.

Richie, L. S. An ether sedimentation technique for routine stool examinations. *Bull. U.S. Med. Dept.* 8:326, 1948.

Sapero, J. J. and Lawless, D. K. The "MIF" stain-preservation technic for the identification of intestinal protozoa. *Am. J. Trop. Med. Hyg.* 2:613–619, 1953.

Sawitz, W. G. and Faust, E. C. The probability of detecting intestinal protozoa by successive stool examinations. *Am. J. Trop. Med.* 22:131–136, 1942.

Scholten, T.H. An improved technique for the recovery of intestinal protozoa. *J. Parasitol.* 58:633–634, 1972.

Schwetz, J. A comparative morphological and biological study of *Schistosoma hematobium, S. bovis, S. intercalatum* Fisher, 1934, *S. mansoni* and *S. rodhaini* Brumpt, 1931. *Ann. Trop. Med. Parasitol.* 45:92–98, 1951.

Senay, H. and MacPherson, D. Parasitology: diagnostic yield of stool examination. *Can. Med. Assoc. J.* 140:1329–1331, 1989.

Stamm, W. P., II. The laboratory diagnosis of clinical amoebiasis. *Trans. Roy. Soc. Trop. Med. Hyg.* 51:306–312, 1957.

Stoll, N. R. and Hausheer, W. C. Concerning two options in dilution egg counting; small drop and displacement. *Am. J. Hyg.* 6(suppl): 134–145, 1926.

Sun, T. *Pathology and Clinical Features of Parasitic Diseases*. New York: Masson Publishing, Inc., 1982.

Swartzwelder, C. Laboratory diagnosis of amebiasis. *Am. J. Clin. Pathol.* 22:379–395, 1952.

Tompkins, V. N. and Miller, J. K. Staining intestinal protozoa with iron-hematoxylin-phosphotungstic acid. *Am. J. Clin. Pathol.* 17:755–757, 1947.

Wenyon, C. M. Protozoology, A Manual for Medical Men, Veterinarians and Zoologists, Vol. I, II. Facsimile, 1926 Ed. New York: Hafner, 1965.

Wheatley, W. B. A rapid staining procedure for intestinal amoebae and flagellates. *Am. J. Clin. Pathol.* 21:990–991, 1951.

Wolfe, M. S. Current concepts in parasitology: Giardiasis. *N. Engl. J. Med.* 298:319–321, 1978.

Young, K. H., Bullock, S. L., Melvin, D. M., and Spruill, C. L. Ethyl acetate as a substitute for diethyl ether in the formalin-ether sedimentation procedure. *J. Clin. Microbiol.* 10:852–853, 1979.

Yang, J. and Scholten, T. A fixative for intestinal parasites permitting the use of concentration and permanent staining procedures. *Am. J. Clin. Pathol.* 67:300–304, 1977.

Yoeli, M., Most, H., Hammond, J., and Scheinesson, G. P. Parasitic infections in a closed community. Results of a 10-year survey in Willowbrooke state school. *Trans. Roy. Soc. Trop. Med. Hyg.* 66:764–776, 1972.

Zierdt, C. H. Pathogenicity of *Blastocystis hominis* [letter]. *J. Clin. Microbiol.* 29:662–663, 1991a.

Zierdt, C. H. *Blastocystis hominis* —past and future. *Clin. Microbiol. Rev.* 4:61–79, 1991b.

Zierdt, C. H., Rude, W. S., and Bull, B. S. Protozoan characteristics of *Blastocystis hominis. Am. J. Clin. Pathol.* 48:495–501,1967.

Index

Page numbers that appear in **boldface** refer to the principal discussion of an organism.

A

B

C

D

E

F

G

H

I

J

K

L

M

N

O

P

Q

R

S

T

U

V

W

X

Y

Z